I0072629

Phenology and Climate Change

Phenology and Climate Change

Edited by **Loren Gilbert**

New York

Published by Callisto Reference,
106 Park Avenue, Suite 200,
New York, NY 10016, USA
www.callistoreference.com

Phenology and Climate Change
Edited by Loren Gilbert

International Standard Book Number: 978-1-63239-510-8 (Hardback)

Printed in the United States of America.

Contents

Preface

A descriptive account based on phenology and climate change has been highlighted in this up-to-date book. Phenology is referred to as the study of animal as well as plant life cycle. It is one of the most direct and obvious phenomena on our planet. The timing of phenological events offers important information for the purpose of natural resource management, carbon sequence analysis, climate change investigation, and crop and forest growth monitoring. This book elucidates current developments in the comprehension of seasonal variation in plants and animals and its interaction with climate variables. Renowned scientists in this field from across the world have contributed in this book. The information provided in this book covers topics like plant seasonality, remote sensing phenology, animal life cycle, and phenology in fruit plants. It provides the readers with an extensive analysis of phenology observations and climate effects. This book aims to stimulate further advancements in the field of phenology; monitoring, modeling as well as predicting.

The information contained in this book is the result of intensive hard work done by researchers in this field. All due efforts have been made to make this book serve as a complete guiding source for students and researchers. The topics in this book have been comprehensively explained to help readers understand the growing trends in the field.

I would like to thank the entire group of writers who made sincere efforts in this book and my family who supported me in my efforts of working on this book. I take this opportunity to thank all those who have been a guiding force throughout my life.

Editor

Part 1

Animal Life Cycle

1

A Review of Phenological Patterns of Amphibians and Reptiles in Central Mediterranean Ecoregion

Pierangelo Crucitti
Società Romana di Scienze Naturali, SRSN, Roma, Italy

1. Introduction

Phenology is consistent with different definitions and topics: (i) the study of seasonal phenomena conditioning lifestyle of organisms; (ii) the science of recording and classifying the remarkable occurrence of animal development, particularly ectothermic ones, e.g. insects and lower vertebrates; (iii) the branch of Biology which studies the relationships between climatic factors and various seasonal events of plant and animal lives; the last one being suitable to the problems discussed in this chapter (Lanza et al., 1982; Brandmayr & Ghirardelli, 1995; Speight et al., 2008). Phenological data are represented by climatic factors - seasonal events and related parameters, especially temperature and rainfall - in their correlation with various aspects of animal lifestyles. So, phenology and bioclimatology are somewhat subsidiary subjects. Biological patterns are represented, in their seasonal context, by a remarkable amount of data concerning different, ecological and ethological, items; 1- reproductive activity; i.e. variation in space and time of reproductive success and its relationships with macroclimatological (regional), meso- and microclimatological (local) patterns; 2- population turnover (population size, sex and sub adult/adult ratios); 3- site fidelity and its variation from year to year; 4- foraging behaviour (diet composition; prey - predator relationships and its seasonal variability). Current problems stimulate some basic questions. Beginning from the predictable assumption that weathering events strongly affect daytime and seasonal activities and, to a certain extent, the distribution, both local and worldwide, of amphibians and reptiles, what contribution may arise from the knowledge of phenological patterns to the dramatic problems of the decline of their populations in disparate regions of the world? Recent changes of yearly reproductive phenology together with the modification of species distributional area might be related to, or induced by, the so called "global warming"? Predictive models and related strategies of conservation measures may stimulate the increase of our knowledge in this field of research. Scientific literature suggests that the Mediterranean area as a whole is one of the most sensitive landscape to current climatic modifications (Gatto et al., 2009). Climatic scenario for the near future shows for the Mediterranean area, and particularly for Italy, an increase of heat waves and natural dramatic events such as inundation and / or dryness with a significant loss of marine, freshwater and terrestrial biodiversity. Global Climatic Changes and its impact events on ecosystems and species may be grouped in the following categories: physiology and behaviour; yearly biological cycles; geographical distribution; composition of ecological communities including interactions between species (Nanni et al., 2009). At the global level, the imminent extinction emergency of Amphibia is a tribute to a number of additional threats such us diseases, invasive species, pollution and over-exploitation (Alford & Richards, 1999; Collins & Storfer, 2003; Stuart et al., 2004; Di Rosa et al., 2007).

2. Amphibians and reptiles in the central mediterranean ecoregion

Climate. It appears as a *petitio principi* that a large part of the Mediterranean area is characterized by a typical Mediterranean climate with a dry season generally as long as the summer period when air temperature is higher. However, climatic scenario is much more complicated, e. g. 15 different bioclimatic units are reported for Latium, Central Italy (Blasi, 1994; Blasi & Michetti, 2007). Climatic scenario of the Italian territory shows, at present, an increase of daily, maximum and minimum, temperatures, coupled with the increase of mid temperature and in accordance with the increase of the frequency of heat waves which have trebled in the last fifty years; besides, all the Italian territory is characterized by a significant decrease of feeble rainy days, contrary to the increase of intense rainy days in some regions of northern Italy (Lionello et al., 2009). The Sardinian climate especially due to the position of this island in the Tyrrhenian Sea, overlapping with its typical mountainous territory. However, desertification, as pointed out by aridity and drought indexes, shows for Sardinia an 88.7% of the regional surface (Sicily 95%; Apulia 72%); for the Italian territory, desertification increases 6.5% between 1951 and 1980, at present being about 18% of the whole surface of Italy (Sciortino et al., 2009).

Biodiversity. The Mediterranean Basin, a hyper-hotspot of biodiversity, is the largest among the five Mediterranean systems of the world and includes at least ten mini-hotspot of animal, plant and landscape diversity (Blasi et al., 2007). It is well known that Italy shows an extraordinary richness in species, communities and landscapes, the significance and state of conservation of this biological diversity is self-evident (Bianchi et al., 2002; Minelli et al., 2002). The position of the Central Mediterranean Ecoregion makes this landscape a remarkable reservoir of biodiversity. As a mere example, in the Region of Latium, a landscape of just 17,227 km^2, 5.7% of the Italian political territory, 3228 species and subspecies of vascular plants (2000 only in the Province of Rome) were recorded together with some 70 terrestrial mammals, 23 of which are bats (Chiroptera) out of 34 species of bats of four families belonging to the Italian fauna (Crucitti, 2011; Veith et al., 2011). Amphibians and Reptiles are represented in the Italian territories considered as a biogeographical unit (mainland and peninsular Italy, Sicily, Sardinia, Corsica and related archipelagos; besides, the Nice area, Canton Ticino, Istria, Quarnaro Islands and neighbouring Dalmatian coastal area; finally, the Maltese Islands) by some 100 species and species complex (Bologna et al., 2000; Sindaco et al., 2006; Lanza et al., 2007; Corsetti & Romano, 2008; Bassu et al., 2008, 2010; Corti et al., 2011: Luiselli, in verbis VIII.2011) (Tab. 1). More generally, Sardinia and Corsica, among the largest islands in the Tyrrhenian Sea, are known to harbour a comparatively high number of endemic species among different taxa, both animals and plants (Grill et al., 2007); recent results of researches in *Pipistrellus* and *Hypsugo* cryptic diversity highlight the importance of Sardinia as a major Mediterranean hotspot for microbat biodiversity (Veith et al., 2011).

Area / Taxon	ITALY	Central Italy	Sardinia
AMPHIBIA	43 (20 + 23)	16 (7 + 9)	10 (6 + 4) (8)
REPTILIA	55 (9 + 27 + 19)	24 (7 + 7 + 10)	20 (6 + 9 + 5) (4)

Table 1. Total numbers, outside parentheses, of Amphibia and Reptilia species in the Italian fauna. Amphibia: the first and second number (inside parentheses) indicates species and species complex of Caudata and Anura, respectively. Reptilia: the first, second and third number (inside parentheses) indicate the species of Chelonii, Sauria and Serpentes, respectively. Sardinia: numbers inside the second parentheses indicate the endemic species for each taxon.

3. Current methodological problems

For the goal of this contribution, Central Italy and Sardinia, belonging to the Central Mediterranean Ecoregion, have been selected as paradigmatic areas, a by-product of our specific interest and field experience. Several methods are available to provide suitable data in the study of vertebrate phenology, especially ectothermic ones, commonly related to the problems investigated in a given research. Some remarkable kinds of methodological approaches as emerging from current literature in the topics mentioned in the previous paragraph are uninterruptedly discussed.

i. Field studies on the reproductive phenology during the annual cycle of a given species in relation to its ecology, particularly population dynamics, sex ratio, average growth rate etc. in other words the study of the influence of environmental conditions on adult body size and other biological parameters. In this field of investigation, particularly intriguing are the numerous cases represented by newt populations living at persistent vs. temporary ponds, i.e. in a predictable / unpredictable environments which require different, counterintuitive, life strategies. In this context, experimental drawings generally require the following steps:
   a. the choice of the species;
   b. the choice of the study area and sites with a concise physiographical description;
   c. the collection of data on the (local) climate, e.g. mean of the minimum air temperature and mean rainfall, recorded monthly during the whole period of observations;
   d. the recurrence (monthly, weekly) of catching individuals and correlated methods of recognition such as belly-pattern recorded photographically and toe-clipping marking (less invasive methods are desirable though recent observations agree with the absence of secondary effects provoked by the removal of a single finger from one leg (see Sacchi & Delaugerre, 2011 for the Gekkonidae *Hemidactylus turcicus* (Linnaeus, 1758)) including sex and secondary sexual characters together with biometrical data;
   e. the adoption of suitable indexes to derive the absolute size of population and its dynamic - variation of size through time, generally one year (Accordi et al., 1990; Angelini et al., 2001, 2010 a, b).

ii. Topical issues in Amphibians. The detrimental role of climate change by itself or in association with other anthropogenic stressors - structural modifications of native habitat, pollutant releases, emerging infectious diseases, invasive species, over-exploitation - as the main causes of amphibian populations' decline in Europe and elsewhere, has been pointed out. Evidence of climate-linked amphibian declines in Italy has been highlighted, first adopting for the presence data of Italian amphibians, the CKmap 5.3.8 software, the largest and most authoritative databank of faunistic knowledge of the Italian fauna, 500,000 records of approximately 10,000 terrestrial and freshwater species (Stoch, 2000-2005; Ruffo & Stoch, 2005). The CKmap database contains 12,585 records representing the current knowledge of the distribution in Italy of the 36 amphibian species of the Italian fauna. Species occurrence within any of the 3557 cells of UTM grid, 10x10 km, that intersect the Italian territory is reported; in each cell one single record is reported. Ancillary information, e.g. the year of last observation is provided for each cell. Nineteen species, out of which 12 Anura and 7 Caudata (salamandrids), have been analyzed according to their present status (D'Amen & Bombi, 2009). Environmental data, i.e. climate change, habitat alteration and solar irradiation incidence have been analyzed for the Italian territories in their possible association with amphibians declines. SCIA databank, the National System for collection, elaboration, and dissemination of Climatologic data of environmental interest, provides annual mean values of annual number of dry days, annual cumulate precipitation, and annual mean temperature for the period 1961-1990. Habitat modifications was obtained by calculating the percentage of agricultural and urban land cover classes (Corine Land Cover 1990 data, provided by the Institute for Environmental Protection and Research, ISPRA). Finally, the map of horizontal irradiation available from Joint Research Centre has been utilized as a proxy of geographic differences in solar irradiation incidence relative to the period 1981 - 1990. In order to measure the influence of these three variables on amphibians' decline, auto covariate models on cell status with a binomial probability distribution ("stable" = 0; "decline" = 1) have been fitted together with a logit link utilizing environmental variables as predictors. Results reveal an alarming situation. The status of "decline" was assigned to 17.04% cells, the status of "stable" to 82.96% cells. A large fraction of amphibian species (52.63%) had 15-20% decline cells. Five species with the highest rate of local decline RLD ("decline" / "stable" + "decline"), *Pelobates fuscus* (Laurenti, 1768), *Bombina pachypus* (Bonaparte, 1838), *Discoglossus pictus* Otth, 1837, *Salamandra atra* Laurenti, 1768 and *Triturus carnifex* have RLD 54.35, 37.54, 32.08, 23.33 and 20.31 respectively, while the lowest RLD' values are typical of some, yet relatively common species, the green frogs *Pelophylax* spp. (D'Amen & Bombi, 2009). The complex influence of several factors associated with amphibian population declines has been demonstrated thanks to these results. Important evidence of global warming's impacts - climatic shifts e.g. the decreasing annual precipitation and increasing temperatures during the warmest months - and its interaction with diseases, for example increased susceptibility to chytrid fungi (Simoncelli et al., 2005), have a strong impacts on amphibians populations, in association with habitat loss too. Human land use patterns, particularly changes linked to urbanization, have substantially contributed to amphibian declines in Italy (Andreone & Luiselli, 2000). Finally, irradiation, though probably not a direct cause of mortality, may have an important role in association with other stressors; the increase in sub-lethal exposures of UV-B radiation and its link with the susceptibility to contaminants and pathogens is paradigmatic (D'Amen & Bombi, 2009).

iii. Topical issues in Reptiles. A meaningful example of methodological protocol applied to chelonians, is a recent study on seasonal changes of the diet in Hermann's Tortoises *Testudo hermanni hermanni* Gmelin, 1789, a threatened species (NT according to the Red List of IUCN). *T. h. hermanni* is a relatively small (800-1000 g on average) terrestrial turtles of the family Testudinidae, endemic to Southern Europe. Terrestrial chelonians mostly belonging to the widespread family Testudinidae display a generalized herbivorous or omnivorous feeding habits contrary to most terrestrial vertebrate ectotherms which are exclusively or predominantly carnivorous. Exactly, 66% of terrestrial chelonians are strictly herbivorous, 33% are omnivorous (but many are also primarily herbivorous) and only one species, *Terrapene carolina* (Linnaeus, 1758), is predominantly carnivorous (Luiselli, 2006 a, b). *T. h. hermanni* occupies a wide range of dry sites along coastal areas of Western Central Italy where it is markedly linked to maquis habitat. Phylopatry together with a detailed knowledge of their home ranges based on vegetation and other landmarks is a peculiar trait of their life history. To understand the habitat requirements is strategic, since this tortoise is declining in most part of its range. At least 132 plant species belonging to 46 families, mainly Asteraceae and Fabaceae, are eaten by Hermann's Tortoise (Cheylan et al., 2011). Field studies have been conducted in the coastal pine forest of Castelfusano near Roma, Central Italy, with the aid of faecal-pellet analysis to determine intersexual and interseasonal variation in the diet of this free-ranging population. A third goal of this study is the identification of the most important food plants for this species, discovering any plant species that might be rare but important diet items, their relative abundance being critical to tortoise conservation. The diet of *T. h. hermanni* has been quantified by analyzing faecal pellets collected during spring and autumn. Tortoises typically defecated during capture and manipulation; on the contrary, individuals may be placed in small cages without any vegetation until defecation occurs, up to 48 h later in some instances; after, they were released at their capture points. The analysis of faecal pellets, placed in 50% ethanol-water solution and stored in plastic vials, is described in detail by Del Vecchio et al. (2011). At the same time, plant sampling at the study area has been performed from April to June, and the relative abundance of each plant species has been estimated. Leaves were the most commonly consumed plant structure contrary to seeds and fruits (Del Vecchio et al., 2011). During spring the most important families were Fabaceae, Asteraceae and Poaceae, in autumn Ruscaceae, Fabaceae and Asteraceae; at least 21 plant species were found in males faecal pellets and 18 in females, the composition of the diets in the two sexes being similar both qualitatively and quantitatively. Interseasonal differences in diets reveal interesting pattern; diet composition differs remarkably among seasons, notwithstanding the quantitative compositions of the diet, as revealed by the frequency of consumption of the various food items, were not significantly different between spring and autumn. *Carduus, Medicago, Melilotus, Trifolium* with some undetermined species of Leguminosae were consumed in spring but not in autumn, contrary to *Ruscus* which was consumed during autumn but never during springtime. Abundances of food availability in the field were strongly uneven, some taxa are extremely abundant e.g. *Coronilla emerus, Ruscus aculeatus* and *Smilax aspera* and others extremely rare, e.g. *Rosa sempervirens* and some Papaveraceae. Five main patterns have been focused; 1 - in the study area, *T. h. hermanni* is entirely herbivorous; 2 - during spring, leaves are consumed more than flowers or fruits; 3 - the sexes do not differ significantly in dietary composition; 4 - dietary composition of spring has been found significantly different from that of autumn, thus indicating a seasonal change in diet (fruits of some species, e.g. *Arbutus unedo,* are available in November-December, too late in the season); 5 - tortoises are mostly generalist, the relative abundance of the eaten species is generally higher than that of the species that were not eaten, the foraging strategy of *T. hermanni* is a mixture between a generalist and a specialist forager. Plants which are preferred because they provide important nutrients or other dietary needs should be carefully managed for the maintenance of this declining species (Del Vecchio et al., 2011). More generally, the genus *Testudo* is a group of conservation concern - highly vulnerable as a result of habitat destruction, commercial exploitation and overharvesting for food.

An other example of appropriate methodological protocol has been recently applied to the endangered species *Hemorrhois* (= *Coluber*) *hippocrepis* (Linnaeus, 1758), the horseshoe whip snake, a medium sized (up to 180 cm in length) quick-running terrestrial Colubridae, which actively forages on lizards and mice. The Italian distribution of this snake is limited to Pantelleria island, south of Sicily, and Sardinia, nearly only in the south-western portion of the island. The morphology of the specimens from both the Pantelleria and Sardinia was found to be different from that of mainland conspecific populations from Iberian Peninsula and North Africa, suggesting their indigenous origin (Cattaneo, 1985; Corti et al., 2000; Luiselli et al., 2010). In Sardinia, *H. hippocrepis* is so rare and localized that this snake was considered to be extinct in 1960. Notwithstanding its occurrence on the island was repeatedly confirmed in recent years, the Sardinian populations have been regarded among the most critically endangered in Europe, this species representing an absolute priority for biodiversity conservation in Italy, low population density, competition with the western whip snake *Hierophis* (= *Coluber*) *viridiflavus* (Lacépède, 1789) and fragmentation of available habitats representing the major threats to its survival (Corti et al., 2000; Filippi & Luiselli, 2000; Luiselli et al., 2010). A goal of pressing realization is planning helpful instruments for the short- and long-term conservation of *H. hippocrepis* in Sardinia. First, habitat suitability of the horseshoe whip snake in Sardinia under current environmental conditions has been modelled; the species ecological requirements is projected in future climatic scenarios in order to evaluate the probability of persistence of the species in the study area. Presence data have been derived from CKmap databank. Data indicates the occurrence of the species within the UTM, 10 x 10 km, grid. Downscaling coarse resolution atlas data is necessary for gaining the spatial accuracy required by conservation initiatives, so 1000 random points all across the study area have been generated and a binary code (0/1) has been assigned to the random points according to their relative location, outside (0) or within (1) the range of *H. hippocrepis*. Environmental predictors - land cover, land morphology and human impact layers - in addition to climatic predictors, have been considered in order to take into account as many aspects of the species environmental requirements as possible. Two different time slices, 2020 and 2050, were utilized for projecting habitat suitability into future climate and, in addition, percentage of 5 different Corine Land Cover classes in 30 arcsec squares were calculated. Land morphology descriptors were represented by using four variables such as elevation, slope, northness and eastness. The Human Impact Index for representing the human impact on environment with a spatial resolution of 1 km, has been utilized. Finally, five different methods, CTA, GBM, GLM, MARS and RF for modelling habitat suitability have been performed; the models fitted on current conditions were projected into future climatic scenarios. Four retaining models agree in considering suitable, under present conditions, large areas of south-western Sardinia, nevertheless they evidenced an alarming trend for *H. hippocrepis* conservation. All but four squares currently occupied by the snake are expected to lose completely their suitability within 2020 and other two squares became totally unsuitable within 2050. The consensus maps indicate a reduction of 86.2% in suitable surface by 2020 and a collapse of 99.2% by 2050, i.e. to survive in one protected area only, disappearing completely from the other protected areas of the Province of Cagliari. Results demonstrate that climate change must be considered one of the most immediately threatening factor. The only area that is expected to maintain suitable conditions for *H. hippocrepis* under future climate change must be considered as a crucial refuge; this is the case of the site of Community Importance named "Stagno di Cagliari, Saline di Macchiareddu, Laguna di Santa Gilla" although in this site the species is uncommon - especially with respect to the sinthopic one *Hierophis viridiflavus*. Most of the sites inhabited by *H. hippocrepis* are strongly anthropized - sea-side tourism areas with busy roads. However, connectivity among source areas is crucial for conservation of this snake. According to this view the systems of natural corridors connecting various sites of actual presence of the species should be implemented and constantly monitored together with restoration of microhabitats and construction of drinking places and laying sites, strategically important for guarantying the persistence of this attractive species (Bombi et al., 2011).

4. Some remarkable case studies in amphibia

A selection of recent Italian literature on this topic is provided. The genus *Euproctus* includes, at present, two species, *E. montanus* (Savi, 1838), a Corsican endemism, and *E. platycephalus* (Gravenhorst, 1829), a Sardinian endemism. Suitable habitats of these newts are the cool and calm waters of streams and rivers (Sardinian species; ponds and pools too) of hilly and mountainous areas. Adults life-cycles is generally characterized by the existence of terrestrial and aquatic phases; the Corsican species may be found in water from March to September, the Sardinian during summer. Contrary to *E. montanus*, the present distributional area of *E. platicephalus* is extremely reduced with respect to the past. *E. platicephalus* is considered as endangered (EN) based on IUCN categories; according to some authors it is, among European newts, the most endangered species. So, this species is a priority for biodiversity conservation in Italy. One of the major threats to the survival of its populations is considered the strong dryness of the last decades and the fragmentation / disappearance of streams, especially during summer. Different altitudinal and optimum ranges of both species indirectly suggests for the Sardinian species a problematic destiny concerning its survival (Andreone & Luiselli, 2000; Lecis, 2004, 2007) (Tab. 2). Serious consequences for conservation and management often came out from the overlapping between anthropogenic factors and eco-ethological constraints of species and environment.

Shifting of habitat suitability with climate change could led to population declines for species that are unable to move to new appropriate conditions (Thomas et al., 2004; Araújo et al., 2006; D'Amen & Bombi, 2009).

Species	Total altitudinal range	Optimal altitudinal range
Euproctus platicephalus	50 - 1800	400 - 800
Euproctus montanus	15 - 2200	700 - 1900

Table 2. Total and optimal altitudinal range (m a.s.l.) for each species of *Euproctus* of the Sardinian - Corsican Massif

Salamandra salamandra (Linnaeus, 1758) is the largest Italian newt with a body length up to 28 cm in Southern Europe. This species occurs mainly in hilly and mountainous areas of the Alpine Arch and the Apennine system, from about the sea level to 1800 m a.s.l., preferably between piedmont areas to about 1000 m a.s.l. with a relatively unbroken distribution in mainland Italy - with the exception for the Po Valley - and Calabria; its scattered distribution along the remnants of Italian territories, especially the Central Apennine, is a debated question (see for instance Caldonazzi et al., 2007). *S. salamandra* occurs mostly in beech, chestnut and oak woods. Adults, strictly terrestrial, are active during the whole year under favourable meteorological conditions. The distribution of *S. salamandra* along the Northern and Central Apennines corroborates the hypothesis of a strong dependence from rainfall, mostly mesophilous sites are inside the area with over 1,000 mm / year of precipitation (Tedaldi, 2001). The species shows a bimodal pattern of activity, courtship and breeding occurs on land during spring and autumn time. Female sperm storage lasts over one year. Births occur during April and larvae of unusual size are lied down in still water of mountain streams; larvae are found at water up to mid October when metamorphosis occurs (Caldonazzi et al., 2007). *Salamandrina* is an endemic genus of peninsular Italy. Two species are acknowledged since 2005, *S. perspicillata* (Savi, 1821) of the Central and Northern Apennine, and *S. terdigitata* (Lacépède, 1788) of the Southern Apennine - a pattern of parapatrical distribution (Lanza et al., 2009). *S. perspicillata* shows a relative continuous distribution from the Genoa Province to the whole Latium, from the sea level to over 1900 m a.s.l. Mesophilous mixed woods and Mediterranean scrub together with caves, debris, pastures, garigues and reforestation areas represents suitable habitats for this, essentially terrestrial, species. Generally, only adult females are found inside water - low running streams or similar bodies, drinking through, stone wells, entrance of caves (Crucitti et al., 2010). Eggs deposition during the aquatic phase is few days vs. few weeks long. A bimodal activity period has been ascertained in Latium in accordance to oviposition time; from September at the onset of January and from February to April, spending unfavourable period inside hypogeal environment. Daily terrestrial activity, during day or night-time, is between sunset and midnight. During light time, activity occurs in moist days and, in this case, shadow sites with litter are preferred (Angelini et al., 2007). *Lissotriton italicus* (Peracca, 1898) and *Lissotriton vulgaris* (Linnaeus, 1758), generally known as "water salamanders", are small newts living, often sintopically, in drinking through, irrigation tanks, pools and ponds of hilly and mountainous areas of Central Italy. Reproductive activity of *L. italicus* is observed mostly between February and May, courtship and eggs deposition during autumn too. This species shows a typical yearly cycle with multiple pattern of four phases - aquatic, emigration, terrestrial, immigration. The reproductive activity is especially concentrated between December and February. Between the end of September and the beginning of November, the majority of newts are still terrestrial; in this period males first enter water, where courtship began. Larvae appear during January and their development is completed before the beginning of terrestrial phase. Variation of spermatogenetic pattern occurs during the whole cycle, for example, during the aquatic phase testis are evident, contrary to the terrestrial phase. In summer, newts are active during the first hours of the day and at sunset, when the soil is moist. Diet is extremely varied including both terrestrial insects, especially at larval stage, and aquatic invertebrates, woodlice of the genus *Asellus* are included in mountain ponds (Scillitani & Tripepi, 2007). Many studies have been carried out on the ecology of *L. vulgaris* especially in Northern and Central Europe. In the last twenty years some studies have focused on the ecology of the Italian species *Lissotriton meridionalis* (Boulenger, 1882). The present distribution of Italian newts have been influenced by historical factors coupled with climatic conditions such as temperature and rainfall (Giacoma, 1988). The annual cycle of a population of this subspecies has been studied at a pond, close to the southern limit of its distributional area, inside the "President's Estate of Castelporziano", few kms south of Rome, one of the last areas of the Tyrrhenian Coast still covered by floodplain forest. A particularly dried climate is typical of Castelporziano and the long summer drought causes ponds to dry up completely. Average water level fluctuates yearly, since water is of rainfall origin. Timing of migration differs significantly from that described in other countries, the Central Italy population is present at water from November to May, much earlier in comparison with Northern populations of *L. vulgaris* - in England immigration starts in February-March and emigration lasts from July to December. The aquatic period is short compared to that of northern populations and activity during the summer terrestrial phase is scanty. Temperature has a strong influence on the annual reproductive cycle, different water seasons have also been observed in Italy according to latitude and altitude of the breeding sites. Immigration and reproduction take place in December, as soon as weather conditions are favourable. Emigration lasts from April-May, ending before pond desiccation in June. Therefore, summer drought is a limiting factor especially for larval survival. The sex ratio is unbalanced because in particularly dry conditions few females may reach the pond; males, which are at ponds earlier than females, often exceed in breeding populations. The mean body size registered at Castelporziano is the smallest among populations of Central and Northern Italy owing to unfavourable conditions, typical of coastal and xeric habitats; food consumption is irregular especially

Phases / **Species**	**AQUATIC**	**EMIGRATION**	**TERRESTRIAL**	**IMMIGRATION**
Lissotriton italicus	October - May	June	Summer	September - November
Lissotriton meridionalis	November - April	April - May	Summer	December

Table 3. A prospect of time - dependent phases in the lifecycle of two species of *Lissotriton* from Central Italy

during the terrestrial juvenile phase. Fecundity is size-specific, larger females of *L. meridionalis* lay more and larger eggs that hatch earlier (Accordi et al., 1990). Nearly similar patterns have been observed in a population of *L. meridionalis* living at temporary ponds in the "Bosco di Palo", a WWF Oasis near Rome, characterized by a meso-mediterranean climate; mild winter, summer dryness three months long coupled with abundant rainfall typical of maritime climate. Mediterranean climate allows to anticipate the eggs layering in late fall. Concentration of newts at the water reaches its maximum between February and April. Activity pattern during aquatic phase has been ascertained with a maximum of 14-17 °C and a minimum of 3-6°C (Pizzuti Piccoli, 2006/07) (Tab. 3).

5. Some remarkable case studies in reptilia

A selection of recent Italian literature on this topic is provided. Among Gekkonidae, four species in Italy, *Euleptes europaea* (Gené, 1839) is considered Near Threatened (NT), particularly the micro insular or satellite populations of Sardinian and Corsican Archipelagos. This species shows a nocturnal activity pattern, store up on heat during the day better than other species of the family living in the same Corsican habitat, *Tarentola mauritanica* (Linnaeus, 1758) and *Hemidactylus turcicus* (Salvidio et al., 2011). *Tarentola mauritanica,* one of the commonest species of the family, is active in the first hours of the day, during the hottest months and, in the central hours, during the coldest one. Thermoregulatory ability has been developed both on islands and mainland - in this last situation, a noticeable density of its predators, typically snakes, has been observed. Foraging and thermoregulatory activities are carried out on rocky substrata. On the island of Lampedusa (Sicily) and during summer, body temperature is positively correlated with air temperature. However, a decrease of nocturnal activity has been observed after midnight; during the last hour of the day, body temperature shows an increase of about 2° C, decreasing in the same way at the beginning of nighttimes. With air temperature below 15° C, the activity is low or scanty, however coastal populations of Tuscany (Central Italy) are active with air temperatures 8-9° C and rocky substratum 10° C or just upper. Hibernation is not evident in the populations of coastal areas of Sicily, their activity is displayed during the whole year (Aprea et al., 2011). Like Gekkonidae, Scincidae is a family of tropical / subtropical origin; only three species occur in Italy, *Chalcides chalcides* (Linnaeus, 1758) being the widespread one. This species is active only during the day both in spring and summer; a bimodal pattern of activity has been ascertained, during the central hours of the day and the first hours of the afternoon. Aestivation occurs in August. Active lizards have been observed during winter month too, hot days raging (Caputo et al., 2011 a). Similarly, *Chalcides ocellatus* (Forskål, 1775) is active from April to October, especially from April to June with a peak of activity during May. Circadian activity pattern during this period appears unimodal, showing a preference for the central and hottest hours of the day; during summer it becomes bimodal, being concentrated during the first hours of the morning and in late afternoon. Activity pattern becomes again unimodal with a peak during the central time of the day in September-October. As a rule, during the second half of October, when thermal fluctuations are especially marked, *C. ocellatus* is at rest at their hibernation roosts. Populations of coastal sites show scattered winter activity under particularly favourable meteorological conditions. In inland and mountainous areas of Sicily, the phase of dormancy lasts up to March. Reproductive season, together with a marked activity pattern, falls between April and May; spermatogenesis is seasonal, spermatogonial proliferation occurs in August and spermiohystogenesis in April-May shortly before mating. Gravid females are found in Sicily beginning from the second half of May. In southern sites, two or three yearly reproductive cycles may be performed (Caputo et al., 2011 b). *Archaeolacerta bedriagae* (Camerano, 1885) is a lizard (Lacertidae) restricted in its distribution to eastern Sardinia and Corsica and considered as "Near Threatened", the extension of crystalline rocky outcrop to which its populations are linked, being strategic for its survival; shelters at risk are satellite islands or small inland areas. Daily activity has been studied in the populations of Corsica between 1750 and 1800 m a.s.l. (however *A. bedriagae* is not exclusively a mountain species as regarded till up to thirty years ago). Activity is marked especially during spring time, being constant between 7:00 and 16:00-16:30 hours at sunset. Males explain a bimodal pattern of activity with the exception for the central hours of the day contrary to adult females which are active in the central hours of the day and at afternoon. The species has been observed on the sunny surface of the rocks; thermoregulation occurs in the hottest hours of the day too, lizards get out at their shelters with clouding over within an hour from passing sun. Average activity temperature is higher in *A. bedriagae* contrary to *Podarcis tiliguerta* (Gmelin, 1789) - the commonest and relatively ubiquitous lizard of the sardo-corso district - living sintopically in the same sites (Sindaco et al., 2011). *Coronella girondica* (Daudin, 1803) is a fine, small, thermo-xerophilous and oviparous snake of the family Colubridae with a Western-Mediterranean distributional range. In Italy, it is a rare species with the exception of few and relatively large areas such as Piedmont, Liguria and Tuscany; besides, many isolated European populations are at higher risk of extinction within few years. This species shows a remarkable ability pattern of adaptation in mediterranean / oceanic bioclimatic regions where it lives on stone areas of rocky slopes. Activity period begin at mid March or at the beginning of April lasting up to September in Central Italy. Males latency period is two weeks earlier than females, on the contrary hibernation is practically simultaneous. *C. girondica* shows an unimodal yearly cycle of activity contrary to many other species which live under Mediterranean bioclimatic conditions. This unusual phenological pattern may be related with the peculiar twilight and nocturnal daily activity of the species (Razzetti & Bernini, 2011). *Zamenis situla* (Linnaeus, 1758), is a mid sized snake (60 - 100 cm) with a distributional Italian area confined to the south east of Apulia, Eastern Sicily and the Maltese Islands, few sites in Basilicata near the Apulia border are also known. Open areas with scrub vegetation and sparse rock surface are elective biotopes, the conservation and proper management of these ecosystems being strategical for the survival of *Z. situla*, considered as LC species according to IUCN categories. As a rule, the activity period extends from March to November; occasionally, individuals have been observed outside their shelters in wintertime during sunny and hot days - winter latency generally lasts from December to February. Activity pattern of this strictly daytime snakes - however some Authors suggest a nocturnal activity during hottest months too - is explained by two peaks, during February - March and during June, linked to breeding behaviour, which lasts up to May, and intense foraging behaviour, respectively. Newborns emerge between the end of August and September. A marked circadian activity rhythm may be observed between 7 to 10 hours during May and September and between 6 to 11 hours during June (Scillitani & Turrisi, 2011). *Vipera ursinii* (Bonaparte, 1835) *sensu lato* has a wide geographic distribution that extends from Southern France to Eastern Asia, being part of a complex of several species or semi species whose origin goes back to the Pliocene (Cheylan et al., 2011). A recent revision changes substantially the validity of many taxa of this polytypic group. Particularly, among the five subspecies up to now considered valid, the nominal one, *V. u. ursinii*, is exclusive of some areas of the Central Apennines and some mountainous areas of south-eastern France (Nilson & Andrén, 2001; Filippi et al., 2011). Orsini's viper is the smallest viper in Europe and the smallest snakes of Italy and France, adults are 40-45 cm long, a very large adult female may reach 50 cm. This snake is associated with open areas such as grasslands and steppes. In the Central Apennines it is almost exclusively linked to mountain areas characterized by the outcrop of fissured limestone on which extensive formations of *Juniperus nana* and *Pinus mugo* - a bearing prostrate shrub - groves. The influence of seasonal conditions of the habitat in which its populations evolve is marked. Phenological pattern are correlated with the existence of two distinct activity phases, a long latency period in autumn-winter and an activity period between late spring and the end of summer. The species shows the shortest cycle of activity among Italian snakes. Males first are vagrant not before the end of April, followed by females two or three weeks after and adults came back inside their winter shelters between mid September - mid October. Winter spermatogenesis is typical of males. Foraging activity falls between June and September, mating activity during the second half of May; ovulation falls at the end of June, birth during late summer. Being ovoviviparous, *V. ursinii* give birth perfectly developed and self-sufficient newborns, 15 cm long. Unfortunately, this small, attractive, viper is a priority for biodiversity conservation in Italy and the whole Europe. Many Apennine populations are more or less critically endangered owing to the following risk factors; 1- mountain overgrazing of domestic and semi domestic cattle; 2- high population density of some ungulate, particularly the wild boar *Sus scrofa* Linnaeus, 1758; 3- intentional killing by man or provoked by car traffic at highest altitude; 4- modification and / or destruction of natural habitats (Cheylan et al., 2011; Filippi et al., 2011).

6. Conclusion

To sum up, it is hardly feasible to overestimate the influence of temperature on reproductive cycle and migration of Italian (and European) amphibians, especially newts, and reptiles. Global Climatic Changes - including high levels of incident solar radiation- in unison with habitat alterations are responsible for the biodiversity crisis of Italian (and European) populations of Amphibia. The following steps attract reader's attention on the landmarks up to now focalized. Ecological modelling approaches to the change of habitat suitability is a powerful tool in the study of short and long-term conservation of endangered species. An up-to-date knowledge of the present distribution of a given species in a certain territory and climatic scenario is required. Such examples are numerous in the various countries; for instance, in Italy, the databank is linked to the so-called CKmap' initiative which is supported financially by the Italian Ministry for the Environment, Land and Sea Protection (Blasi et al., 2007). Environmental factors related to species occurrence and present suitable areas, clearly represent the rudiments of this approach. Further on, both land cover and morphology together with human impact layers are considered as covariates to climatic predictors. Different methods for modelling habitat suitability and binary predictions (presence / absence of the species) are performed. Models may be applied for some time slices, present and future - at selected intervals, e.g. 2030, 2050, 2070. Results indicate which squares of the CKmap loose completely their suitability within a certain period. Present conditions with currently known distribution and different (hypothetical) climatic scenarios and environmental changes may be compared. Existing controversies in the field of climatic (and related) phenomena are pointed out. The efficiency of predictions is influenced by many factors; the simplistic assumption that some factors remain unchanged drives to under / over estimation of the persistence probability of the species under the spotlight. A dramatic perspective is the complete extinction from the Central Mediterranean Ecoregion of *H. hippocrepis* during the next decades; immediate actions for habitat conservation should be organized by the pertinent authorities (Corti et al., 2000). However, the near future will reserve to mankind enormous challenges in the light of supporting and preserving both geodiversity and biodiversity of the Mediterranean Ecoregion.

7. Acknowledgments

The Author is indebted to all those provide useful suggestions, first with Luca Luiselli for the critical review of the manuscript and Giovanni Amori which provides bibliographical sources. Francesca Santamaria improves the English style. Peoples of the Società Romana di Scienze Naturali provide helpful assistance, particularly Chiara Bufalieri and Federica Emiliani. Finally, the Author is really thankful to anonymous reviewers for their extremely careful comments.

8. References

Accordi, F., Massarek, A. & Nobili, G. (1990). Ecological responses in a population of smooth newts (*Triturus vulgaris meridionalis*) in an unpredictable environment. *Herpetological Journal*, Vol. 1, pp. 509-513.

Alford, R. A. & Richards, S. J. (1999). Global amphibian declines: a problem in applied ecology. *Annual Review of Ecology and Systematic*, Vol. 30, pp. 133 - 165.

Andreone, F. & Luiselli, L. (2000). The Italian batrachofauna and its conservation status: a statistical assessment. *Biological Conservation*, Vol. 96, pp. 197-208.

Angelini, C., Antonelli, D. & Utzeri, C. (2001). Aspetti della fenologia riproduttiva di *Salamandrina terdigitata* (Lacépède, 1788). *Pianura*, Vol. 13, pp. 105-108.

Angelini, C., Vanni, S. & Vignoli L. (2007). *Salamandrina terdigitata* (Bonnaterre, 1789) - *Salamandrina perspicillata* (Savi, 1821), pp. 228-237. In: Lanza, B., Andreone, F., Bologna, M. A., Corti, C. & Razzetti, E. (eds.) (2007). *Fauna d'Italia. XLII. Amphibia.*, "Calderini - Edizioni Calderini de Il Sole 24 ORE Editoria Specializzata S. r. l." Bologna, ISBN -978-88-506-5256-3.

Angelini, C., Antonelli, D. & Utzeri, C. (2010 a). Capture-mark-recapture analysis reveal survival correlates in *Salamandrina perspicillata, Amphibia-Reptilia*, Vol. 31, pp. 21-26.

Angelini, C., Cari, B. & Utzeri, C. (2010 b). Studio della dinamica di una popolazione di *Lissotriton italicus* dei Monti Ausoni (Lazio), pp. 195-198. In: Di Tizio, L., Di Cerbo, A. R., Di Francesco, N. & Cameli, A. (eds.) (2010). Atti VIII Congresso Nazionale Societas Herpetologica Italica, Chieti 22-26 settembre 2010, Ianieri Edizioni, Pescara.

Aprea, G., Lo Cascio, P., Corti, C. & Zuffi, M.A.L. (2011). *Tarentola mauritanica* (Linnaeus, 1758), pp. 277-285. In: Corti, C.; Capula, M. Luiselli, L. Razzetti, E. & Sindaco, R. (2011). *Fauna d'Italia. XLV. Reptilia,* "Calderini - Edizioni Calderini de Il Sole 24 ORE S.p.A." Milano, ISBN -978-88-506-5390-4.

Araújo, M.B., Thuiller, W. & Pearson, R.G. (2006). Climate warming and the decline of amphibians and reptiles in Europe. *Journal of Biogeography,* Vol. 33, pp. 1712-1728.

Bassu, L., Nulchis, V., Satta, M. G., Fresi, C. & Corti, C. (2008). Atlas of amphibians and reptiles of Sardinia - state of the art and general considerations, pp. 52-58. In: Corti C. (ed.) (2008). *Herpetologia Sardiniae.* Societas Herpetologica Italica / Edizioni Belvedere, Latina "le scienze" (8).

Bassu, L., Nulchis, V., Satta, M. G., Fresi, C. & Corti, C. (2010), Anfibi e Rettili di Sardegna II, brevi considerazioni sulla loro distribuzione, pp. 19-26. In: Di Tizio, L., Di Cerbo, A. R., Di Francesco, N. & Cameli, A. (eds.) (2010). Atti VIII Congresso Nazionale Societas Herpetologica Italica, Chieti 22-26 settembre 2010, Ianieri Edizioni, Pescara.

Bianchi, C. N., Boero, F., Fraschetti, S. & Morri, C. (2002). La fauna del Mediterraneo, pp. 248-335. In: Minelli, A., Chemini, C., Argano, R. & Ruffo, S. (eds.) (2002). *La fauna in Italia.* Touring Editore, Milano e Ministero dell'Ambiente e della Tutela del Territorio, Roma.

Blasi, C. (1994). Fitoclimatologia del Lazio. Università "La Sapienza" e Regione Lazio, Assessorato Agricoltura-Foreste Caccia e Pesca, Usi Civici.

Blasi, C. & Michetti, L. (2007). Biodiversity and Climate, pp. 57-66. In: Blasi, C. (ed. in chief), Boitani, L., La Posta, S., Manes, F. & Marchetti, M. (2007). Biodiversity in Italy. Contribution to the National Biodiversity Strategy. Palombi & Partner s. r. l. , Roma, ISBN 978-88- 6060-041-7.

Blasi, C. (ed. in chief), Boitani, L., La Posta, S., Manes, F. & Marchetti, M. (2007). *Biodiversity in Italy. Contribution to the National Biodiversity Strategy.* Palombi & Partner s. r. l. , Roma, ISBN 978-88- 6060-041-7.

Bologna, M. A., Capula, M. & Carpaneto, G. M. (eds.) (2000). *Anfibi e Rettili del Lazio.* Fratelli Palombi Editori, Roma.

Bombi, P., Capula, M., D'Amen, M. & Luiselli, L. (2011). Climate change threatens the survival of highly endangered Sardinian populations of the snake *Hemorrhois hippocrepis. Animal Biology,* Vol. 61, pp. 239-248.

Brandmayr, P. & Ghirardelli, E. (1995). *Gli animali e l'ambiente: ecologia,* pp. 469 - 603. In: Baccetti, B., Baldaccini, N. E., Bedini, C., Brandmayr, P., Capanna, E., Chieffi, G., Cobolli, M., Ferraguti, M., Ghirardelli, E., Ghiretti, F., Giusti, F., Grigolo, A., Mainardi, D., Minelli, A., Papi, F., Parrinello, N., Ricci, N., Ruffo, S., Sarà, M., Scali, V. & Zullini, A. (1985). *Zoologia. Trattato Italiano.* Zanichelli Editore S.p.A., Bologna.

Caldonazzi, M., Nistri, A. & Tripepi, S. (2007). *Salamandra salamandra* (Linnaeus, 1758), pp. 221-227. In: Lanza, B., Andreone, F., Bologna, M. A., Corti, C. & Razzetti, E. (eds.) (2007). *Fauna d'Italia. XLII. Amphibia.,* "Calderini - Edizioni Calderini de Il Sole 24 ORE Editoria Specializzata S. r. l.." Bologna, ISBN -978-88-506-5256-3.

Caputo, V., Guarino, F.M. & Giovannotti, M. (2011 a). *Chalcides chalcides* (Linnaeus, 1758), pp. 298-310. In: Corti, C.; Capula, M. Luiselli, L. Razzetti, E. & Sindaco, R. (2011). *Fauna d'Italia. XLV. Reptilia,* "Calderini - Edizioni Calderini de Il Sole 24 ORE S.p.A." Milano, ISBN -978-88-506-5390-4.

Caputo, V., Lo Cascio, P. Turrisi, G.F. & Vaccaro, A. (2011 b). *Chalcides ocellatus* (Forskål, 1775), pp. 310-322. Corti, C.; Capula, M. Luiselli, L. Razzetti, E. & Sindaco, R. (2011). *Fauna d'Italia. XLV. Reptilia,* "Calderini - Edizioni Calderini de Il Sole 24 ORE S.p.A." Milano, ISBN -978-88-506-5390-4.

Cattaneo, A. (1985). Il colubro ferro di cavallo dell'isola di Pantelleria: *Coluber hippocrepis nigrescens* subsp. nova (Reptilia Squamata Colubridae). *Atti della Società italiana di Scienze naturali e del Museo civico di Storia naturale di Milano,* Vol. 126: 165-184.

Cheylan, M., Corti, C., Carpaneto, G.M., Mazzotti, S. & Zuffi, M.A.L. (2011). *Testudo hermanni* Gmelin, 1789, pp. 188-199. In: Corti, C.; Capula, M. Luiselli, L. Razzetti, E. & Sindaco, R. (2011). *Fauna d'Italia. XLV. Reptilia,* "Calderini - Edizioni Calderini de Il Sole 24 ORE S.p.A." Milano, ISBN -978-88-506-5390-4.

Cheylan, M., Croquet, V., Dragone, C., Ferchaud, A-L. Garcia, A., Lisse, H., Lyet, A., Reboul, D. & Reyna K. (2011). Technical guide to manage and monitor populations of Orsini's viper. ARPE-Corinne Dragone; CEEP-Hèlène Lisse.

Collins, J. P. & Storfer, A. (2003). Global amphibian declines: sorting the hypotheses. *Diversity and Distributions,* Vol. 9, pp. 89 -98.

Corsetti, L. & Romano, A. (2008). On the occurrence of the Italian Aesculapian snake, *Zamenis lineatus* (Camerano, 1891), in Latium (Central Italy). *Acta Herpetologica,* Vol. 3, pp. 179-183.

Corti, C., Luiselli, L., Filippi, E. & Capula, M. (2000). Distribution, natural history and morphometrics of the critically endangered *Coluber hippocrepis* populations of Sardinia: a review, with additional data and conservation implications. *Amphibia-Reptilia,* Vol. 21, pp. 279-287.

Corti, C., Capula, M. Luiselli, L. Razzetti, E. & Sindaco, R. (2011). *Fauna d'Italia. XLV. Reptilia,* "Calderini - Edizioni Calderini de Il Sole 24 ORE S.p.A." Milano, ISBN -978-88-506-5390-4.

Crucitti, P. (2011). Bats of Latium: A review of past and recent studies and the contribution of B. Lanza. *Hystrix It. J. Mamm.* (n. s.), Vol. 22, pp. 15-22.

Crucitti, P, Brocchieri, D, Emiliani, F., Malori, M., Pernice, S., Tringali, L. & Welby, C. (2010). Amphibians of the Simbruini Mountains (Latium, Central Italy). *Acta Herpetologica,* Vol. 5, pp. 91-101.

D'Amen, M. & Bombi, P. (2009). Global warming and biodiversity: Evidence of climate-linked amphibian declines in Italy. *Biological Conservation,* Vol. 142, pp. 3060-3067.

Del Vecchio, S., Burke, R.L., Rugiero, L., Capula, M. & Luiselli, L. (2011). Seasonal changes in the diet of *Testudo hermanni hermanni* in Central Italy. *Herpetologica,* Vol. 67, pp. 236-249.

Di Rosa, I., Simoncelli, F., Fagotti, A. & Pascolini, R. (2007). The proximate cause of frog decline ?. *Nature,* Vol. 447, E4-E5.

Filippi, E. & Luiselli, L. (2000). Status of the Italian snake-fauna and assessment for conservation threats. *Biological Conservation,* Vol. 93, pp. 219-225.

Filippi, E., Capula, M., Luiselli, L. & Rugiero, L. (2011). *Vipera ursinii* (Bonaparte, 1835), pp. 623-627. In: Corti, C., Capula, M. Luiselli, L. Razzetti, E. & Sindaco, R. (2011). *Fauna d'Italia. XLV. Reptilia,* "Calderini - Edizioni Calderini de Il Sole 24 ORE S.p.A." Milano, ISBN -978-88-506-5390-4.

Gatto, M., Zingone, A., Fiorese, G. & De Leo G. A. (2009). La biodiversità nell'era dei cambiamenti climatici: un'eredità da salvaguardare, pp. 295 – 335, In: Castellari, S. & Artale, V. (Ed.) (2009). *I cambiamenti climatici in Italia: evidenze, vulnerabilità e impatti*. Bononia University Press, Bologna. ISBN: 978-88-7395-484-2.

Giacoma, C. (1988). The ecology and distribution of newts in Italy. *Ann. Ist. Mus. Zool., Univ. Napoli*, Vol. 26, pp. 49-84.

Grill, A., Casula, P., Lecis, R. & Menken, S. (2007). Endemism in Sardinia, pp. 273-296. In: Weiss, S. & Ferrand, N. (eds.) (2007). *Phylogeography of southern refugia*. Springer, Dordrecht.

Lanza, B.; Azzaroli, M. L., Borri, M., Poggesi M. & Vanni, S. (1982). *Dizionario del Regno Animale*. Arnoldo Mondadori Editore S.p.A., Milano.

Lanza, B.; Andreone, F., Bologna, M. A., Corti, C. & Razzetti, E. (eds.) (2007). *Fauna d'Italia. XLII. Amphibia.*, "Calderini - Edizioni Calderini de Il Sole 24 ORE Editoria Specializzata S. r. l.." Bologna, ISBN -978-88-506-5256-3.

Lanza, B.; Nistri, A. & Vanni, S. (2009). *Anfibi d'Italia*. Quaderni di Conservazione della Natura; numero 29. Ministero dell'Ambiente e della Tutela del Territorio e del Mare, I.S.P.R.A., Grandi & Grandi Editori.

Lecis, R. (2004). The endemic Sardinian newt *Euproctus platicephalus*: local threats and population decline. *Italian Journal of Zoology*, suppl. Vol. 2, pp. 195-198.

Lecis, R. (2007). *Euproctus platicephalus* (Gravenhorst, 1829), pp. 192-197. In: Lanza, B., Andreone, F., Bologna, M. A., Corti, C. & Razzetti, E. (eds.) (2007). *Fauna d'Italia. XLII. Amphibia.*, "Calderini - Edizioni Calderini de Il Sole 24 ORE Editoria Specializzata S. r. l.." Bologna, ISBN -978-88-506-5256-3.

Lionello, P., Baldi, M., Brunetti, M., Cacciamani, C., Maugeri, M., Nanni, T., Pavan, V. & Tomozeiu R. (2009). Eventi climatici estremi: tendenze attuali e clima futuro sull'Italia, pp. 81-106. In: Castellari S. & Artale, V. (Ed.) (2009). *I cambiamenti climatici in Italia: evidenze, vulnerabilità e impatti*. Bononia University Press, Bologna. ISBN: 978-88-7395-484-2.

Luiselli, L. (2006 a). Resource partitioning in the communities of terrestrial turtles: A review of the evidences. *Revue d' Ecologie (Terre et Vie)*, Vol. 61, pp. 353-365.

Luiselli, L. (2006 b). Food niche overlap between sympatric potential competitors increases with habitat alteration at different trophic levels in rain-forest reptiles (omnivorous turtles and carnivorous vipers). *Journal of Tropical Ecology*, Cambridge, Vol. 22, pp. 695-704.

Luiselli, L., Corti, C., Salvi, D. & Capula, M. *Hemorrhois hippocrepis* (Linnaeus, 1758), pp. 500-506. In: Corti, C., Capula, M. Luiselli, L. Razzetti, E. & Sindaco, R. (2011). *Fauna d'Italia. XLV. Reptilia*, "Calderini - Edizioni Calderini de Il Sole 24 ORE S.p.A." Milano, ISBN -978-88-506-5390-4.

Minelli, A., Chemini, C., Argano, R. & Ruffo, S. (eds.) (2002). *La fauna in Italia*. Touring Editore, Milano e Ministero dell'Ambiente e della Tutela del Territorio, Roma.

Nanni, T., Maugeri, M. & Brunetti, M. (2009). La variabilità e le tendenze del clima in Italia nel corso degli ultimi due secoli, pp. 11-45. In: Castellari S. & Artale, V. (Ed.) (2009). *I cambiamenti climatici in Italia: evidenze, vulnerabilità e impatti*. Bononia University Press, Bologna. ISBN: 978-88-7395-484-2.

Nilson, G. & Andrén, C. (2001). The meadow and steppe vipers of Europe and Asia - the *Vipera* (*Acridophaga*) *ursinii* complex. *Acta Zoologica Academiae Scientiarum Hungaricae*, Vol. 47, pp. 87-267.

Pizzuti Piccoli, A. (2006/07). Fenologia riproduttiva del tritone punteggiato *Lissotriton vulgaris meridionalis* (Boulenger, 1882), (Amphibia, Urodela) nel Bosco di Palo (Roma). *Ann. Mus. civ. St. nat. Ferrara*, Vol. 9/10, pp. 99-110.

Razzetti, E. & Bernini, F. (2011). *Coronella girondica* (Daudin, 1803), pp. 481-487. In: Corti, C.; Capula, M. Luiselli, L. Razzetti, E. & Sindaco, R. (2011). *Fauna d'Italia. XLV. Reptilia*, "Calderini - Edizioni Calderini de Il Sole 24 ORE S.p.A." Milano, ISBN -978-88-506-5390-4.

Ruffo, S. & Stoch, F. (eds.) (2005). Checklist e distribuzione della fauna italiana . *Memorie del Museo Civico di Storia Naturale di Verona*, Verona.

Sacchi, R. & Delaugerre, M. (2011). *Hemidactylus turcicus* (Linnaeus, 1758), pp. 271-276. In: Corti, C.; Capula, M. Luiselli, L. Razzetti, E. & Sindaco, R. (2011). *Fauna d'Italia. XLV. Reptilia*, "Calderini - Edizioni Calderini de Il Sole 24 ORE S.p.A." Milano, ISBN -978-88-506-5390-4.

Salvidio, S., Lanza, B. & Delaugerre, M. (2011). *Euleptes europaea* (Gené, 1839), pp. 258-270. In: Corti, C., Capula, M. Luiselli, L. Razzetti, E. & Sindaco, R. (2011). *Fauna d'Italia. XLV. Reptilia*, "Calderini - Edizioni Calderini de Il Sole 24 ORE S.p.A." Milano, ISBN -978-88-506-5390-4.

Scillitani, G. & Tripepi, S. (2007). *Lissotriton italicus* (Peracca, 1898), pp. 239-246. In: Lanza, B., Andreone, F., Bologna, M. A., Corti, C. & Razzetti, E. (eds.) (2007). *Fauna d'Italia. XLII. Amphibia.*, "Calderini - Edizioni Calderini de Il Sole 24 ORE Editoria Specializzata S. r. l.." Bologna, ISBN -978-88-506-5256-3.

Scillitani, G. & Turrisi, G. F. (2011). *Zamenis situla* (Linnaeus, 1758), pp. 591-599. In: Corti, C.; Capula, M. Luiselli, L. Razzetti, E. & Sindaco, R. (2011). *Fauna d'Italia. XLV. Reptilia*, "Calderini - Edizioni Calderini de Il Sole 24 ORE S.p.A." Milano, ISBN -978-88-506-5390-4.

Sciortino, M., Luise, A. & Genesio, L. (2009). La desertificazione e il degrado del territorio, pp. 160-174. In: Castellari S. & Artale, V. (Ed.) (2009). *I cambiamenti climatici in Italia: evidenze, vulnerabilità e impatti.* Bononia University Press, Bologna. ISBN: 978-88-7395-484-2.

Simoncelli, F., Fagotti, A., Dall'Olio, R., Vagnetti, D., Pascolini, R. & Di Rosa, I. (2005). Evidence of *Batrachochytrium dendrobatidis* infection in Water frogs of *Rana esculenta* complex in Central Italy. *Ecohealth*, Vol. 2, pp. 307-312.

Sindaco, R.; Doria, G., Razzetti, E. & Bernini, F. (eds.).(2006). *Atlante degli Anfibi e dei Rettili d'Italia. Atlas of Italian Amphibians and Reptiles.* Edizioni Polistampa, Firenze, ISBN 88-8304-941-1.

Sindaco, R., Bombi, P., Salvi, D. & Corti, C. (2011). *Archaeolacerta bedriagae* (Camerano, 1885), pp. 341-348. In: Corti, C.; Capula, M. Luiselli, L. Razzetti, E. & Sindaco, R. (2011). *Fauna d'Italia. XLV. Reptilia*, "Calderini - Edizioni Calderini de Il Sole 24 ORE S.p.A." Milano, ISBN -978-88-506-5390-4.

Speight, M.R., Hunter, M.D. & Watt, A.D. 2008. *Ecology of Insects. Concepts and Application*, Wiley-Blackwell, Oxford, UK, ISBN 978-1-4051-3114-8.

Stoch, F. (2000-2005). CKmap 5.3.8. Ministero dell'Ambiente e della Tutela del Territorio, Dir. Prot. Nat. http://www.minambiente.it/index.php?id_sezione=1930 (downloaded on January 2008).

Stuart, S.N., Chanson, J.S., Cox, N.A., Young, B.E., Rodrigues, A.S.L., Fischman, D.L. & Waller, R.W. (2004). Status and trends of amphibian declines and extinctions worldwide. *Science*, Vol. 306, pp. 1783 - 1786.

Tedaldi, G. (2001). La salamandra pezzata nell'Appennino tosco-romagnolo e in alcune zone limitrofe dell'Emilia, della Toscana e delle Marche: stato attuale delle conoscenze e problemi di conservazione, pp. 129-132. In: Barbieri, F., Bernini, F. & Fasola, M. (eds.) (2001). Atti 3° Congresso Nazionale Societas Herpetologica Italica, Pavia (2000). *Pianura*, Cremona.

Thomas, C.D., Cameron, A., Green, R.E., Bakkenes, M., Beaumont, L.J., Collingham, Y.C., Erasmus, B.F. N., de Siqueira, M.F., Grainger, A., Hannah, L., Huges, L., Huntley, B., van Jaarsveld, A.S. Midgley, G.F., Miles, L., Ortega-Huerta, M.A., Peterson, A.T., Phillips, O.L. & Williams, S.E. (2004). Extinction risk from climate change. *Nature*, Vol. 427, pp. 145-148.

Veith, M., Mucedda, M., Kiefer, A. & Pidinchedda, E. (2011). On the presence of pipistrelle bats (*Pipistrellus* and *Hypsugo*; Chiroptera Vespertilionidae) in Sardinia. *Acta Chiropterologica*, Vol. 13, pp. 89-99.

2

Breeding Phenology and Reproductive Strategies in Terrestrial Isopod Species (Crustacea Isopoda Oniscidea) from Tunisia

Mohamed Sghaïer Achouri
Unité de Recherche de Bioécologie Animale et Systématique évolutive, Faculté des Sciences de Tunis, Campus Universitaire Manar, Université Tunis El Manar, Tunis

1. Introduction

Reproduction in terrestrial isopods has been dealt with in numerous studies (summarized in Dangerfield and Telford, 1995). There are remarkable variations in breeding phenology between species (Sutton et al., 1984, Warburg, 1987) and between different populations of the same species (Souty- Grosset et al., 1988). Even within the same population, annual variations have been described (Zimmer and Brauckmann, 1997). A characteristic of many populations, especially those in seasonal environments, is that reproduction is frequently confined to a distinct breeding season or period. These periods are usually located in those months when physical conditions are less severe. It is often an implicit assumption that there is an optimal time of year for release of an individual's offspring, when the physical and biotic environment is most suitable (Willows, 1984).

Theory of life history evolution predicts specific patterns of growth, survivorship and reproduction based on a genotypic strategy (Pianka 1970; Greenslade, 1983). Models based on optimality arguments take into account environmental effects but often predict an optimal strategy or tactic for a given set of environmental conditions (Sibly and Calow, 1986). The breeding phenology of woodlice is influenced by both abiotic and biotic environmental factors, resulting in temporal coincidence of the release of offspring with favourable conditions for growth and survival and thus in discrete annual breeding events in temperate climates (Lawlor 1976; Sutton et al., 1984; Souty-Grosset et al., 1988). However, breeding phenology can vary annually within these constraints (Dangerfield and Hassall, 1992; Zimmer and Brauckmann, 1997). More generally speaking, the optimal time for breeding may be more or less predictable from year to year, but those animals which reproduce at that time may be viewed as the most successful (Willows, 1984).

In arid zones as in temperate ones, there are remarkable variations in breeding phenology. Isopods breed mostly during spring and early summer (Aljetlawi and Nair, 1994; Nair, 1984;

Warburg, 1994; Achouri and Charfi-Cheikhrouha, 2001; 2002), but variations in breeding phenology among both species (Sutton et al., 1984; Warburg, 1987) and different populations of the same species (Souty-Grosset et al., 1988; Achouri and Charfi-Cheikhrouha, 2005) are common. Even within the same population, annual variations have been described (Dangerfield and Hassall, 1992; Zimmer and Brauckmann, 1997). Female fecundity is proportional to female size (Sutton et al., 1984). Thus, population dynamics depend on both the number and size of reproducing females (Rushton and Hassall, 1987; Hassall and Dangerfield, 1990). Due to variable individual growth rates (Hubbell, 1971; "cohort splitting": Sunderland et al., 1976), even a cohort of similarly aged females exhibits a considerable range in size (Zimmer and Kautz, 1997; Achouri and Charfi-Cheikhrouha, 2002; Achouri et al., 2003). Within given species-specific constraints, it is thus advantageous to grow large before reproducing, while, on the other hand, reproductive investment constrains increase in body size (for discussion, see Zimmer and Kautz, 1997). As for intraspecific comparison, we hypothesise that, on an interspecific scale, (1) individuals of larger species are able to invest more of the available energy in reproduction than individuals of small species (cf. Glazier et al., 2003).

In most species of the Porcellionidae, two or more brood waves per year have been observed (Warburg, 1987). Thus, these species have been considered iteroparous. However, it has rarely been demonstrated unambiguously that a single female was capable of breeding more than once in its lifetime (iteroparous) or whether it would not breed again (semelparous) (Cole, 1954); from more than one brood wave per year in a given population, it does not necessarily follow that each individual female breeds more than once (Warburg, 1994). Knowing whether a particular species is semelparous or iteroparious, however, is essential for a comparative interspecific analysis of breeding strategies, since reproductive investment in current reproduction varies inversely with expectation of future offspring. Thus, semelparous females should invest significantly more in their first and only brood than potentially iteroparous females. In iteroparouas species, females may tend to invest less into the first brood and allocate more energy into growth, in order to grow larger for a second brood (cf. Zimmer and Kautz, 1997). We hypothesise that (2) females of semelparous species exhibit high reproductive investment as compared to iteroparous species. Accordingly [see hypothesis (1)], we expect small species to be iteroparous, enabling them to spread their reproductive effort over several breeding events. On the other hand, it may -under certain conditions- be advantageous for large species to be semelparous, concentrating their reproductive effort on a single event, while at the same time reducing the risk of dying prior to a second brood (cf. Zimmer and Kautz, 1997).

As reproductive activity and investment may vary with environmental conditions, interspecific comparisons should be restricted to syntopic species that are exposed to similar environments. We performed a long-term study on interspecific differences in breeding phenology and investment in syntopic isopod species, predicting that large isopod species invest high proportions of their energy in reproduction and tend to be semelparous, while the opposite is true for small species. The main aim of this study is to compare the reproductive patterns of seven species of Oniscidea collected in Chambi area, to test whether these syntopic species exhibit interspecific differences in breeding phenology (cf. Zimmer, 2003) and hypothesise that differently sized syntopic species differ in terms of female reproductive investment and to compare their reproductive phenology with others populations from the Mediterranean areas.

2. Material and methods

2.1 Study site and sampling programme

The study area is the Mt. Chambi or Djebel Chambi which includes the highest point of Tunisia (1544m/5066ft). It is accepted as a biosphere reserve by UNESCO in 1977 and declared a National Park in 1981. This mountain is Located in center-oust of Tunisia at 17 km from the city of Kasserine and a few kilometers from the Algerian border.

Several plant species are identified (262). They are distributed in different stages of the mountain. The first stage (up to approximately 900 m) is the area of the alpha, then above, the Aleppo pine which grows in the Phoenician juniper and rosemary. Still above, beyond 1100 m are the oaks.

Populations of *Armadillidium sulcatum* A. Milne-Edwards, 1840, *Leptotrichus panzerii* (Audouin, 1826), *Hemilepistus reaumuri, Porcellio djahizi* n. sp. (Medini & Charfi-Cheikhrouha, 2001), *Porcellio laevis,* (Latreille 1804), *Porcellio variablis* Lucas, 1846 and *Porcellionides pruinosus* Brandt, 1833, were hand-collected in the Chambi area. Data were obtained from a field survey of twenty six months (October 2008-December 2010).

2.2 Laboratory procedures

Individuals collected were taken to the laboratory alive, the majority of them are preserved in 70% ethanol and the others are raised for further investigation. Subsequently, we counted and sexed all individuals, and measured their total length (distance between the anterior edge of the head and the end of the pleotelson).

The animals were kept in translucent plastic boxes (20 x 35 x 10cm) containing a loamy soil about 3cm height, under natural conditions, with temperature mostly fluctuating between 20 and 33°C, and the daily photoperiod was 12h. They were fed on disks of carrot, and sporadically, on some fresh linden leaves. To guarantee the detection of all females with brood pouches before manca release, females were examined for the occurrence of brood pouches once a week. Based on the detection of ovigerous females and the young in laboratory populations, we were able to distinguish between seasonal reproduction and continuously breeding populations. Moreover, the appearance of gravid females and juveniles served as a measure for the onset and the duration of the reproductive period and sexual rest.

During the annual reproductive period, 20 ovigerous females of each species were sized using a stereomicroscope (Leica MS5), then they were dissected and marsupial eggs or embryos were counted to estimate fecundity (defined here as the number of eggs for a female for one brood); this should not be confused with fertility, which is the number of young leaving the brood pouch. To find a relation between brood size and body length, studies were conducted on gravid females of varying body dimensions, and simultaneously, it was possible to examine the ovaries after the discharge of the large oocytes. At that point the ovary was largely an empty sleeve-like structure, except for the smaller oocytes (Warburg, 1994). By examining the ovaries, it was possible to check for future generations of oocytes, indicating the potential for iteroparity; their absence was interpreted as indicator of semelparity.

In every reproductive period, 15 to 20 additional gravid females were maintained individually in cylindrical plastic tubes (7x4cm) and fed on disks of carrot until manca release. Using a Mettler AB22204-S balance (±0.1 mg accuracy), each female was weighed weekly during the breeding season, both before (W_0) and immediately after releasing mancae (W_1). The weight loss of the female corresponds to both reproductive allocation and parental investment. Upon hatching, mancae were counted, and they were separated from their mother and kept separately in Petri dishes under the same laboratory conditions (temperature, photoperiod, food) as the adult individuals.

2.3 Data analysis

Since recruitment was discontinuous we estimated field growth rates by tracking recognizable cohorts from size-frequency distribution (0.2mm length classes) on successive sample dates. Size frequency analysis was achieved by using the probability paper method (Harding, 1949), as performed by Cassie (Cassie, 1954; 1963). To test the reliability of the method, we employed the χ^2 and G tests ($P \geq 0.05$) (Fisher, 1950; Sokal and Rohlf, 1981). Computations were performed using ANAMOD software (Nogueira, 1992).

Usually, field growth rates are not constant throughout the year. Thus, results from field growth estimations were fitted to the Von Bertalanffy growth function (VBGF) with seasonal oscillation, according to a model proposed by Gaschütz and co-workers (Gaschütz et al., 1980), and previously applied (Marques et al., 1994; Pardal et al., 2000) with satisfactory results. Length is expressed as:

$$L_t = L_\infty \{1 - e^{-[kD(t-t_0) + C(kD/2\pi)\sin 2\pi(t-ts)]}\}^{1/D}$$

L_t= length of the organism at a given moment t;
L_∞ = maximum possible length of the organism;
t = time instant;
t_0 = time instant when the organism would have a length = 0;
t_s = time interval between start of growth (when t=0) and the first growth oscillation; growthis expressed by a sinus curve, with a period of 1year;
k= intrinsic growth rate;
C= constant, ranging from 0 to 1;
D = parameter that expresses metabolic deviations from the Von Bertalanffy 2/3 rule.

Egg loss, brood pouch mortality, number of broods per female, reproductive allocation and the potential of reproduction were quantified:

Reproductive allocation (R) was calculated as the percentage of weight lost by the female during the process of producing a brood of young:

$$R = \frac{(W_0 - W_1)}{W_0} \times 100$$

where W_0 is the initial weight before the mancae release and W_1 is the weight immediately following mancae release (Warburg, 1994).

R estimates a female's reproductive investment, while P quantifies the weight loss of a female in order to produce a single manca: P = R/N, where N is the number of mancae per brood (Warburg, 1994).

The estimation of the reproductive potential was based on the number of juveniles released, the proportion of females in the sample, the life span of a female, and the number of broods during a female's lifetime. The reproductive potential (R') was estimated using the Wildish equation R' = bnp. (Wildish, 1971, 1979, 1982 *in* El Kaïm et al., 1985), where b represented the mean number of young per brood, n the annual number of broods per female, and p the average number of ovigerous females.

The sex ratio was estimated by the ratio of males to females. The observed and expected values were compared using Chi-square (χ^2) test.

3. Results

3.1 Breeding phenology

Breeding season is defined as the time from the first appearance of gravid females to the time of the last appearance of empty brood pouches in the population. Figure 1 shows that the duration of the reproductive period varied considerably among species, ranging from 3 months in the largest species *Hemilepistus reaumuri* and the two smallest species *Armadillidium sulcatum* and *Leptotrichus panzerii*, to 6 months in *Porcellio variabilis* and *Porcellionides pruinosus*. Marsupial eggs were detected in *Hemilepistus reaumuri*, *Armadillidium sulcatum* and *Leptotrichus panzerii* in May. Mancae emerged during June and July.

Porcellio variabilis carried mature oocytes in February; marsupial eggs were found in March and April. Mancae were released from May to September with a sexual rest during July and August. In *Porcellionides pruinosus*, ovigerous females and juveniles were present throughout the period extended from March to October, however their percentage decreased in June and July (9.43%). For the other species (*Porcellio djahizi* and *Porcellio laevis*), the reproductive period was 5 months with a sexual rest from June to August for the first species, and during July and August for the second one.

These latter species of the genus *Porcellio* and *Porcellionides pruinosus* breed twice a year, in spring and during monsoon in the period extending from the fourth week of August until the second week of October. Females contained mature oocytes during March and April. Egg release into the marsupium began in May, and mancae were released from June to September. In all studied species, the date of the onset of breeding appeared to be correlated to female size. Overall, in species with seasonal reproduction behaviour, larger females started reproducing earlier in the year than smaller ones, except in *Hemilepistus reaumuri* (Fig. 1). Further, on an interspecific scale, larger species started reproducing earlier in the year than smaller ones.

The marsupial period starts with the parturial moult which is followed a few days later by the movement of mature oocytes into the marsupium. The marsupial eggs then develop into embryos and finally mature to hatch as mancae (Holdich et al., 1984; Warburg, 1994). The marsupial development, defined in this case as the period from the appearance of marsupial eggs to the emergence of mancae, was spread over a period of 22-49 days. It was longest in *Porcellio laevis*, one of the largest species, and *P. variabilis*, and shortest in the smallest species, *Armadillidium sulcatum* (Table 1).

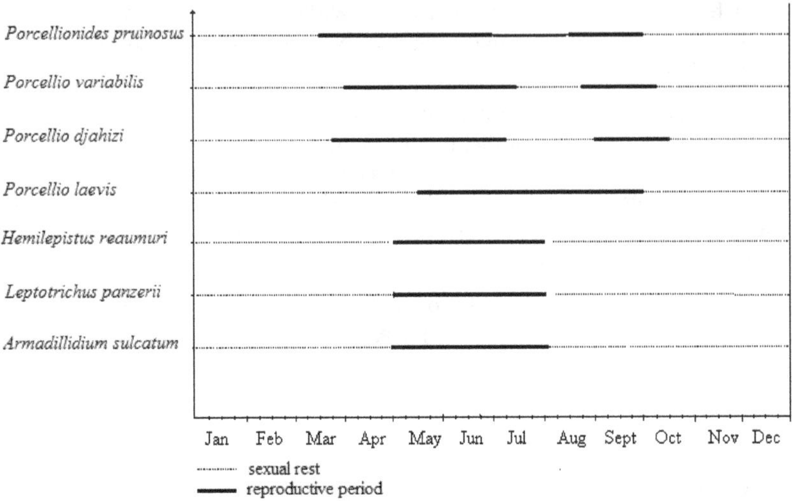

Fig. 1. Onset of reproduction and duration of the reproductive period.

The sexual rest lasted for 9 months in Hemilepistus reaumuri, Armadillidium sulcatum and Leptotrichus panzerii, and about 6 to 7 months in Porcellio variabilis, Porcellionides pruinosus, Porcellio djahizi, and Porcellio laevis (Fig. 1).

Species	Duration of reproductive period	Gravidity (days) [M ± S.D. (N)]	Avg. number Eggs [M ± S.D. (N)]	Avg. number mancae [M ± S.D. (N)]
A. sulcatum	May-June- July	22 ± 2 (31)	14 ± 0.9 (29)	9 ± 2.1 (25)
H. reaumuri	May-June- July	43 ± 4 (45)	68 ± 0.7 (79)	47 ± 1.9 (70)
L. panzerii	May-June- July	23 ± 6 (27)	11 ± 0.5 (25)	9 ± 0.78 (37)
P. djahizi	Mar-May[]Sept-October	39 ± 2 (50)	48 ± 0.6 (57)	38 ± 1.81 (50)
P. laevis	April-June[]Sept-October	49 ± 5 (42)	52 ± 0.8 (48)	41 ± 0.91 (41)
P. variabilis	Feb-May[]Sept-October	46 ± 3 (36)	51 ± 0.5 (54)	43 ± 0.73 (45)
P. pruinosus	Mar- may///August-October	34 ± 5 (38)	26 ± 1.3 (37)	19 ± 1.2 (37)

(N): number of used females, [] sexual rest /// only a few number of ovigerous females was found

Table 1. Marsupial eggs, mancae released numbers and duration of reproductive period in seven terrestrial isopod species from Chambi area.

3.2 Fecundity, fertility and egg losses

Corrected for female size, small species (e.g., *Armadillidium sulcatum* and *Leptotrichus panzerii*) released less mancae than did large species (e.g., *Hemilepistus reaumuri, Porcellio laevis* and *Porcellio variabilis*) (Table 1).

In all species, the number of released mancae was lower than the number of eggs released into the marsupium ("Egg loss" in Table 1). This loss in marsupial eggs was low in

Leptotrichus panzerii but relatively high in *Hemilepistus reaumuri*, ranging in total from 2 to 21 (Table 1). Otherwise, both fecundity and fertility was positively correlated with female size and weight (Table 2), but we observed significant interspecific differences in size/fecundity relationships.

Species	Fecundity		Fertility	
	Equation	R^2	Equation	R^2
A. granulatum	Y = 7.1 X -18.7	**0.81**	Y = 5.4 X - 18.9	**0.82**
H. reaumuri	Y = 5.3 X - 13.7	**0.79**	Y = 4.3 X - 12.3	**0.53**
L. panzerii	Y = 13.2 X - 71.1	**0.91**	Y = 8.2 X - 39.2	**0.89**
P. djahizi	Y = 13.6 X - 89.1	**0.88**	Y = 12.1 X - 69.4	**0.91**
P. laevis	Y= 6.9 X - 26.9	**0.89**	Y= 5.1 X - 17.3	**0.73**
P. variabilis	Y = 7. 3 X -35.8	**0.79**	Y = 4.9 X - 15.9	**0.82**
P. pruinosus	Y = 5.8 X - 35.6	**0.92**	Y = 6.5 X - 18.1	**0. 87**

Y is the total number of eggs produced per brood pouch; X is the body length of gravid females

Table 2. Equations and R^2 of relationship between female body length and fecundity and fertility.

3.3 Growth and life span

We analysed size frequency distributions for recognisable cohorts. Based on data from the first sampling results, one cohort could be identified for these species, *Armadillidium sulcatum*, *Hemilepistus reaumuri*, and *Leptotrichus panzerii*, and three to four cohorts in *Porcellio variabilis*, *Porcellio djahizi*, *Porcellio laevis* and *Porcellionides pruinosus*. Additionally, new cohorts appeared, respectively, in June and july for *A. sulcatum*, *H. reaumuri*, and *L. panzerii*, (one cohort), in May and October for the other species (two cohorts). The minimum average length of newly recognized cohorts ranged from 3.5 ± 0.136mm to 4.87 ± 0.163mm.

Growth was approximately continuous throughout life, but field growth rates were clearly higher in early life phases. Growth rates were also higher during spring (Fig. 2), but showed, on the other hand, a decline during autumn and winter (Fig. 2). We could follow new cohorts (appeared in our sampling) from detection up to their extinction, which allowed estimating the average life span at 14 to 25 months. In *Porcellio variabilis*, *Porcellio djahizi*, *Porcellio laevis* and *Porcellionides pruinosus* cohorts born by the end of the reproductive period had longer life span (19 to 25 months) than those (born in spring; 14 to 18 months). Data on field growth rates from these cohorts (followed in sampling period) were fitted to the Von Bertalanffy growth function. Parameters were estimated from each cohort data set, and the resulting model was used to simulate the expected growth in each case (Fig. 2). The degree of fit was assessed from correlation (r^2) between predicted values and observations.

3.4 Life cycle

Size frequency analysis allowed identifying the cohorts to which males and ovigerous females belonged, and thus we could evaluate the possible contribution of each cohort to recruitment. These results suggest that cohorts born in early spring for *P. djahizi*, *P. laevis*, *P. variabilis*, and *P. pruinosus*, will in principle be able to reproduce in the same reproductive period, and most probably also in the beginning of the next. Cohorts born late in the reproductive period will grow slower during winter and not reproduce until the next spring, (the next reproductive period). However, in *A. sulcatum*, *H. reaumuri*, and *L. panzerii*, from July to October, this was a period of rapid growth in all size classes, and the bimodality in size class distribution observed during this period disappeared from November to next July.

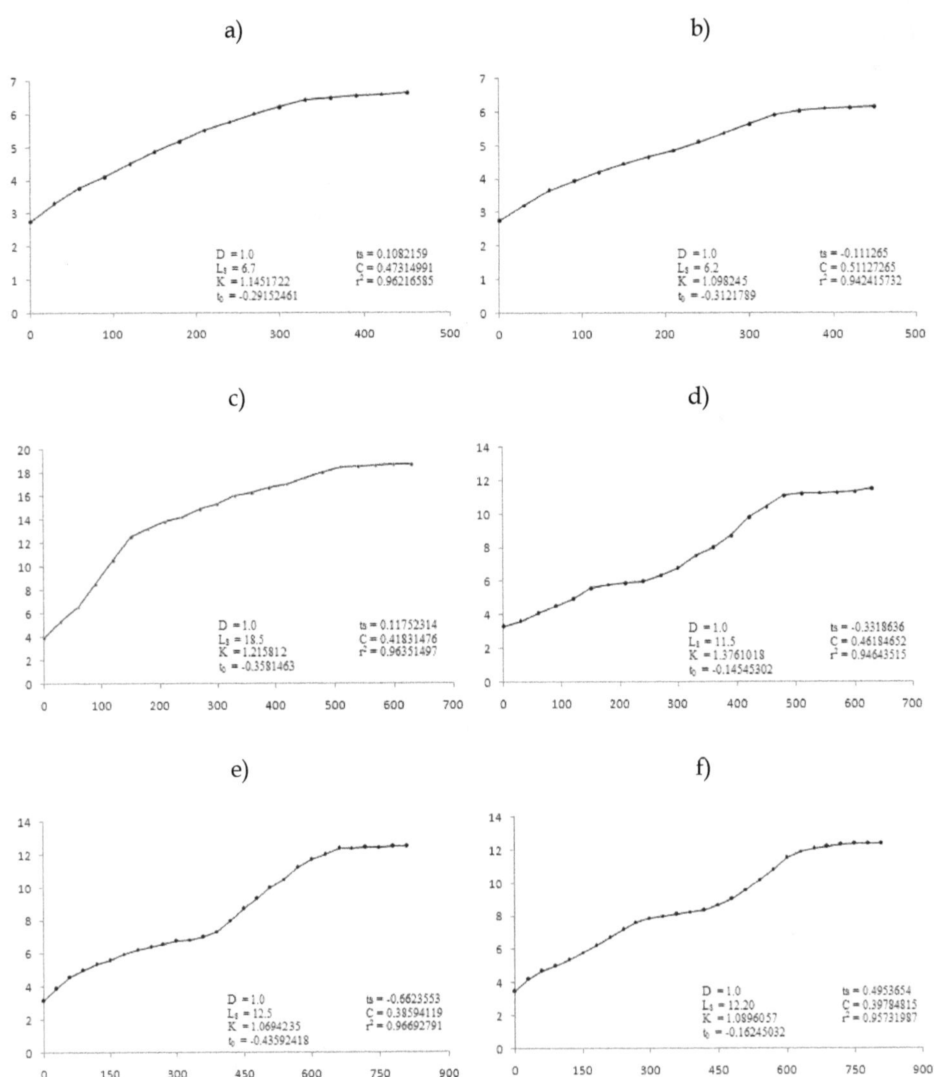

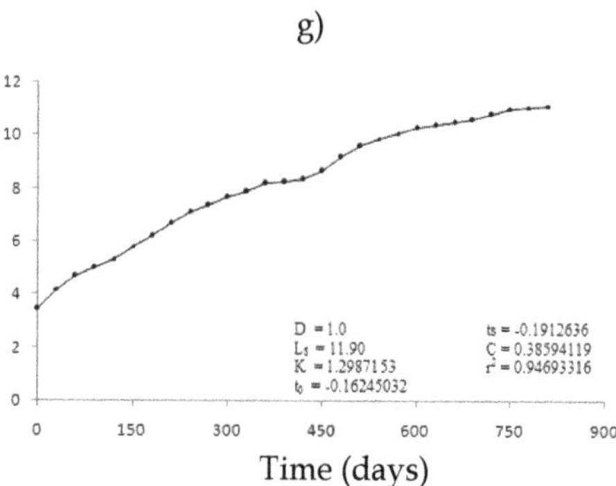

Fig. 2. Growth model of studies species at Chambi area. Graphic adjustment of the growth curves of cohorts: a. *A. sulcatum*, b. *L. panzerii*, c. *H. reaumuri*, d. *P. pruinosus*, e. *P. variabilis*, f. *P. laevis* and g. *P. djahizi* is given. Values of mean length estimated for each given instant are also plotted; r^2, correlation between adjusted curves and plotted values.

3.5 Reproductive strategies

Four out of seven species were iteroparous, with single females reproducing repeatedly. On the other hand, twenty females, each, of *Armadillidium sulcatum*, *Leptotrichus panzerii* and *Hemilepistus reaumuri*, were undoubtedly semelparous. In all other species, we found immature oocytes inside the ovary after mature eggs had been released into the marsupium upon mating, indicating the potential for a subsequent brood. Contrary to our hypothesis (2), the semelparous species are among the two smallest *Armadillidium sulcatum*, *Leptotrichus panzerii*, and the largest *Hemilepistus reaumuri*, in the Chambi area.

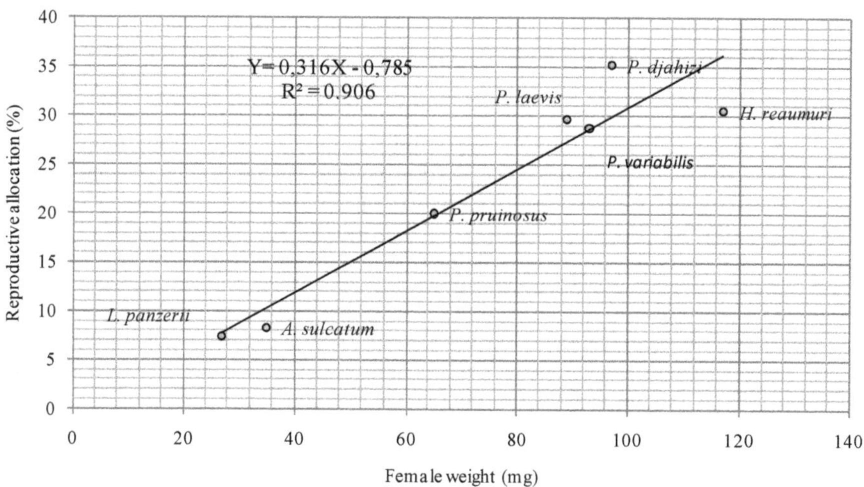

Fig. 3. Relationship between mean female weight and mean reproductive allocation of seven isopod species.

The two smallest species exhibited the lowest reproductive allocation, corroborating our hypothesis (1) (Table 3). The largest species, *Hemilepistus reaumuri*, *Porcellio djahizi* and *Porcellio laevis* were among the three species, with the highest reproductive allocation, and the numbers of both marsupial eggs and mancae released were highest in these species. Although *Porcellio variabilis* females on average were heavier than those of the other species, they produced much smaller numbers of both marsupial eggs and mancae than the other species; despite their large mancae, reproductive allocation was intermediate. Overall, however, reproductive allocation was positively correlated with female weight (Fig. 3; $R^2 = 0.906$).

Parental investment in producing a single manca varied between 0.61 (*L. panzerii*) and 0.89 (*P. djahizi*) mg per manca. Parental investment was not significantly correlated with female weight ($R^2 = 0.1313$).

The reproductive potential differed greatly between the small and large species. *H. reaumuri* (2386 juveniles released) *P. variabilis* (2456) *P. laevis* (2284) and *P. dalensis* (2284) exhibited the highest reproductive potential. These species produced six times more offspring than did *L. panzerii* (154) (176) and *A. sulcatum* (Table 4).

4. Conclusion

Various parameters used to define reproductive strategies herein covaried with species-specific female size. Overall, small isopod species from the Chambi area tend to be semelparous [contrary to hypothesis (2)], to allocate less energy in reproduction [corroborating hypothesis (1)], to produce less mancae (corrected for female size), to start breeding later in the year, and to exhibit a shorter reproductive period and a shorter marsupial development than large species.

Overall, the isopod species studied herein confirmed several aspects of breeding phenology that had been described in other syntopic woodlice species in the Mediterranean area (Warburg 1994; Medini et al., 2000; Hamaied et al., 2004; Achouri and Charfi-Cheikhrouha, 2006; Achouri et al., 2008). However, no clear relationship emerges between the reproductive pattern, on the one hand, and either the taxonomic status or the environment, on the other hand; rather the period from April (spring) to October (autumn) can generally be considered favourable for reproduction in the Mediterranean, irrespective of species identity. Thus, the reproductive pattern of terrestrial isopods appear to be independent, to a certain extent, of the habitat condition (cf. Warburg, 1994). Based on our present results, we hold that it is rather the species-specific body size of females that determines reproductive strategies. It remains, however, unclear why - in contrast to our hypothesis (2) - the smallest species rather than the larger ones are semelparous in Chambi area.

The loss in marsupial eggs did not covary with female size in the present study, and was in the same range of magnitude as values found by Warburg (1994) in species inhabiting the Negev desert and xeric and mesic habitats in the Mediterranean region (ca 26 %). Reproductive allocation to produce one brood pouch ranged between 6 and 35%, depending on female size. Warburg (1994) reported reproductive allocation of 7-52 % in seven species from Palestine, but in all but one species it was less than 22 %. The maternal investment in producing offspring ranged between 0.6 and 1.6 mg per manca, covering the range of 0.6-1.4 Warburg (1994) reported for different species in the Negev desert and the Mediterranean. Females of the small species, *Armadillidium sulcatum*, and *Leptotrochus panzerii* invested less than did females of large species [hypothesis (1)], and overall, semelparous species exhibited a lower reproductive potential than iteroparous species, although the latter showed high interspecific variation in this respect. Reproductive investment and output of the species in this study were positively correlated with female size and weight, both within and among species. Intraspecific variation in size-adjusted reproductive output frequently occurs (e.g., Dangerfield and Telford, 1990; Nair, 1984) that may be explained by individual differences in the ability to accrue resources, the timing of allocation of resources to reproduction, genetic determinants of reproductive characteristics, or seasonal variation in environmental conditions during individual development and reproduction. This may also hold true for interspecific comparison, but further studies are needed to unambiguously decide upon this issue.

According to the present results, iteroparous species *P. variabilis*, *P. djahizi*, *P. laevis*, and *Porcellionides pruinosus* were reproductively active from February/March to August/October with a cessation during June and July, when semelparous species *H. reaumuri*, *A. sulcatum* and *L.panzerii* engaged in reproduction. Although we did not test for competitive exclusion or character displacement in the presence of a superior competitor, it is interesting to speculate that small species in the Chambi area may avoid competition for their young by the larger offspring of larger species through a shift in time of reproduction (cf. Zimmer, 2003, Achouri et al., 2008).

Interspecific variation in the onset and the duration of reproduction prevailed in the studied species despite the imposed equal conditions of temperature, photoperiod and nutrition. Since interspecific interactions were absent in our experimental design, we hold that individual genotypes dictate individual phenology. Direct genetic control (De Jong, 1991), or a genetically determined norm reaction (Stearns and Koella, 1986), exist for reproduction traits such as the timing of reproduction (Dangerfield and Telford, 1995). However, flexibility in female reproductive tactics is apparent in many isopod species. Lifetime brood production, phenology of brood production, fecundity and offspring size all differ to some degree between individuals. The theoretical concepts of bet-hedging (Slatkin, 1974) and phenotypic plasticity (Bradshaw, 1965; Caswell, 1989) consider such variation in reproductive traits as strategies capable for maximising female reproductive success under changeable environmental conditions (Houston and McNamara, 1992). In the long run, flexibility in phenology in response to environmental conditions (cf. Dangerfield and Hassall, 1992; Zimmer and Brauckmann, 1997) will determine the potential of a species to adapt to changing climatic conditions.

The outputs from our calibrated growth model showed that growth rates are always higher in the initial life cycle stages, slowing down when individuals become mature, viz., during the reproductive phase. This is consistent with observations carried out on *Porcellionides sexfasciatus*, as well as other Oniscidea species (Matsakis, 1955). Cohorts born in late summer and autumn may in principle live 19 to 25 months, while cohorts born in spring will live only 14 to 18 months. Since cohorts born in spring will be able to produce offspring in the same reproductive period, differences observed might be explained by the fact that females from these cohorts invest in reproduction in an early phase of their lives, and face the harsh winter conditions immediately after. On the contrary, cohorts born later in the season will live through the winter and reproduce only in the next year, at a later stage in their life. These will face more favourable conditions after their reproductive effort and consequently will tend to survive longer.

Next to effects on fecundity, variable growth rates appear to be important as determinants of the maturation of young individuals. In combination with discrete breeding events, growth rates also control lifetime brood production. For instance, in a population of *Armadillidium vulgare* relaxation of intraspecific competition and favourable growth conditions allowed young born in July to grow rapidly and reproduce the following summer, thus breaking the normal sequence of two years maturation and semelparity (Dangerfield and Hassall, 1992).

Combining biological observations with the growth model output makes it possible characterize the life cycle of the populations studied of *Porcellio djahizi, P. laevis, P. variabilis,* and *Porcellionides pruinosus,* as that of a semi-annual species, with iteroparous females and bivoltine life cycle. However, *H. reaumuri, A. sulcatum* and *L. panzerii* could be a semi-annual species, with semelparous females and univoltine life cycle.

Reproductive allocation to produce one brood pouch ranged between 7.46 and 35.21% and thus showed marked interspecific differences. Overall, reproductive allocation was positively correlated with female weight. For seven isopod species in England, the reproductive allocation ranged from 24 to 47 % (Sutton et al., 1984), and Warburg (1994) reported reproductive allocation of 7-52 % in seven species from Palestine, but in all but one species it was less than 22 %. The maternal investment in producing offspring ranged between 0.61 and 0.89 mg per manca, covering the range of 0.6-1.4 Warburg (1994) reported for different species in the Negev desert and the Mediterranean. Females of the small species, *Armadillidium sulcatum,* and *Leptotrichus panzerii* invested less than did females of large species, and overall, semelparous species exhibited a lower reproductive potential than iteroparous species, although the latter showed high interspecific variation in this respect.

5. References

Achouri, M.S., Charfi-Cheikhrouha, F., 2001. Biologie de la reproduction de *Porcellionides pruinosus* (Brandt, 1833) Isopode Terrestre de Garat Nâam, Tunisie. *Crustaceana,* 74 (1): 11-26.

Achouri, M.S., Charfi-Cheikhrouha, F., 2002. Breeding and population dynamic of *Porcellionides sexfasciatus* (Crustacea, Isopoda, Oniscidea) *C.R. Acad. Sci. Paris,* 325 (1):1-12.

Achouri, M.S., Charfi-Cheikhrouha, F., 2005. Morphological and reproductive variability in *Porcellionides pruinosus* (Brandt, 1833) in Tunisia (Crustacea Isopoda, Oniscidea). *Crustaceana,* 75 (10):1241-1262.

Achouri, M.S., Charfi-Cheikhrouha, F., 2006. Stratégie de reproduction et de croissance de deux espèces sympatriques du genre *Porcellionides* (Crustacea Isopoda, Oniscidea) de Tunisie, *Crustaceana* 79 (7) : 843-864.

Achouri, M.S., Charfi-Cheikhrouha, F., Marques J.C., 2003. Biology, population and field-growth rates of *Porcellionides pruinosus* (Isopoda, Onicidea) at Garat Nâam (Kasserine, Tunisia). *Crustaceana,* 75 (10): 1241-1262.

Achouri, M.S., Charfi-Cheikhrouha, F., Zimmer M., 2008. Reproductive patterns in syntopic Terrestrial Isopod species (Crustacea, Isopoda, Oniscidea) from Morocco. *Pedobiologia,* 52 : 127—137.

Aljetlawi, A. A., Nair, G. A., 1994. Breeding and population of *Armadillo officinalis* (Isopoda: Oniscidea) in Benghazi, Libya. *J. Arid. Environment* 27: 241-255.

Bradshaw, A. D.,1965. Evolutionary significance of phenotypic plasticity in plants. *Adv. Genet.* 13: 115-155.

Cassie, R.M., 1954. Some uses of probability paper in the analysis of size-frequency distributions, *Aust. J. Mar. Freshwater Res.* 3: 513–522.

Cassie, R.M., 1963.Tests of significance for probability paper analysis, *N. Z. Sci. Rev.* 6: 474–482.

Caswell, H., 1989. Life history strategies. In. J.M. Cherret (ed.), Ecological concepts: The contribution of ecology to an understanding of the natural world. *Symposia of the Britsh Ecological Society*. 29. Oxford: Blackwell.

Cole, L. C., 1954. The populations consequences of life history phenomena. *Ouart. Rev. Biol.*, 29: 103-137.

Dangerfield, J.M., Hassall, M., 1992. Phenotypic variation in the breeding phenology of the woodlouse *Armadillidium vulgare* (Latr.). *Oecologia*, 89:140-146.

Dangerfield, J.M., Telford, S.R., 1990. Breeding phenology, variation in reproductive effort and offspring size in a tropical population of the woodlouse *Porcellionides pruinosus*. *Oecologia*, 82: 251-258.

Dangerfield, J.M., Telford, S.R., 1995. Reproduction in woodlice: Flexibility to maximise individual fitness. In Alikhan M. A. Ed., Terrestrial isopod biology (Crustacean issues 9). *Rotterdam: Balkema*, 69-82.

De Jong, G., 1991. Genotype-by-environment interaction and the genetic covariance between environments: multilocus genetics. *Genetica*, 81: 171-177.

El Kaïm, B., Irlinger, J.P., Pichard, S., 1985. Dynamique de la population d'*Orchestia mediterranea* L. (Crustacé, Amphipode) dans de l'estuaire Bou Regreg (Maroc). *Can. J. Zool.*, 63: 2800-2809.

Fisher, R.A., 1950. The significance of deviation from expectation in a Poisson series, *Biometrics* 6: 17–24.

Gaschütz, G., Pauly, D., David, N., 1980. A versatile BASIC program for fitting weight and seasonally oscillating growth data. *ICES CM*, G24: 123.

Glazier, D.S., Wolf, J.F. and Kelly, C.J., 2003. Reproductive investment of aquatic and terrestrial isopods in central Pennsylvania (U.S.A.). *Crustaceana Monographs*, 2: 151-179.

Greenslade, P.J.M., 1983. Adversity selection and the habitat template. *Am Nat* 122: 352-365.

Hamaied, S., Nasri-Ammar, K., Charfi-Cheikhrouha F, 2004. Phénologie de la reproduction d'une population naturelle de Armadillidium pelagicum Arcangeli, 1955 (Isopoda, Oniscidea). *C. R. Acad. Sci. Paris*, 327: 335-342.

Harding, J.P., 1949. The use of probability paper for the graphical analysis of polymodal frequency distributions, *J. Mol. Biol. Ass. UK* 28: 141–153.

Hassell, M., Dangerfield, J. M., 1990. Effect of habitats and climate on the population dynamics of Armadillidium vulgare Latr. (Isopoda: Oniscoidea). *Journal of Animal Ecology*, 55: 941-958.

Holdich, D. M., Lincoln, R. J., Ellis, J. P., 1984. The biology of terrestrial isopods: Terminology and classification. *Symp. Zool. Soc. Lond.* 53: 1-6.

Houston, A. I., McNamara, J. M. 1992. phenotypic plasticity as a state-dependent life-history decision. *Evol. Ecol.* 6: 243-253.

Hubbell, S.P., 1971. Of sowbugs and systems: the ecological bioenergitics of a terrestrial isopod. In:B.C. Pattern (ed.), Systems Analysis and simulation in Ecology. *New York: Academic Press.*

Lawlor, L.R., 1976. Molting, growth and reproductive strategies in the terrestrial isopod *Armadillidium vulgare*. *Ecology* 57: 1179-1194.

Marques, J.C., Martins, I., Teles-Ferreira, C.,Cruz, S., 1994. Population dynamics, life history, and production of Cyathura carinata (KrØyer) (Isopoda: Anthuridae) in the Mondego estuary (Portugal). *Journal of Crustacean Biology*, 14:258-272.

Matsakis, J., 1955. Contribution à l'étude du développement post-embryonnaire et de la croissance des Oniscoïdes. *Bulletin de la Société Zoologique de France*, 80 : 52-65.

Medini, L., Nasri-Ammar, K., Charfi-Cheikhrouha, F.,2000. Reproduction saisonnière de Porcellio variabilis (Lucas 1846) (Crustacea, Isopoda Oniscidea). *C. R. A. Sci. Paris*, 323: 689–695.

Nair, G. A., 1984. Breeding and population biology of the terrestrial isopod Porcellio laevis in the Delhi region. *Symp. Zool. Soc. Lond.* 53: 315-337.

Nogueira, A.J.A., 1992. ANAMOD - Extracção das componentes modais de distribuições de frequências de variáveis biométricas, Trabalho de Síntese, Provas de Aptidão Pedagógica e de Capacidade Científica, Universidade de Coimbra, pp. 67.

Pardal, M.A., Marques, J.C., METELO, I. LibellebØ,A.I., Flindt,M.R., 2000. Impact of eutrophisation on the life cycle, population dynamics and production of Amphithoe valida (Amphipoda) along an estuarine spatial gradient (Mondego estuary, Portugal). *Mar. Ecol. Progr. Ser.*, 196: 207-219

Pianka, E.R., 1970. On r- and k- selection. *Am Nat* 104: 592-597.

Rushton, S.P., Hassall M., 1987. Effects of food quality on isopod population dynamics. *Func. Ecol.* 1: 359-368.

Sibly R.M., Calow P., 1986. Physiological ecology of animals. An evolutionary approach. *Blackwell Scientific Publications, Oxford.*

Slatkin, M.,1974. Hedging one's evolutionary bets. *Nature*, 250: 704-705.

Sokal, R.R., Rohlf, F.J., 1981. Biometry, W.H. Freeman & Co., San Francisco, pp. 859.

Souty-Grosset, C., Chentoufi, A, Mocquard, J.-P., Juchault, P., 1988. Seasonal reproduction in the terrestrial isopod Armadillidium vulgare (Latreille): Geographical variability and genetic control of the response to photoperiod and temperature. *Inv. Rep. Dev.* 14: 131-151.

Stearns , S.C., Koella, J. C., 1986. The evolution of phenotypic plasticity in life-history traits: predictions of reaction norms for age and size at maturity. *Evolution*, 40: 893-913;

Sutton, S. L., Hassall, M., Willows, R., Davis, R.C., Grundy, A., Sunderland, K.D., 1984. Life histories of terrestrial isopods. In: S.L. Sutton and Holdich (eds). The Biology to Terrestrial Isopods. *Symposium of the Zoological Society of London*, 53, Oxford; Clarendon.

Warburg, M.R., 1987. Isopods and their terrestrial environment. *Adv. Ecol. Res.* 17: 187-242.

Warburg, M.R., 1994. Review of recent studies on reproduction in terrestrial isopods, *Inv. Rep. Dev.* 26 (1): 45-62.

Willows, R. 1984. Breeding phenology of *Ligia oceanica*. *Symp. Zool. Soc. Lond.* 53: 469-485.

Zimmer, M., 2003. Habitat and resource use by terrestrial isopods (Isopoda, Oniscidea). Biology of terrestrial Isopods, V: 243-261.

Zimmer, M., Brauckmann H.-J., 1997. Geographical and annual variations in the phenology of some terrestrial isopods (Isopoda: Oniscidea), *Biologia*, 52: 281-289.

Zimmer, M., Kautz, G., 1997. Breeding phenological strategies of the common woodlouse, *Porcellio scaber* (Isopoda: Oniscidea). *Eur. J. Soil Biol.*, 33 (2): 67-73.

3

Using Geographically Robust Models of Insect Phenology in Forestry

David R. Gray
Natural Resources Canada, Canadian Forest Service, Canada

1. Introduction

The history of observations of phenology dates to at least the 11th C BCE (Chen, 2003) when phenological events (commonly the flowering of woody plants) were recorded together with descriptions of weather and farming events. The principle objective of these early unsystematic observations was to improve agriculture. This general objective continued into the 13th C when it became accepted that repeated observations could provide insight into natural processes (Demarée & Rutishauser, 2011). In the 18th C Carl Linnaeus made special note of anomalous phenological events and inter-species synchrony. The principle objective remained the improvement of agriculture, and phenological calendars were compiled for individual regions in recognition that phenology varied geographically. Phenological calendars of the modern era include attempts to predict phenology based on longitude, latitude and elevation (Chen, 2003). Hopkins "Bioclimatic Law" (Hopkins, 1938) predicts the progression of spring phenological events based on these three geographic coordinates.

Despite the long history of phenological observations, it was not until 1849 that the term "phenology" first appeared in print (Demarée & Rutishauser, 2011) in an article by Charles-François-Antoine Morren (1807–1858). In 1735 Réne Antoine Ferchault Réaumur suggested that differences between locations in phenological events could be explained by differences in daily temperatures summed from some arbitrary date to the date of interest (Réaumur, 1735). Réaumur is now commonly credited with the concept of the degree-day sum model that continues to be widely used today.

Integrated pest management (IPM) has been one of the strategies to improve agricultural (including forestry) productivity. The prediction of pest phenology is a critical component of an IPM strategy (Damos & Savopoulou-Soultani, 2010) whether it is for the efficacious application of pesticides or for optimizing biological control tactics (Moerkens et al., 2011). Insect phenology models have been developed in response to this need, particularly within the agricultural industry. Insect phenology models are less common within forestry probably because there are fewer insect pest species and options for pest control are more limited so a model is not needed. Insect phenology models may be less common in forestry also because major insect pest species occur over a larger spatial scale in forestry than in agricultural, the accompanying topographic and climatic complexity is therefore greater, and the ability to accurately predict phenological events has been more difficult (Schaub et al., 1995).

The ability of a phenology model to perform satisfactorily over a broad, and climatically variable geographic range has been termed geographic robustness (Gray, 2004). This chapter examines the development of two geographically robust models of insect phenology in forestry, and the emerging application of geographically robust phenology models to investigate basic ecological questions such as population establishment and stability under changing climatic conditions, and risk assessment of alien pest introduction.

2. Insect developmental rates and temperature

Among the requisites of every insect is the availability of food during its feeding stage[s] and suitable temperatures to complete its life cycle. These needs can be summarized as a requisite seasonality. Seasonality is the predictable "occurrence of [a life stage event] within a definite limited period or periods of the astronomic (solar, calendar) year" (Lieth, 1974: 5). Implicit in this definition for temperate forest insects is the attainment of a cold-hardy stage before the onset of cold winter temperatures, that the emergence of the post-winter feeding stage will coincide with the availability of suitable food (which is often newly emerged foliage, hence the requisite of phenological synchrony between host trees and its insect herbivores), and that these events will coincide sufficiently each year for the continual survival of the population. Temperature, photoperiod, moisture and nutrition all influence insect phenology. Of these factors, temperature is the most spatially and temporally variable, is the strongest determinant of poikilotherm phenology, and is perhaps the best understood.

Insects progress through a life stage at a rate (*R*) that is dependent on the temperature (*T*) of that time (*t*):

$$R(t) = f\left(T[t]\right) \tag{1}$$

The developmental rate at a constant temperature *T* is calculated as the inverse of the time (*d*) needed to complete the life stage ($1/d$). The physiological age (*A*) of an individual at time *t* is given by the integral of the function

$$A(t) = \int_0^t f\left(T[t]\right)\,dt \tag{2}$$

and physiological age can be thought of as the state of the individual (i.e., the proportion of the life stage completed). The life stage is completed (and the subsequent life stage is begun) when physiological age reaches 1. In a computational environment (i.e., a simulation of insect phenology) developmental rates are calculated for each small time step Δt, and the stage is completed at $t = t_c$ when

$$\sum_{t=0}^{t=t_c} f\left(T[t]\right) = 1 \tag{3}$$

This generalized construct assumes that the response to a given temperature (*T*) is uniform for the duration of the life stage (i.e., $A = 0$ to 1). As we shall see (below), this assumption is infrequently tested, and at least in some cases where tested has been shown to be invalid, and developmental rate (*R*) is temperature and age dependent:

$$R(T[t]) = f(T[t], A[t]) \tag{4}$$

The earliest numerical relationship between developmental rate of insects and temperature used the linear heat summation model of Réaumur (1735). The general form of the model is

$$d = DD/(T - T_{\min}) \tag{5}$$

where d is the duration of the phenophase at temperature T, T_{min} is the temperature below which no development occurs, and DD is the thermal constant (i.e., the number of degree-days above T_{min} during the phenophase). Estimates of the thermal constant (DD) of a species can be obtained in the laboratory under a range of constant temperature regimes by conducting a simple linear regression between temperature (T) and the inverse of the duration of the life stage at constant T:

$$\frac{1}{d} = \frac{1}{DD}T + m \tag{6}$$

A minimum temperature threshold (T_{min}) for development is the x-intercept of regression equation (6), and is equal to

$$T_{\min} = -m \times DD \tag{7}$$

Field estimates of DD are obtained by assuming a reasonable T_{min} and summing the number of degrees by which the daily mean temperature exceeds T_{min} until the phenophase is completed. The ease with which parameters can be estimated, and more importantly, the ease with which phenological predictions can be made, have made linear heat summation (i.e., degree-day) models very popular.

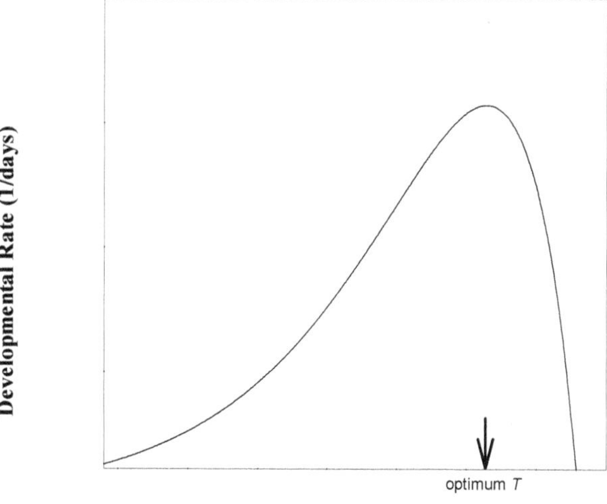

Fig. 1. Generalized developmental rate curve.

As early as 1932, the non-linear nature of the developmental rate function in insects was observed (Janisch, 1932) and various non-linear functions have been proposed and adopted (see (Wagner et al., 1984) for an early review). It is now widely accepted that developmental rates increase, with a positive first derivative, at low temperatures and decrease abruptly above some optimum temperature (Fig. 1). Logan et al. (Logan et al., 1976) used the technique of matched asymptotic expansion to capture this dual and asymmetric nature of the developmental rate function; their two formulas (with an exponential or a sigmoidal behavior for the low temperature segment) are now widely used. But non-linear models of phenological development did not become common until parameter estimation and summation of non-linear functions were simplified by the ready availability of digital computers.

3. Variability in developmental rate response

Part of the inherent variability within a population of insects includes developmental responses to temperature. Estimating this variability, and including it, in phenology models is important: pest management decisions are often made on the basis of the timing of early emerging insects (i.e., those with faster than average developmental responses); estimates of population stability depend more on the responses of the population as a whole than the response of just the average (or median) individual in the population (see below for a discussion of estimating population stability from phenological predictions). Population variability has been included in phenology models most often by one of three methodologies. The first is by using distributed delays (Manetsch, 1976) wherein individuals pass through a number of discrete "boxes" within each life stage in such a way that output from the life stage (and progression to the next life stage) depends on the number of boxes and transition rates between boxes. The second, and admittedly less abstract method, is to partition the population into subsets that each display a different developmental response (Stinner et al., 1975); the combination of the subsets reflects the population variability. This can be termed the "same-post" construct, in that all individuals develop to the same end post, but at different rates. The third methodology involves calculating normalized developmental times (τ) from a set of constant temperature observations:

$$\tau = \frac{time@T}{\text{median } time@T} \tag{8}$$

and describing the cumulative probability distribution of normalized times with an appropriate function (Sharpe & Demichele, 1977; Sharpe et al., 1977). This method is the most computationally efficient because the entire population is simulated using only the median developmental rates ($\tilde{R}(T)$ with the result that the proportion of the population (p) that has completed the life stage at time t is equal to the value of the cumulative probability distribution function for the summed developmental rates at time t

$$p(t) = \int_0^t \tilde{R}(T)\, dt \tag{9}$$

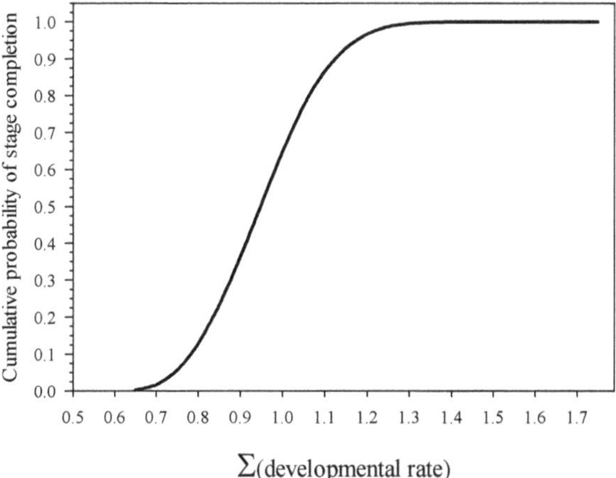

Fig. 2. Probability of stage completion with same-rate construct.

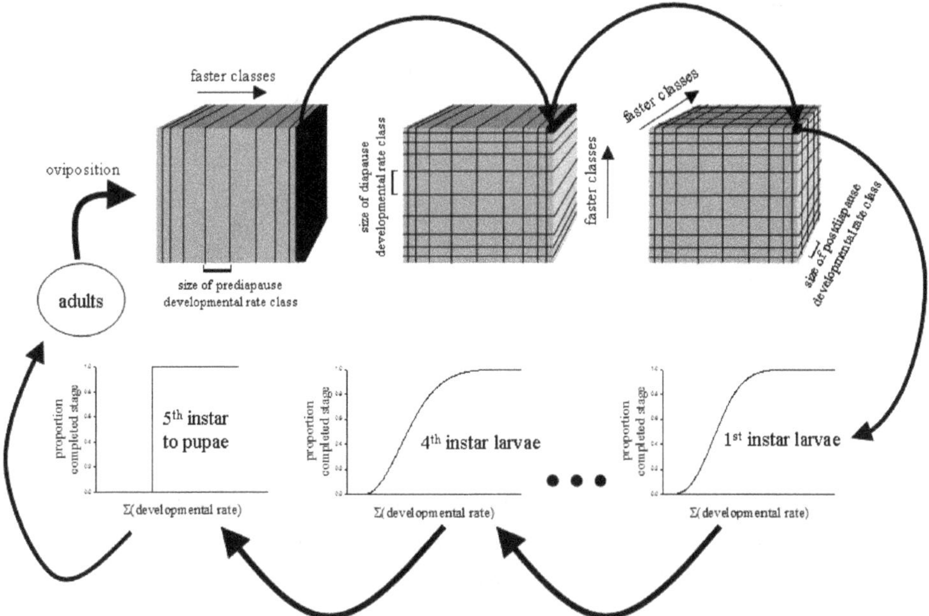

Fig. 3. A mixture of same-post and same-rate construction.

(Fig. 2). This can be termed the "same-rate" construct, in that all individuals develop at the same rate, but to different end posts. It has the obvious drawback that the whole population is assumed to respond according to the same developmental rate function; and therefore, the whole population fails to age when temperature (T_i) is below the developmental threshold, whereas it would be more accurate to age the portion of the population whose developmental threshold is lower than T_i. Gilbert et al. (2004) expanded an age-structured partial differential model to account for variability over the full range of the rate curve. Gray (2004) used a combination of the same-post and the same-rate methods (Fig. 3) to model gypsy moth (*Lymantria dispar* (L.)) egg development after showing that developmental responses during diapause (Gray et al., 2001) and postdiapause (Gray et al., 1995; Gray, 2009) are dependent on physiological age and temperature—this violation of independence on physiological age meant that age variability had to be specifically modeled in the population.

4. Geographic robustness of phenology models

High populations of forest insects often exist over very large geographic areas: during periods of outbreaks the spruce budworm (*Choristoneura fumiferana* Clem.) can occur in great abundance over an area of several hundreds of thousand km^2 (Hardy et al., 1986); a recent outbreak of the mountain pine beetle (*Dendroctonus ponderosae* Hopkins) occurred in over 90×10^3 km^2 (Aukema et al., 2008) of topographically diverse western Canada; the range of gypsy moth (*Lymantria dispar* L.) in North America extends from the Atlantic coast to the western Great Lakes, south to North Carolina and north into New Brunswick, and isolated populations have been detected in British Columbia, California, Utah, and many other

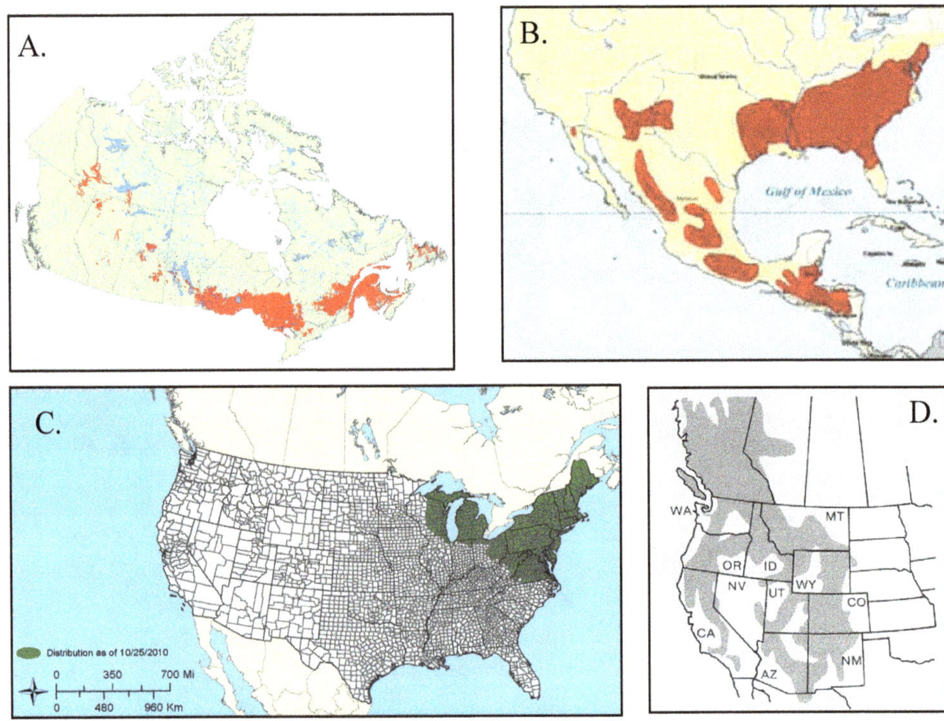

Fig. 4. Outbreak ranges of four dominant forest insect pests in North America
A) spruce budworm (Gray, unpublished; USA range not shown);
B) southern pine beetle (Salinas-Moreno et al., 2004);
C) gypsy moth (Canadian range not shown) (USDA: http://nrs.fs.fed.us/tools/afpe/?src=);
D) mountain pine beetle (Amman et al., 1990 (reprinted)).

locales; the range of the southern pine beetle (*Dendroctonus frontalis* Zimmermann) extends from Pennsylvania to Nicaragua (Fig. 4). Temperatures vary greatly over such large areas, and, therefore, insect development will vary also. The expected accuracy of a phenology model, measured on a large landscape scale is a function of the geographic robustness of the model—the ability to perform satisfactorily over a broad, and climatically variable geographic range (Gray, 2004). That few phenology models exhibit satisfactory geographic robustness can probably be attributed to at least one factor, and the implications that arise from that factor: an accurate description of the relationship between temperature and insect development has been elusive for the diapause developmental phase that is common among temperate insects.

The circular nature of insect life-cycles means that there is no default life stage at which to initiate a phenology model; instead an arbitrary "biofix" is chosen at which all individuals begin development (Logan & Bentz, 1999). The combination of the difficulty in modelling diapause development, and the greater interest in the feeding life stages (i.e., the larvae) has resulted in the construction of many phenology models that use diapause completion as the biofix (see as examples Moerkens et (2011) and many others). Here, I will describe early models of gypsy moth egg phenology to illustrate the likely reason for a lack of geographic robustness. The larvae within newly oviposited eggs begin development in a prediapause phase that is characterized by abundant morphological development (Leonard, 1968) and developmental rates that are favored by high temperatures (Gray et al., 1991). After spending 25 days in prediapause (assuming a constant temperature regime of 20°C), the gypsy moth enters the diapause phase, which is characterized by a virtual absence of morphological development and developmental rates that are favored by low temperatures (Gray et al., 2001). During the postdiapause phase, developmental rate is again favored by high temperatures (Gray et al., 1995). Eggs hatch upon completion of the postdiapause phase. Due largely to the difficulty in observing, either directly or indirectly, the transition between successive phases, model developers have modeled only the postdiapause phase after assuming that diapause is completed by the arbitrary biofix (Johnson et al., 1983; Waggoner, 1984; Lyons & Lysyk, 1989; Hunter, 1993). However, any error in choosing the date of diapause completion/postdiapause initiation (the biofix) results in an error in the amount of phenological development still to be completed for egg hatch, and a concomitant error in the estimate of developmental rate response to temperature. A date that precedes the true date of diapause completion will underestimate postdiapause developmental rates because *d*, the number of days to complete postdiapause was inflated by the portion of diapause not yet completed. A date that succeeds the true date of diapause completion will overestimate postdiapause developmental rates because *d* was deflated by the portion of postdiapause already completed. When such a model is run only in the locale where it was developed, the error in the date of diapause completion will tend to be compensated for by the error in developmental rates because the climatic regime within the locale tends to be consistent from year to year. However, when such a model is run in a locale with a markedly different climatic regime, the errors will lead to a bias in egg hatch prediction. Phenology models of larval development (Logan et al., 1991) will then commence with biased predictions of egg hatch. In two examinations of gypsy moth egg phenology models (Nealis et al., 1999; Régnière & Nealis, 2002), models that had been developed in NE United States (Johnson et al., 1983) and central Canada (Lyons & Lysyk, 1989) performed very poorly when tested against observed egg hatch in western Canada (Victoria, British Columbia)—they lacked geographic robustness. The model of Sawyer et al. (1993) includes the diapause phase and was developed using a combination of geographically diverse laboratory and field observations of egg hatch. It performed better than the former two models in the Victoria test; however, it performed poorly against a small subset of the same observations used in model construction.

The gypsy moth egg phenology model of Gray includes the prediapause (Gray et al., 1991), diapause (Gray et al., 2001), and postdiapause phases (Gray et al., 1995) of egg development. They used detailed measurements of the respiration rates of individual eggs to monitor phase transition (and therefore phase duration) under controlled temperatures. Their observations of postdiapause development demonstrated that the classic assumption that developmental response to a given temperature is uniform for the duration of the life stage (or phase, in this case) is invalid (Gray et al., 1995). They estimated "instantaneous" developmental rates at five temperatures and constructed an age- and temperature-dependent response curve for the postdiapause phase (Fig. 5). Estimates of "instantaneous" developmental rates during the diapause phase could only be modeled assuming control by two simultaneous temperature-dependent processes: (1) a typical developmental response to temperature that is inhibited by a temperature-activated biochemical agent; and (2) the temperature-dependent removal of the inhibiting agent (Fig. 6). Diapause is a critical developmental phase in maintaining seasonal development within a locale; and an accurate estimation of the day of diapause termination is a critical component of a geographically robust phenology model. The egg hatch model of Gray showed the highest level of geographic robustness in the Victoria test (Nealis et al., 1999; Régnière & Nealis, 2002); its geographic robustness has been improved more recently by a newer postdiapause submodel (Gray, 2009).

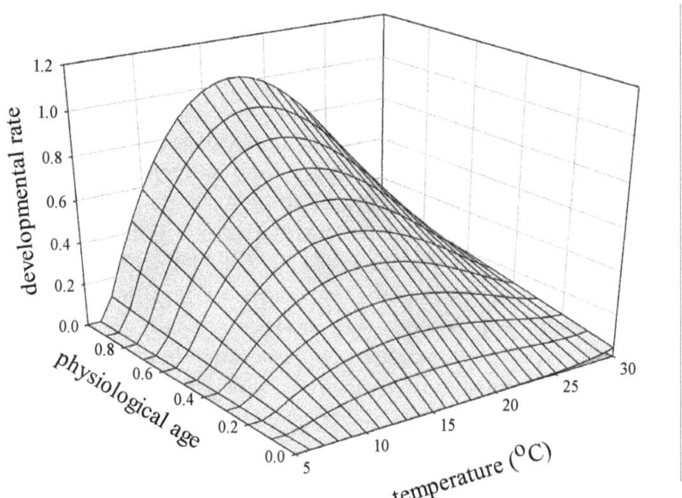

Fig. 5. Postdiapause developmental rates of gypsy moth are dependent on temperature and physiological age.

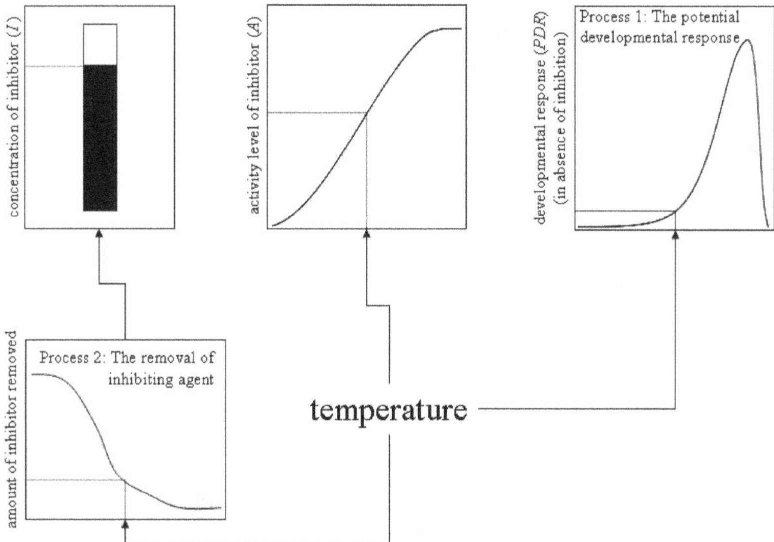

Fig. 6. Schematic representation of the dual-process diapause model. In the absence of an inhibiting agent, exposure to a given temperature (T) results in a developmental response (PDR) in the first process. However, the developmental response is inhibited at the onset of diapause by the presence of an inhibiting agent. In the second process a portion of the inhibiting agent is removed by exposure to T. Temperature also determines the level of activity ($A(T)$) of the remaining inhibitor (I).

Armed with a geographically robust model of egg hatch, a life-cycle phenology model was constructed by submitting a cohort of first instar larvae on each day that egg hatch occurred to the larval model of Logan et al. (1991). The size of each first-instar cohort is equal to the number of eggs that hatch on that day. Each day, the individuals that complete a larval life stage constitute a new cohort in the subsequent life stage. Individuals completing the final larval life stage constitute a new cohort in the pupal/adult model described by Sheehan (1992). Oviposited eggs constitute new egg cohorts of the next generation (Fig. 3). This composite, multi-generational phenology model was named the *GLS* model (=Gray/Logan/Sheehan, or Gypsy moth Life Stage model) (Gray, 2004).

5. Using phenology models to estimate population establishment and stability

Waggoner (1974) listed four requisites for the practical use of models of seasonality:

i. Considerable importance to mankind (usually, but not restricted to, economic importance).
ii. Variability in the phenophase—if inter-annual variation in the phenophase is relatively minor, the information probably already exists.
iii. Accuracy.
iv. Possibility of action; this may not be an absolute requisite, but it "certainly helps."

Forestry is not a classic example of crop production; the combination of the need for accuracy over extremely large, topographically complicated landscapes (Fig. 4) and the limited options for actions that depend on a phenological prediction may largely explain why insect phenology models are far fewer in forestry than in agriculture. However, climate has been linked to shifts in spring phenological events (Bradley et al., 1999), and one of the natural effects of climate change is to alter the spatial distribution of insect species (i.e., their natural range). Numerous examples already exist of a climate-mediated modification, in latitude and elevation, of historical ranges of insects (Parmesan & Yohe, 2003; Crozier, 2004). Perhaps the most interesting recent applications of phenology models within forestry have been to address the question of potential shifts in, or expansions of, the range of forest pest insects. Two prime examples are now discussed.

Given the existence of suitable host species, the ability to maintain seasonal development (defined above) is among the strongest determinant of the likelihood that an insect population can maintain seasonal development in a location (i.e., for that location to be part of the natural range of the insect). A geographically robust phenology model can estimate the probability of population establishment by quantifying the likelihood that temperature regimes in the location will consistently produce seasonal development (Gray, 2004). Gypsy moth and mountain pine beetle phenology modeling will illustrate this concept.

5.1 Mountain pine beetle

The mountain pine beetle is a native insect that constitutes a valuable link in the basic ecological cycle of western pine forests (Schmidt, 1988). Nonetheless, periodic, widespread, and severe outbreaks of mountain pine beetle pose a serious problem for the forest industry by killing trees, reducing wood supply, forcing changes to harvesting schedules, and increasing the fire hazard in areas of an outbreak. The most recent outbreak in the western province of British Columbia, which began in approx. 1994, eventually covered almost 14 million ha in 2008 (Safranyik et al., 2010). Previous outbreaks in British Columbia have normally collapsed following a severe, although not uncommon, cold weather event, and the Rocky Mountains have posed an impenetrable barrier to eastward expansion of outbreaks. However, the absence of a collapse-causing cold weather event, coupled with an abundance of highly suitable, even-aged host material, created an outbreak of historic proportions, and significant beetle populations appeared as far north in Alberta as Slave Lake (Safranyik et al., 2010), and as far east as the Alberta-Saskatchewan border, where the principle host, lodgepole pine (*Pinus. contorta* Dougl. *ex* Loud. var. *latifolia* Engelm. *ex* S. Watson), mixes with the closely related, potential host jack pine (*P. banksiana* Lamb.), which extends all the way to the Atlantic coast.

The mountain pine beetle phenology model (see Logan & Powell, 2001 for a description of rate curves), hereafter called the Logan et al. model, was used to evaluate mountain pine beetle seasonality in central Idaho under historic climate conditions and under a simulated climate regime with temperatures elevated by 2.5°C (Logan & Bentz, 1999) by testing for two requisite conditions of seasonality: (1) adult emergence from the host tree late enough to avoid lethal freezing temperatures but early enough to achieve full ovipositional potential before the onset of winter temperatures; and (2) synchronous adult emergence that leads to mass attack to overcome host defences. They demonstrated that historic temperatures satisfy the requisites of seasonality in central Idaho; that an increase in future temperatures would cause previously inhospitable locations (higher elevation) to become hospitable; but that the same warming of temperatures can have deleterious effects on the necessary synchrony of adult emergence. Logan & Powell (2001) used the phenology model to demonstrate how synchronous adult emergence within narrow bands of climatic regime are separated by asynchronous emergence within warmer or cooler bands of climatic regimes. The results suggest that warmer than average temperatures during the 1930s may have, for a short period, made high-elevation white bark pine (*P. albicaulus* Engelmann) forests susceptible to mountain pine beetle, and that mortality during this period was, in fact, due to a self-sustaining mountain pine beetle outbreak.

The determination that the distribution and abundance of mountain pine beetle is highly sensitive to variation in mean annual temperature in climatically marginal habitats led to consideration of the potential shift or expansion of its range in Canada under simulated climate-change temperatures (Safranyik et al., 2010). The phenology model estimated regions of adaptive seasonality in the recent past that reached the Atlantic coast and that coincide very well with the current distribution of pine species. However, under their climate-warmed scenario, the regions east of the Rocky Mountains that will produce adaptive seasonality are predicted to shift northward and beyond the region with high-volume boreal pine forests.

5.2 Gypsy moth

Since the accidental escape of a purposeful introduction near Boston, Massachusetts in 1869 (Liebhold et al., 1989), gypsy moth populations have spread to the north, south, and west. Populations were first reported in the eastern province of New Brunswick in 1936. However, since that time, populations in New Brunswick have never reached the sustained level observed further south and frequently decline without intervention. Populations were detected in the province of Quebec in 1924, but northward expansion has not continued for much of the past 90 years. In the western province of British Columbia, there have been repeated introductions of gypsy moth since 1978. However, in the year following a detected introduction, intensive pheromone trapping programs have many times failed to capture a single moth (Humble & Stewart, 1994). Similarly, numerous introductions of gypsy moth occur in the western United States every year, and many fail to lead to established populations even without intervention (Logan et al., 2007). This apparent failure in New Brunswick, Quebec, and British Columbia and in the western United States for populations to either establish, expand, or grow to outbreak proportions, is likely linked to the climatic conditions in each locale.

The geographically robust *GLS* model has been used several times to assess the potential range of gypsy moth within North America (Gray, 2004; Régnière et al., 2009), or sections thereof (Régnière & Nealis, 2002; Logan et al., 2007), New Zealand (Pitt et al., 2007), and trans-nationally to estimate the risk of introduction via international trade routes (Gray, 2010) under current climate conditions and under projected climate conditions of the future. In addition, *GLS* is used in the large-scale multi-year effort to slow the spread of gypsy moth in the USA.[1] (Roberts & Ziegler, 2007). However, despite using the same basic phenology

[1] Slow the Spread of the Gypsy Moth Project; http://www.gmsts.org/operations/index.htm

model (*GLS*), certain variations exist among these investigations in how model outputs were managed and interpreted.

Four of the preceding six examples (Régnière & Nealis, 2002; Logan et al., 2007; Pitt et al., 2007; Régnière et al., 2009) chose to run *GLS* within the BioSIM© system (Régnière et al., 1995). BioSIM© is primarily a weather generator and can be used to generate daily temperatures for specific locales where temperature records do not exist or to generate a gridded network of weather traces (Régnière & St-Amant, 2007). Unfortunately, BioSIM© also imposes two limitations on *GLS* that weaken its geographic robustness and influence the estimates of climatic suitability. These impositions are discussed below.

5.2.1 An arbitrary and globally defined date before which a specific phenological milestone must be achieved for generation success

In brief, the central criterion in the four examples for generational success in a location is that a phenological event (e.g., oviposition) must be able to achieve a stable equilibrium (=same date in successive generations) under the climatic conditions of the location (Fig. 7). Régnière et al (2002) stipulate that generation success in the western province of British Columbia can occur only if oviposition occurs before 25 October each year. Logan et al. (2007) require median oviposition to occur by 2 October in Utah, USA. Pitt et al. (2007) require all immature life stages to be in the egg stage by 31 December in the eastern Canadian province of New Brunswick and by 30 June in New Zealand. Régnière et al. (2009) make the same stipulation as Pitt et al. (2007) in their Canada-wide estimation of where generation success will be climatically possible.

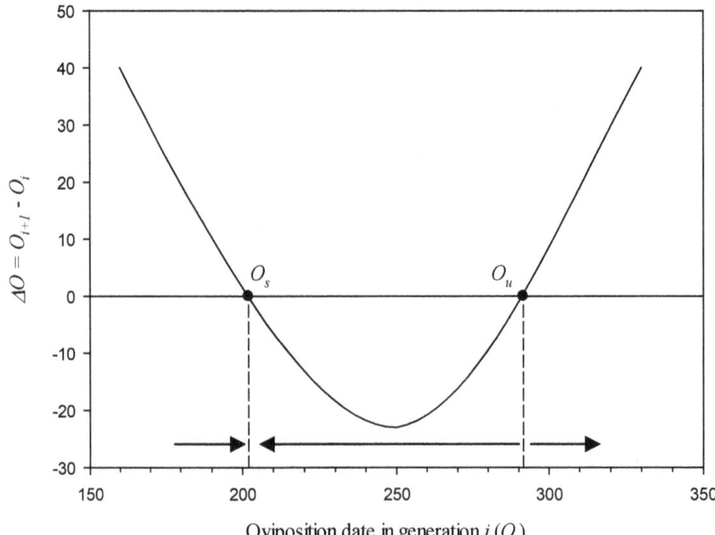

Fig. 7. Seasonal stability is possible in a location when the phenological event (e.g., oviposition) occurs earlier than O_u, an unstable equilibrium (identical dates in generation i and $i+1$). Oviposition later than O_u in generation i will result in ever-later oviposition dates. Oviposition dates earlier than O_u will result in a convergence to a stable equilibrium at O_s.

The rationale for the deadlines is that embryogenesis must reach the cold-hardy diapause phase (Leonard, 1968) before the onset of winter temperatures. However, these dates seem liberal (i.e., late), given that much of British Columbia and Utah will experience sub-zero temperatures by October, and sub-zero temperatures are extremely common throughout New Brunswick prior to 31 December. At the same time, daily maximum temperatures above the minimum prediapause developmental threshold 4°C (Gray et al., 1991) will still be common. Thus, these formulations of *GLS*, with arbitrary dates, allow the simulated population to escape cold-induced mortality while continuing prediapause development.

5.2.2 Removal of all population variability in oviposition date

In each location of interest, the authors (Régnière & Nealis, 2002; Logan et al., 2007; Pitt et al., 2007; Régnière et al., 2009) ran N independent simulations. Each simulation was initiated with a population oviposited on an arbitrary date; a stability flag (1 or 0) was assigned to the simulation based on the success or failure of the population to consistently achieve one or more specified phenological milestones by an arbitrary date during n generations under a climate regime typical of the location. All oviposition in generation $i+1$ occurred on the day of peak female abundance (or median female emergence) in generation i. The probability of stability was then calculated from the N stability flags.

Although the egg hatch model within *GLS* has been shown to be relatively insensitive to oviposition date, it is not entirely insensitive. Within the northern hemisphere, hatch success becomes strongly influenced by oviposition date in northern locations (Gray, 2004; Gray, 2010); presumably this is also the case in southern locations within the southern hemisphere, or at higher elevations regardless of the latitude.

Furthermore, initiating each generation with 100% of the oviposition on the single day of peak female abundance (or median female emergence) of the previous generation in essence "saves" a portion of the population whose true simulated oviposition dates were later than the peak (or median) date and, therefore, too late for survival of the oviposited eggs (Gray, 2004; Gray, 2010).

And finally, the combination of a binomial flag and the removal of population variability in oviposition date has the effect of estimating identical risk for locations where the real risk should be judged to be different. A location in which the median date of oviposition meets the arbitrary deadline in each of n generations—but 49% of oviposition each generation occurs too late—has an estimated p (viability: (Pitt et al., 2007), or probability of establishment (Régnière & Nealis, 2002; Logan et al., 2007; Régnière et al., 2009), equal to a location where 100% of the oviposition meets the deadline each generation. Conversely, a location in which the median date of oviposition fails to meet the deadline—but 49% of oviposition does meet the deadline—has a p equal to a location in which none of the oviposition meets the deadline.

Although it is correct that seasonal stability (Fig. 7) is a requisite for population establishment, it is not known what proportion of a population must survive each generation in order to maintain a population. Thus, it may be the case that less than 50% intergenerational survival is sufficient for gypsy moth establishment; or it may be the case that more than 50% is necessary. But until such a proportion is known, it is much preferable to estimate probability of establishment (or viability) in a location by the intergenerational survival that a geographically robust model such as *GLS* permits. Contrary to the claim of Pitt et al. (Pitt et al., 2007), variability in oviposition dates within a population is a critical component in estimating the potential for seasonal development and establishment within a location. Within the broad geographic areas that are either highly climatically suitable or highly unsuitable, estimates of probability of establishment will be little affected (or unaffected) by the removal of oviposition variability. However, where climatic suitability is marginal, it is recommended that the risk of establishment be estimated from the intergenerational survival of a model that maintains population variability in all its life stages.

6. Conclusion

Although insect phenology models may be less common in forestry than in agricultural systems, at least two examples in forestry illustrate the potential of geographically robust phenology models to address questions of considerable ecological importance: the potential range of insect species under climate change scenarios.

There can be little doubt that establishment of a stable population of poikilothermic organisms can only occur where the climatic regime leads to adaptive seasonality. A stage-specific, geographically robust phenology model is ideally suited to determining where adaptive seasonality is possible, and therefore, the potential range of an insect species under current climatic regimes, and under the climate regimes that may exist in the future. However, it should be emphasized that imposing an arbitrary biofix on a geographically robustness model, and/or failing to maintain population variability in key phenological events (e.g., oviposition date) will negatively impact the accuracy of the predictions of even geographically robust models where climatic regimes are marginally suitable for adaptive seasonality. These marginal regions should be of greater interest, not less, than the broad regions where the climatic regime is highly suitable.

7. References

Amman, G.D., McGregor, M.D., & Dolph, R.E.J. (1990 (reprinted)). Mountain pine beetle. Forest Insect & Disease Leaflet 2. United States Department of Agriculture, Forest Service, Washington, DC, USA.

Aukema, B.H., Carroll, A.L., Zhen, Y., Zhu, J., Raffa, K.F., Moore, R.D., Stahl, K., & Taylor, S.W. (2008). Movement of outbreak populations of mountain pine beetle: influences of spatiotemporal patterns and climate. *Ecography*, Vol. 31, pp. 348–358

Bradley, N.L., Leopold, A.C., Ross, J., & Huffaker, W. (1999). Phenological changes reflect climate change in Wisconsin. *Proceedings of the National Academy of Sciences of the United States of America*, Vol. 96, No. 17, pp. 9701–9704

Chen, X. (2003). Phenological data, networks, and research: East Asia. In: *Phenology: an integrative environmental science*. M.D. Schwartz. pp. 11–25, Kluwer Academic Publishers, ISBN 1-4020-1580-1, London, UK

Crozier, L. (2004). Warmer winters drive butterfly range expansion by increasing survivorship. *Ecology*, Vol. 85, No. 1, pp. 231–241

Damos, P.T., & Savopoulou-Soultani, M. (2010). Development and statistical evaluation of models in forecasting moth phenology of major lepidopterous peach pest complex for Integrated Pest Management programs. *Crop Protection,* Vol. 29, No. 10, pp. 1190-1199

Demarée, G.R., & Rutishauser, T. (2011). From "Periodical Observations" to "Anthochronology" and "Phenology" - the scientific debate between Adolphe Quetelet and Charles Morren on the origin of the word "Phenology". *International Journal of Biometeorology,* Vol. pp. 1-9

Gilbert, E., Powell, J.A., Logan, J.A., & Bentz, B.J. (2004). Comparison of three models predicting developmental milestones given environmental and individual variation. *Bulletin of Mathematical Biology,* Vol. 66, No. 6, pp. 1821-1850

Gray, D.R., Logan, J.A., Ravlin, F.W., & Carlson, J.A. (1991). Toward a model of gypsy moth egg phenology: using respiration rates of individual eggs to determine temperature-time requirements of prediapause development. *Environmental Entomology,* Vol. 20, No. 6, pp. 1645-1652

Gray, D.R., Ravlin, F.W., Régnière, J., & Logan, J.A. (1995). Further advances toward a model of gypsy moth (*Lymantria dispar* (L.)) egg phenology: respiration rates and thermal responsiveness during diapause, and age-dependent developmental rates in postdiapause. *Journal of Insect Physiology,* Vol. 41, No. 3, pp. 247-256

Gray, D.R., Ravlin, F.W., & Braine, J.A. (2001). Diapause in the gypsy moth: a model of inhibition and development. *Journal of Insect Physiology,* Vol. 47, No. 2, pp. 173-184

Gray, D.R. (2004). The gypsy moth life stage model: landscape-wide estimates of gypsy moth establishment using a multi-generational phenology model. *Ecological Modelling,* Vol. 176, pp. 155-171

Gray, D.R. (2009). Age-dependent postdiapause development in the Gypsy Moth (Lepidoptera: Lymantriidae) Life Stage Model. *Environmental Entomology,* Vol. 38, No. 1, pp. 18-25

Gray, D.R. (2010). Hitchhikers on trade routes: a phenology model estimates the probabilities of gypsy moth introduction and establishment. *Ecological Applications,* Vol. 20, No. 8, pp. 2300-2309

Hardy, Y., Mainville, M., & Schmitt, D.M. (1986). An atlas of spruce budworm defoliation in eastern North America, 1938-1980. Miscellaneous Publication 1449. United States Department of Agriculture, United States Forest Service, Washington, DC, USA.

Hopkins, A.D. (1938). Bioclimatics—a science of life and climate relations. Miscellaneous Publication 280. United States Department of Agriculture, US Forest Service, Washington, DC, USA. p. 188.

Humble, L., & Stewart, A.J. (1994). Gypsy moth. Forest Pest Leaflet 75. Natural Resources Canada, Canadian Forest Service - Pacific Forestry Centre, Victoria, BC, Canada. p. 8.

Hunter, A.F. (1993). Gypsy moth population sizes and the window of opportunity in spring. *Oikos,* Vol. 68, No. x, pp. 531-538

Janisch, E. (1932). The influence of temperature on the life-history of insects. *Transactions of the Entomological Society of London,* Vol. 80, pp. 137-168

Johnson, P.C., Mason, D.P., Radke, S.L., & Tracewski, K.T. (1983). Gypsy moth, *Lymantria dispar* (L.) (Lepidoptera: Lymantriidae), egg eclosion: degree-day accumulation. *Environmental Entomology,* Vol. 12, No. 3, pp. 929-932

Leonard, D.E. (1968). Diapause in the gypsy moth. *Journal of Economic Entomology*, Vol. 61, No. 3, pp. 596–598

Liebhold, A.M., Mastro, V.C., & Schaefer, P. (1989). Learning from the legacy of Leopold Trouvelot. *Bulletin of the Entomological Society of America*, Vol. 35, No. 2, pp. 20–22

Lieth, H. (1974). *Phenology and seasonality modeling*. Springer-Verlag, ISBN 0-387-06524-5, New York, New York, USA

Logan, J.A., Wollkind, D.J., Hoyt, S.C., & Tanigoshi, L.K. (1976). An analytic model for description of temperature dependent rate phenomena in arthropods. *Environmental Entomology*, Vol. 5, No. x, pp. 1133–1140

Logan, J.A., Casagrande, P.A., & Liebhold, A.M. (1991). Modeling environment for simulation of gypsy moth (Lepidoptera: Lymantriidae) larval phenology. *Environmental Entomology*, Vol. 20, No. 6, pp. 1516–1525

Logan, J.A., & Bentz, B.J. (1999). Model analysis of mountain pine beetle (Coleoptera: Scolytidae) seasonality. *Environmental Entomology*, Vol. 28, No. 6, pp. 924–934

Logan, J.A., & Powell, J.A. (2001). Ghost forests, global warming, and the mountain pine beetle (Coleoptera: Scolytidae). *American Entomologist*, Vol. 47, No. 3, pp. 160–173

Logan, J.A., Régnière, J., Gray, D.R., & Munson, A.S. (2007). Risk assessment in face of a changing environment: gypsy moth and climate change in Utah. *Ecological Applications*, Vol. 17, No. 1, pp. 101–117

Lyons, D.B., & Lysyk, T.J. 1989. Development and phenology of eggs of gypsy moth *Lymantria dispar* (Lepidoptera: Lymantriidae) in Ontario. *In* Proceedings, Lymantriidae: a comparison of features of new and old world tussock moths. USDA Gen. Tech. Rep. NE-123:351-365. *Edited by* W.E. Wallner, &K.A. McManus. United States Department of Agriculture, Forest Service, Radnor, PA, USA. pp. 351–365.

Manetsch, T.J. (1976). Time-varying distributed delays and their use in aggregative models of large systems. *IEEE Transactions on systems, and and cybernetics*, Vol. SMC-6, No. 8, pp. 547–553

Moerkens, R., Gobin, B., Peusens, G., Helsen, H., Hilton, R., Dib, H., Suckling, D.M., & Leirs, H. (2011). Optimizing biocontrol using phenological day degree models: The European earwig in pipfruit orchards. *Agricultural and Forest Entomology*, Vol. 13, pp. 301–312

Nealis, V., Régnière, J., & Gray, D. (1999). Modeling seasonal development of gypsy moth in a novel environment for decision-support of an eradication program. Gen. Tech. Report NE-277. *Edited by* A.M. Liebhold, M.L. McManus, I.S. Otvos, &S.L.C. Fosbroke. United States Department of Agriculture, Forest Service, Washington, DC, USA. pp. 124–132.

Parmesan, C., & Yohe, G. (2003). A globally coherent fingerprint of climate change impacts across natural systems. *Nature*, Vol. 421, No. 6918, pp. 37–42

Pitt, J.P.W., Régnière, J., & Worner, S. (2007). Risk assessment of the gypsy moth, *Lymantria dispar* (L), in New Zealand based on phenology modellng. *International Journal of Biometeorology*, Vol. 51, pp. 295–305

Réaumur, R.A.F. (1735). *Obserations du thermomètre, faites à Paris pendant l'année 1735, comparées avec celles qui ont été faites sous la ligne, à l'isle de France, à Alger et quelques unes de nos isles de l'Amérique*. Académie Royale des Sciences, Paris, France

Régnière, J., Cooke, B.J., & Bergeron, V. (1995). BioSIM: a computer-based decision supprt tool for sseasonal planning of pest management activities. Information Report LAU-X-116. Natural Resources Canada, Canadian Forest Service - Laurentian Forestry Centre, Ste-Foy, QC, Canada. p. 35.

Régnière, J., & Nealis, V. (2002). Modelling seasonality of gypsy moth, *Lymantria dispar* (Lepidoptera: Lymantriidae), to evaluate probability of its persistence in novel environments. *The Canadian Entomologist*, Vol. 134, No. 6, pp. 805–824

Régnière, J., & St-Amant, R. (2007). Stochastic simulation of daily air temperature and precipitation from monthly normals in North America north of Mexico. *International Journal of Biometeorology*, Vol. 51, No. 5, pp. 415–430

Régnière, J., Nealis, V.G., & Porter, K.B. (2009). Climate suitability and management of the gypsy moth invasion into Canada. *Biological Invasions*, Vol. 11, pp. 135–148

Roberts, E.A., & Ziegler, A.H. (2007). Gypsy moth population monitoring and data collection. General Technical Report NRS-6. *Edited by* P.C. Tobin, &L.M. Blackburn. United States Department of Agriculture, Forest Service, Washington, DC. pp. 33–46.

Safranyik, L., Carroll, A.L., Régnière, J., Langor, D.W., Riel, W.G., Shore, T.L., Peter, B., Cooke, B.J., Nealis, V.G., & Taylor, S.W. (2010). Potential for range expansion of mountain pine beetle into the boreal forest of North America. *The Canadian Entomologist*, Vol. 142, No. 5, pp. 415–442

Salinas-Moreno, Y., Mendoza, M.G., Barrios, M.A., Cisneros, R., Macías-Sámano, J., & Zúñiga, G. (2004). Areography of the genus *Dendroctonus* (Coleoptera: Curculionidae: Scolytidae) in Mexico. *Journal of Biogeography*, Vol. 31, pp. 1163–1177

Sawyer, A.J., Tauber, M.J., Tauber, C.A., & Ruberson, J.R. (1993). Gypsy moth (Lepidoptera: Lymantriidae) egg development: a simulation analysis of laboratory and field data. *Ecological Modelling*, Vol. 66, No. x, pp. 121–155

Schaub, L.P., Ravlin, F.W., Gray, D.R., & Logan, J.A. (1995). Landscape framework to predict phenological events for gypsy moth (Lepidoptera: Lymantriidae) management programs. *Environmental Entomology*, Vol. 24, No. 1, pp. 10–18

Schmidt, W.C. (1988). Lodgepole pine: and ecological opportunist. GTR-INT-262. *Edited by* G.D. Amman. United States Dept. of Agriculture-Forest Service, Kalispell, MT, USA. pp. 14–20.

Sharpe, P.J.H., & Demichele, D.W. (1977). Reaction kinetics of poililotherm development. *Journal of Theoretical Biology*, Vol. 64, No. 4, pp. 649-670, 0022-5193

Sharpe, P.J.H., Curry, G.L., Demichele, D.W., & Cole, C.L. (1977). Distribution model of organism development times. *Journal of Theoretical Biology*, Vol. 66, No. 1, pp. 21–38, 0022-5193

Sheehan, K.A. (1992). User's guide for GMPHEN: gypsy moth phenology model. General Technical Report NE-158. United States Department of Agriculture, Forest Service, Washington, DC. p. 29.

Stinner, R.E., Butler, G.D., Jr., Bacheler, J.S., & Tuttle, C. (1975). Simulation of temperature-dependent development in population dynamcis models. *The Canadian Entomologist*, Vol. 107, pp. 1167–1174

Waggoner, P.E. (1974). Using models of seasonality. In: *Phenology and seasonality modeling*. H. Lieth. pp. 401–405, Springer-Verlag, New York, NY, USA

Waggoner, P.E. (1984). The hatching of gypsy moth eggs, a phenological model. *Agricultural and Forest Meteorology*, Vol. 33, pp. 53–65

Wagner, T.L., Wu, H.-I., Sharpe, P.J.H., Schoofield, R.M., & Coulson, R.N. (1984). Modeling insect development rates: a literature review and application of a biophysical model. *Annals of the Entomological Society of America*, Vol. 77, pp. 208–225

Relationships Between the Fruiting Phenology of *Prunus jamasakura* and Timing of Visits by Mammals – Estimation of the Feeding Period Using Camera Traps

Shinsuke Koike et al.*
Tokyo University of Agriculture and Technology, Tokyo, Japan

1. Introduction

Fleshy fruits have an endozoochory seed dispersal system. Fleshy fruits provide food to dispersal agents in the form of nutritious accessory structures closely associated with the seeds. The exact percentage of tree species dispersed by endozoochory varies from place to place, but can exceed 50% in some temperate deciduous forests (Howe & Smallwood, 1982). So, fruits support a high percentage of a forest's frugivore biomass and the majority of woody plant species are dispersed by frugivores. Seed dispersal by frugivores in temperate deciduous forests has received considerable attention over time, and this interaction is considered an important force shaping the ecology and evolution of frugivore and tree populations (Estrada & Fleming, 1986). Also, many studies report endozoochory generally coevolved as a mutualistic relationship (e.g., Estrada & Fleming, 1986; Howe, 1986; Fleming & Estrada, 1993; Levey et al., 2001; Dennis et al., 2007).

Birds and mammals are the most important seed dispersers, but a wide variety of other animals, including lizards and fish can transport viable seeds (e.g., Corlett, 1998; Willson, 1993; Jordano, 1995; Castilla, 2000; Howe & Westley, 1988; Gottsberger, 1978). However, seed dispersal by birds has been examined most frequently (e.g., Herrera, 2002). Monkeys (e.g., Howe, 1980; Estrada & Coates-Estrada, 1994; Yumoto et al., 1998) and bats (e.g., Fleming, 1981; Gribel, 1988; Izhaki et al., 1995) have long been recognized as important seed dispersers and have been studied intensively, because they can observed directly. Some recent studies have documented the importance of carnivorous mammals in seed dispersal (Willson, 1993; Traveset & Willson, 1997; Hickey et al., 1999). However, the relationship between carnivorous mammals as seed dispersal agents and fleshy-fruit producing plants

* Hideto Morimoto[1], Shinsuke Kasai[1,2], Yusuke Goto[1,3], Chinatsu Kozakai[1], Isao Arimoto[1], and Koji Yamazaki[4]
[1]*Tokyo University of Agriculture and Technology, Tokyo, Japan*
[2]*Present address: Shireto Foundation, Hokkaido, Japan*
[3]*Present address: Tateyama Caldera Sabo Museum, Toyama, Japan*
[4]*Ibaraki Nature Museum, Ibaraki, Japan*

has been somewhat understated, compared with the role of other frugivorous mammals (Herrera, 1989). There are few reports of seed dispersal by other orders of mammals, especially for forest inhabitants, because they are mostly nocturnal or difficult to observe directly.

In Japan, seed dispersal by **Japanese macaque (*Macaca fuscata*)** has been studied intensively among the mammals and shows macaques are important seed dispersers (e.g., Otani, 2003, 2004; Tsuji et al., 2011a). However, the importance of carnivorous mammals in seed dispersal is poorly documented in Japan (e.g., Otani, 2002). Some studies have shown carnivores have features which make them effective seed dispersers. For instance, the sectorial or bunodont cheek teeth of carnivores minimize the damage to seeds (Koike et al., 2008a, 2008b). Also, carnivores disperse seeds longer distances than other animals (Koike et al., 2011). This characteristic would allow carnivores to disperse large numbers of seeds throughout a large area.

Basic data on the number of seeds per fecal sample, which serves as an index of the quantitative effectiveness of seed dispersal (Schupp, 1993), as well as basic information on the physical damage to seeds that pass through a mammal's gastrointestinal tract, which is an index of qualitative effectiveness of seed disperal, are known in Japan for carnivores (Koike et al., 2008b) and macaques (Tsuji et al., 2011a). Koike et al., (2008b) studied five carnivores in the Okutama Mountains (Asiatic black bear (*Ursus thibetanus*), Japanese marten (*Martes melampus*), Japanese badger (*Meles meles*), red fox (*Vulpes vulpes*), and raccoon dog (*Nyctereutes procyonoides*)) and recovered seeds from 17 plant species, representing about 50% of the fleshy-fruited plants occurring in the same study forest, from the carnivore fecal samples. Large numbers (nine to 10,256) of seeds were present in those fecal samples. Almost all the seeds of fleshy fruits retrieved from the fecal samples were undamaged, while no intact acorns or nuts were recovered. These findings suggest all five carnivores can act as seed dispersers for some fleshy-fruited plants in cool-temperate deciduous forests. On the Kinkazan Island in northern Japan, Tsuji et al., (2011a) suggest macaque acts as s seed dispersers in summer and fall. The percentage of intact seeds found in macaques fecal samples was significantly negatively correlated with the seed's mean diameter, and this relationship was strengthened for non-fleshy fruits. This suggests the composition of seeds dispersed by macaques is not rigid, but is determined instead by the vegetation found in a given habitat.

To understand systems of endozoochory, it is also important to consider the feeding period in terms of fruiting phenology. For example, previous studies found a close relationship between the maturation periods of many fleshy-fruited plants and the movement of frugivorous birds in several forest ecosystems (e.g., Thompson & Willson, 1979; Stiles, 1980; Stapanian, 1982; Hererra, 1984; Noma & Yumoto, 1997). Also, some carnivores such as bears, martens, and masked palm civet (*Paguma larvata*) can climb trees to feed on fruits before the fruits drop to the ground. If fruits are ingested after seed maturation, dispersed seeds could germinate. Therefore, whether or not these frugivores are effective seed dispersers can be judged by examining the relationship between the fruit-feeding period of frugivores and the phenology of fruits.

Previous studies indicate bears fed only on mature fruits based on the claw marks left on the stems and the presence of a "bear shelf" (Koike et al., 2008a; Koike, 2009). However, other mammals do not necessary clearly create feeding sign when they consume the ripe fruit. Recently, camera traps have proved useful in the study wildlife ecology. Camera traps are a non-invasive method of monitoring wildlife which generally cause minimal disturbance to the target species. We can observe nocturnal mammals or mammals sensitive to disturbance by using camera traps.

We chose the fruit of *Prunus jamasakura* Siebold ex Koidz. to answer the questions posed below for three reasons. First, *P. jamasakura* is a common species in the deciduous forest in central Japan (Oba, 1989). Second, *Prunus* is one of the rare woody species bearing fleshy fruits during June–July (Koike, 2009) and, so is probably a common food item for mammals during that time (e.g., Koike, 2010; Tsuji et al., 2011b). Third, the fruiting phenology of *P. jamasakura* is well known (Koike et al., 2008a).

Our objective was to evaluate the effectiveness of temperate zone mammals as potential seed dispersers by estimating the timing of their feeding on the fleshy fruit of *P. jamasakura*. We intended to answer three questions with camera trapping: (1) When do mammals visit *P. jamasakura* trees in relation to the tree's fruiting phenology? (2) Does fruiting phenology influence the mammal's behavior? (3) Which mammals potentially disperse the *P. jamasakura* seeds? We discuss the significance of the mammals as potential seed dispersers for *P. jamasakura* based on our results. We also discuss the relationship between mammalian behavior and fruiting phenology.

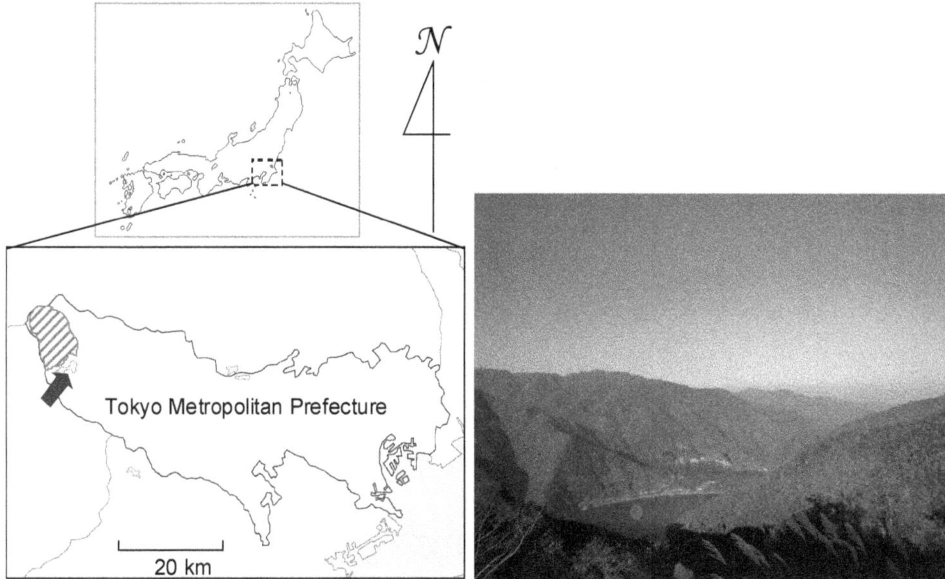

Fig. 1. The study area in an inset map taken from an outline map of Japan (left).
A photograph taken at the position of the arrow in the map (right).

2. Method

2.1 Study area

The study was conducted in the Mine area of the Okutama Mountains, approximately 100 km west of Tokyo central (Fig. 1). The climate of the study area is the coastal Pacific Ocean type, with heavy rainfall in summer and little snow in winter. The mean annual precipitation is 1,586 mm; mean snowfall ranges from 2 to 4 cm; the mean annual temperature is 11.8 °C, and mean temperatures range from 0.6 °C in January to 24.2 °C in August (Tokyo Environmental Office, 2000).

Forest vegetation covers most of the study area. Natural forests cover 41.3% of the area, and conifer plantations (*Cryptomeria japonica* D. Don or *Chamaecyparis obtusa* (Zieb. et Zucc.) Endlicher) cover 50.3% (Koike, unpubl. data). Natural forests in the lower mountain zone (400–500 m a.s.l.) are dominated by *Castanea crenata* Zieb. et Zucc. and *Quercus serrata* Thunb.; in the middle zone (500–1,500 m a.s.l.) by *Quercus crispula* Blume, *C. crenata*, and *Fagus crenata* Blume; and in the upper zone (1,500–1,800 m a.s.l.) by *Abies homolepis* Zieb. et Zucc. and *Tsuga diversifolia* Masters. The study area was located in a *Q. crispula–C. crenata* community that ranged from 800 to 1,000 m a.s.l. in the middle zone.

2.2 General information

We summarized the information about *P. jamasakura* from Oba, (1989), Mogi et al., (2000), Takeuchi et al., (2005), and Koike, (unpublished data). Known in Japanese as *Yama-zakura, P. jamasakura* is a species of cherry native to Japan and Korea, occurring at medium altitudes of up to about 1,000 m in the temperate zone. It prefers sunshine and well-drained soil. This deciduous tree reaches a height of 20–25 m, with a trunk up to 1 m diameter. Its crown spreads to a width of 10–20 m. The flowers are produced in early spring and pollinated by bees.

Fig. 2. The matured fruit of *Prunus jamasakura*. Photo by Motoki Kato.

The fruit is a small drupe, about 10-15 mm in diameter, green at first, then red and finally ripening to black in early summer (Fig. 2). The edible fruit is variably sweet to somewhat astringent and bitter to eat fresh. It contains a single hard-shelled stone 5–12 mm in diameter and 4–8 mm thick with a seed (kernel) inside the stone about 5-8 mm long. The fruits are readily eaten by numerous birds and mammals (Koike & Masaki, 2008; Yoshikawa et al., 2009; Tsuji et al., 2011b), which digest the fruit flesh and disperse the seeds in their droppings. Some rodents and a few birds (notably the *Streptopelia orienta,* Eastern turtle dove) also crack open the stones to eat the kernel (Yagihashi, 2001; Koike, unpublished data).

About 30 species of terrestrial mammals inhabit the Okutama Mountains. The larger ones include Asiatic black bear, Sika deer (*Cervus nippon*), Japanese serow (*Capricornis crispus*), wild boar (*Sus scrofa*), Japanese macaque, red fox, Japanese marten, Japanese badger, masked palm civet, and raccoon dog (Tokyo Environmental Office, 2010). Most of these mammals are carnivores (bears, martens, badgers, foxes, and raccoon dogs); macaques have been recognized as seed dispersers and have been studied in Tokyo (e.g., Koike et al., 2008b; Tsuji 2011).

2.3 Fruiting phenology and defining the flowering date

The fruiting phenology of *P. jamasakura* is known in detail (Ishii, 1991; Koike et al., 2008a) including fruit color, the number of fruits persisting, the fruit and seed size, seed maturation, and the sugar concentration. Ishii, (1991) calculated the *P. jamasakura* fruit maturation period from the flowering date. Below, 'day *n*' represents the date *n* days after the flowering date.

We summarized the information about the fruiting phenology of *P. jamasakura* from Ishii, (1991) and Koike et al., (2008a). All fruits were recorded as green until day 38 for fruit color and bearing number. Red-green fruit then appeared on day 40, followed by red on day 46 and black on day 54. The proportion of persisting fruit decreased gradually until day 60, and then it decreased rapidly. For fruit size and seed size, means of fruit size were not different among fruits of different colors. Fruit size increased from day 10 to 20, from day 34 to 36, and from day 50 to 58. No change in diameter was seen after day 58. Fruit diameter increased in a double sigmoid growth curve and showed three phenological stages. Seed size became measurable beyond day 28, when the endocarp hardened. There was no distinct variation in diameter among fruit colors at the same stage. Seed size was almost constant at all stages. Sugar concentration varied by fruit color and increased as the fruit matured. For seed maturation, no germination was observed until day 44. The percentage rose rapidly on day 50 to 46.1%, and then gradually until day 66. Germination percentages were not different between the differently colored fruit.

We used the method described by Koike et al., (2008a) to determine the flowering day of *P. jamasakura* trees where we set up the camera traps to clarify the fruiting phenology.

In the study area, we chose six *P. jamasakura* trees (DBH > 30 cm) to monitor for flowering once every two days in April 2003 and 2004. We sampled ten branches at random from each sample tree and counted the flowering buds from the tip of each 50 cm long branch while measuring 20 cm wide, 10cm on each side of the branch. The flowering ratio was defined as the number of flowering buds divided by the total number of buds. When the average flowering ratio reached 80%, that day was designated as the "date of flowering" for each tree. After flowering we checked the fruit color ratio of bearing fruit once every ten days to compare this data with existing data on fruiting phenology (Koike et al., 2008a).

2.4 Estimate of fruit production

In our study area, few suitable trees were available for observing fruit production in detail, because *P. jamasakura* trees form part of the upper tree layer in the forest and the trees grow on steep slopes. Therefore, we used the simpler method of estimating fruit production in the study areas even though it was only an indicator of fruit production. We used the methods described by Mizui, (1993). We used binoculars to sample five randomly selected branches from each sample tree and counted the fruits on each branch within a range of about 50 cm length and about 20 cm width (10cm each side of the branch) from the tip of the branch. We counted the fruits once between day 40 and day 50.

2.5 Camera trap survey

While investigating fruiting phenology, we recorded the frequency of visits by mammals under the crown of each target tree. An automatic camera trap (Fieldnote, Marifu Co. Iwakuni City, Japan) was set up 1.5 m above the ground (Fig. 3). One camera was set up next to each tree using the tree as a target in as flat an area as possible. We set five cameras in total. Cameras operated 24 hr per day. To avoid running out of film, we checked the film twice a week from late April to the middle of August 2003 and 2004. To avoid duplicate counts, we defined conspecific animals recorded within a 30 minute period as the same individual (O'Brien et al., 2003).

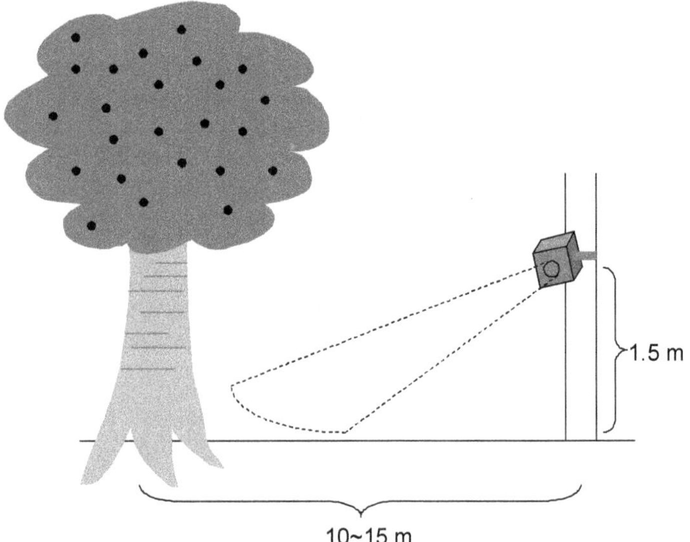

Fig. 3. The setup of an automatic camera trap.

3. Result

3.1 Fruiting phenology and estimate of fruit production

The flowering dates of the sample trees were 24 to 30 April, 2003, and were 27 to 30 April, 2004. There were no differences in the fruit color ratio of bearing fruit after the day of flowering between this study and a previous study (Koike et al., 2008a). We used the date starting at the flowering day to count to the fruit maturation date. The period of fruit maturation was between day 50 and day 66 was after 14 June in 2003 and after 17 June in 2004.

Year	Tree Number	*Fruit production		Number of photographs					
		2003	2004	Bear	Marten	Badger	Fox	Raccoon dog	Civet
2003 - 2004	1	30.0 ± 7.9	3.0 ± 2.0	0,0	3,1	0,0	0,0	0,0	1,0
	2	75.6 ± 9.1	31.2 ± 19.2	2,2	5,1	1,0	1,0	0,0	2,1
	3	70.8 ± 13.8	11.8 ± 3.8	2,1	4,1	0,0	2,0	0,0	2,0
	4	69.6 ± 18.3	2.6 ± 0.8	2,0	4,0	0,0	1,1	0,0	2,0
	5	53.6 ± 18.6	21.4 ± 12.7	1,2	5,1	0,0	0,0	0,0	3,2
	6	46.2 ± 18.9	1.4 ± 0.8	0,0	1,0	0,0	0,0	0,0	1,0
	Total			7,5	22,4	1,0	3,2	0,0	11,3

Tree Number	Number of photographs					
	Boar	Deer	Serow	Macaque	Squirrel	Rodent
1	0,0	5,4	0,1	0,0	7,0	5,2
2	1,1	4,4	1,0	1,0	7,0	3,4
3	0,0	4,5	0,1	0,0	11,0	3,1
4	1,0	4,5	0,2	1,0	8,0	5,3
5	1,0	3,4	0,0	0,0	12,0	4,4
6	0,0	1,3	0,0	0,0	8,0	5,4
Total	3,1	21,25	1,4	2,0	53,0	25,18

Tree Number	Number of photographs					
	Pheasant	Jay	Thrush	Unknown	Total: 2003	Total: 2004
1	1,0	1,0	2,0	0,0	25	8
2	0,0	2,0	2,0	0,0	29	14
3	0,0	0,0	3,0	1,0	30	9
4	1,0	1,0	3,0	0,0	31	11
5	0,0	0,0	1,0	1,0	30	13
6	0,0	1,0	1,0	0,0	18	7
Total	2,0	4,0	12,0	2,0	169	62

Bear, *Ursus thibetanus;* marten, *Martes melampus*; badger, *Meles meles:* fox, *Vulpes vulpes*; raccoon dog, *Nyctereutes procyonoides*; civet, *Paguma larvata*; boar, *Sus scrofa*; deer, *Cervus nippon*; serow, *Capricornis crispus*; squirrel, *Sciurus lis*; macaque, *Macaca fuscata*; rodent, *Apodemus speciosus*; pheasant, *Syrmaticus soemmerringii*; jay, *Garrulus glandarius*; and thrush, *Turdus cardis*.

Table 1. Estimated fruit production of each target tree and the number of photographs of each vertebrate species photographed by automatic cameras under the crown of each target tree, 2003 and 2004. *See text for fruit production calculation.

The average fruit production (number of fruits per tip of the branch) of target trees were 57.6±17.6 (SD) in 2003, and 11.9±12.2 (SD) in 2004 (Table 1).

3.2 Camera trap survey

During two year study, the camera traps took the total of 915 photographs. We could identify mammals in 204 photographs and birds in 18 photographs during the four-month period after the flowering. Table 1 provides a summary of the species seen each year (Fig. 4).

The cameras captured a smaller number of photographs in 2004 than in 2003. We calculated the correlations between fruit production of each tree and number of photographs for some mammal species at each tree, and we can recognize the correlation between fruit production and number of photographs for some species: values are R^2; bears, 0.40 (P=0.02); martens, 0.75 (P<0.01); civets, 0.65 (P=0.01): boars, 0.36 (P=0.03); squirrels (*Sciurus lis*), 0.75 (P<0.01), and macaques, 0.40 (P=0.02). A poor correlation exists between fruit production and number of photographs for some species: values are R^2: foxes, 0.12 (P=0.26); deer, 0.05 (P=0.46); and rodents (*Apodemus* spp), 0.11 (P=0.27). Also a poor correlation exists between fruit production and number of the photographs of deer after day 50 (R^2 = 0.10, P=0.29).

Next, we compared the timing of the sightings of bears, martens, civets, foxes, boars, and deer with the number of days after the initiation of flowering (Fig. 5). Over 90% of the photographs of bears, martens, and civets were taken between day 50 and day 66 (Fig. 5). For foxes and boars, most photographs were taken after day 60 (Fig. 5).

Deer sightings peaked between day 70 and day 80. However, deer were documented immediately after flowering and were recorded periodically throughout the study (Fig. 5). We could not confirm whether each vertebrate species was photographed with fruits in its mouth.

4. Discussion

In this study, the photographs did not necessarily confirm the mammals actually ate fruits. However, because there are correlations between fruit production and the number of photographs for some mammal species, we can conclude these mammals visited the trees to eat the *P. jamasakura* fruit.

In this study, the cameras captured a smaller number of photographs in 2004 than in 2003. One reason for this may be the difference of fruit production. In 2004, the target trees produced a smaller amount of fruit than in 2003. Also, a correlation exists between fruit production and number of photographs for bears, martens, civets, squirrels, and macaques. These results indicate these mammal species moved and selected trees depending on the fruit production of each tree. For pine martens (*Martes martes*) in Mediterranean area, crop size was correlated with the frequency of occurrence and the proportion of fruits by volume in feces. Pine martens feed on fallen fruit and appeared to choose trees with a higher density of fruits under the canopy, that is, trees that produced larger crops (Guitian & Munilla, 2010). Tsuji, (2010) reported Japanese macaques change their home range, move, or select trees depending on the availability of food generally and on the fruit production of each tree. There are no reports on the feeding behavior of bears that indicating bears select trees depending on the food availability of each tree. However, it is known bears can change their diet and movement behavior frequently depending on fruit production (Hashimoto, 2002; Hashimoto et al., 2003; Koike, 2010; Kozakai et al., 2011; Arimoto et al., in press).Thus bears may select trees based on the difference in fruit production for each individual tree. Our results indicate for the first time that the amount of fruit production of each tree may

Fig. 4. Sample camera trap photographs of verteblates as they visit under the crowns of *Prunus jamasakura* trees.(a) deer (*Cervus nippon*); (b) serow (*Capricornis crispus*); (c) bear with cub (*Ursus thibetanus*); (d) marten (*Martes melampus*); (e) is civet (*Paguma larvata*); (f) pheasant (*Syrmaticus soemmerringii*).

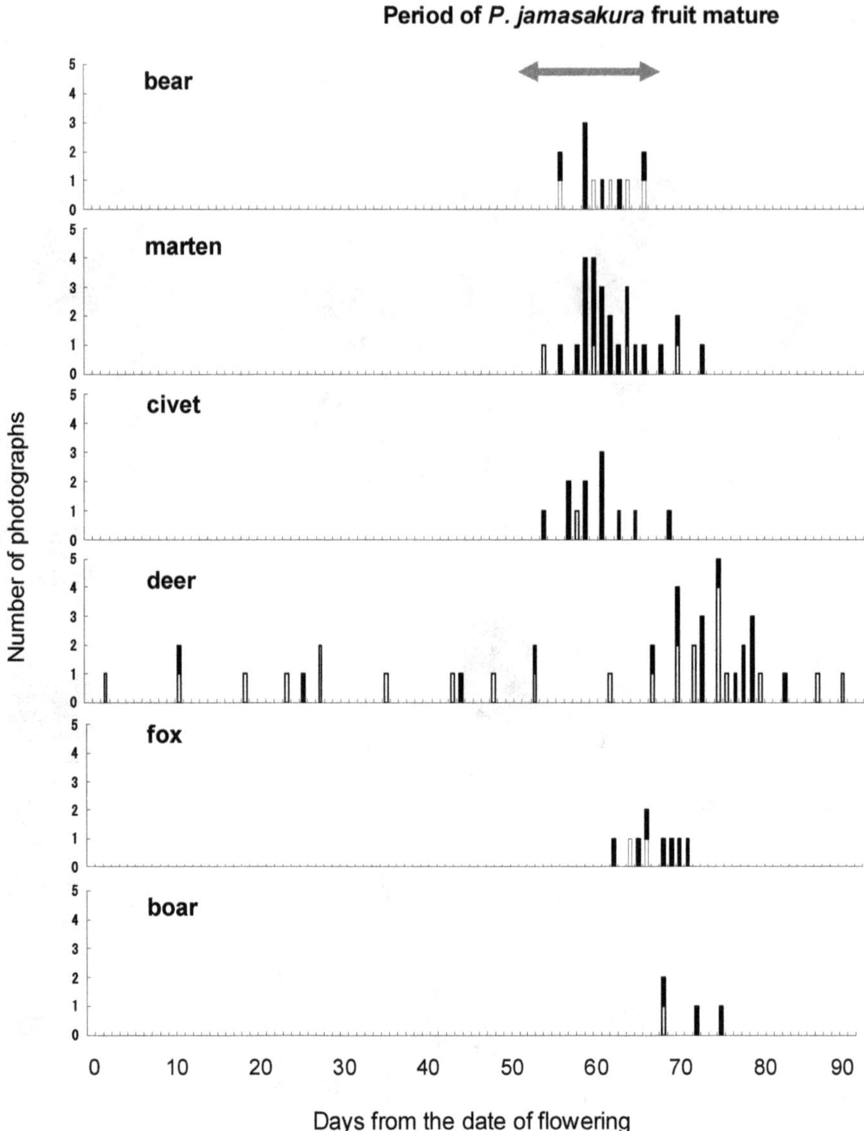

Fig. 5. Relationships between the number of days after the initiation flowering and number of photographs of major mammals under the crowns of *Prunus jamasakura* trees. The period of *P. jamasakura* fruit maturation (maximum of fruit size, sugar concentration, and seed germination) is based on Koike et al., (2008a). Black bar is data of 2003, white bar is data of 2004.

influence individual tree selection by bears; although this result doesn't necessarily clearly demonstrate bears have selective feeding behavior. The ecology of civets in Japan is poorly known (Ohdachi et al., 2009). However, in Southeast Asia, civets are also known to change food items frequently depending on fruit production (Zhou et al., 2008) although the level of fruit production may influence the feeding behavior of civets in Japan. Because some reports indicate squirrels consume the seeds (Ohdachi et al., 2009), squirrels may feed on the fruit of *P. jamasakura*, especially seeds, although documentation of food selection habits of squirrels is lacking. In this study, we recorded many squirrel photographs in 2003, but none in 2004. Because there are no other factors of environmental change such as increases in logging in the study area, fluctuations in the availability of mast may seriously influence the behavior of squirrels.

A poor correlation exists between fruit production and number of photographs for foxes, deer, and rodents. In general, foxes do not inhabit only forest, and they depend strongly on animal food sources (Ohdachi et al., 2009). Also, foxes were not frequently photographed, so the amount of fruit might not affect their behavior much. It is unknown whether deer eat the fruit of *P. jamasakura* or not. Because deer primarily fed on herbaceous plants (Ohdachi et al., 2009), and the presence of deer was recorded before the fruit matured, the fluctuation of fruit production might not affect their behavior. There was also poor correlation between rodent sightings and fruit production. Because rodents eat a variety of foods (Ohdachi et al., 2009), fluctuations in fruit production in one tree species might not affect their behavior. Future research is needed to document the importance of fleshy fruit such as *P. jamasakura* fruit as a food resource for deer and rodents.

Most mammals except deer were photographed after day 50. Koike et al., (2008a) indicated bears consume fruit with high energy content around day 60. The peak for photographs of bears, martens, and civets was between day 50 and day 70. This was earlier than photographs of deer, boars, and foxes. Previous studies indicated the proportion of persisting fruit decreased gradually until day 60, then decreased rapidly by 64.7% in the next 10 days (Koike et al., 2008a). Because bears, martens, and civets can climb tall trees (Hosoda & Tatara, 1996; Huygens et al., 2003; Ohdachi et al., 2009), they can feed on the fruit before it falls. On the other hand, because deer, boars, and foxes cannot climb trees well they feed mainly on fallen fruit. These results indicate the frequency or the peak of photographs of mammalian species reflects their ecological characteristics such as climbing ability and habitat preference.

We show mammals, aside from deer, were recorded after day 50 and they have potential to feed on fruit including mature seed. Previous studies indicate carnivores such as bears, martens foxes, and civets can act as seed disperses for some fleshy-fruited plants (Koike et al., 2008b; Nakashima et al., 2010). This paper clarifies these mammals are not seed eaters in that they don't eat fruit with immature seeds. Although few studies of seed dispersal by ungulates (endozoochory) have been made in Britain, deer have been shown to be effective seed dispersers for a number of plant species. Recent research shows deer do function as seed dispersal agents (Malo et al., 2000; 2001; Myers et al., 2004; Brathen et al., 2007).

5. Conclusions

We investigated the relationship between the fruit phenology of *Prunus jamasakura* and the fruit feeding period of mammals. Our research cannot confirm actual feeding behavior.

However, because the most of photographs for some mammal species were recorded during fruit mature period, we can conclude these mammals visited the trees to eat the *P. jamasakura* fruit. And, these results indicate that some mammal species are potentially effective seed dispersers. We suggest the combination of information gained from camera trap data with information related to fruiting phenology and fruit production will make more intelligent discussion of these relationships possible. Future similar research on other tree species and in other regions will facilitate research related to vertebrate forest ecology by obtaining data that is difficult to observe directly.

6. Acknowledgments

We thank members of the Forest Conservation Biology Laboratory of Tokyo University of Agriculture and Technology for their field assistance. The critical comments on an earlier draft by Dr. K. Furubayashi were helpful in improving this manuscript. We wish to thank M. Soga for its helpful information.This study was partly funded by the Tokyu Foundation for Better Environment and Grant-in Aid for JSPS Fellows (22780141, 22651086).

7. References

Arimoto, I., Goto, Y., Nagai, C., & Furubayashi, K. (in press). Autumn food habits and home-range elevations of Japanese black bears in relation to hard mast production in the beech family in Toyama Prefecture. *Mammal study*

Brathen, KA., Gonzalez, VT., Iversen, M., Killengreen, S., Ravolainen, VT., Ims, RA., & Yoccoz, NG. (2007). Endozoochory varies with ecological scale and context. *Ecography*, 30, 308–320.

Castilla, AM. (2000). Does passage time through the lizard *Podarcis lilfordi*'s gut affect germination performance in the plant *Withania frutescens*?. *Acta Oecologia,* 21, 119–124.

Corlett, RT. (1998). Frugivory and seed dispersal by vertebrates in the Oriental (Indomalayan) Region. *Biological Reviews*, 73, 413-448.

Dennis, A., Schupp, E., Green, R., & Westcott, D. (eds) (2007). *Seed dispersal: theory and its application in a changing world.* CAB International Publishing, UK, Wallingford.

Estrada, A., & Coates-Estrada, R. (1984). Fruit eating and seed dispersal by howling monkeys (*Alouatta palliata*) in the tropical rain forest of Los Tuxtlas, Mexico. *American Journal of Primatology*, 6, 77–91.

Estrada, A., & Fleming, T. (eds) (1986). *Frugivores seed dispersal.* Dr. W. Junk Publishers, The Netherlands, Dordrecht.

Fleming, TH. (1981). Fecundity, fruiting pattern, and seed dispersal in *Piper amalago*, a bat-dispersed tropical shrub. *Oecologia*, 16, 42–46.

Fleming, T., & Estrada, A. (eds) (1993). *Frugivory and seed dispersal: ecological and evolutionary aspects.* Kluwer Academic Publishers, The Netherlands, Dordrecht.

Gottsberger, G. (1978). Seed dispersal by fish in the inundated regions of Humaita, Amazonia. *Biotropica*, 10, 170–183.

Guitián, J., & Munilla, I. (2010). Responses of mammal dispersers to fruit availability: Rowan (*Sorbus aucuparia*) and carnivores in mountain habitats of northern Spain. *Acta Oecologia,* 36, 242–247.

Gribel, R. (1988). The shape of bird- and bat-generated seed shadows around a tropical fruiting tree. *Biotropica,* 20, 347-348.

Hashimoto, Y. (2002). Seasonal food habits of the Asiatic black bear (*Ursus thibetanus*) in the Chichibu Mountains, Japan. *Mammal Study,* 27, 65–72.

Hashimoto, Y., Kaji, M., Sawada, H., & Takatsuki, S. (2003). A five year study on fall food habits of the Asiatic black bear in relation to nut production. *Ecological Research,* 18, 485-492.

Herrera, CM. (1984). A study of avian frugivores, birddispersed plants, and their interaction in Mediterranean scrublands. *Ecological Monograph,* 54, 1–23.

Herrera, CM. (1989). Frugivory and seed dispersal by carnivorous mammals, and associated fruit characteristics, in undisturbed Mediterranean habitats. *Oikos,* 55, 250–262.

Herrera, CM. (2002). Seed dispersal by vertebrates. In: *Plant-animal interactions: an evolutionary approach,* Herrera, CM., & Pellmyr, O. (eds), pp. 185–208, Blackwell Science, Oxford.

Hickey, JR., Flynn, RW., Buskirk, SW., Gerow, KG., & Willson, MF. (1999). An evaluation of mammalian predator, *Martes americana,* as a disperser of seed. *Oikos,* 87, 499–508.

Hosoda, T., & Tatara, M. (1996). Japanese marten and sable. In: *The encyclopaedia of animals in Japan, Vol. 2 Mammals, Part 1,* Izawa, K., Kasuya, T., & Kawamichi, T. (eds), pp. 136–139, Heibonsya, Tokyo. (In Japanese)

Howe, HF. (1980). Monkey dispersal and waste of a Neotropical fruit. *Ecology,* 61, 944–959.

Howe, HF., & Smallwood, J. (1982). Ecology of seed dispersal. *Annual Review of Ecology and Systematics,* 13, 201-228.

Howe, HF. (1986). Seed dispersal by fruit-eating birds and mammals. In: *Seed Dispersal,* Murray, DR. (ed), pp. 123–189, Academic Press, New York.

Howe, HF., & Westley, LC. (1988). Ecological relationships of plants and animals. Oxford University Press, New York.

Huygens, OC., Miyashita, T., Dahle, B., Carr, M., Izumiyama, S., Sugawara., T., & Hayashi, T. (2003). Diet and feeding habits of Asiatic black bears in the Northern Japanese Alps. *Ursus,* 14, 236–245.

Ishii, Y. (1991). Methods of seed handling in cherry trees, *Prunus* species. *Forest Tree Breeding,* 161, 7–12. (In Japanese)

Izhaki, I., Korine, C., & Arad, Z. (1995). The effect of bat (*Rousettus aegyptiacus*) dispersal on seed germination in eastern Mediterranean habitats. *Oecologia,* 101, 335–342.

Jordano, P. (1995). Angiosperm fleshy fruits and seed dispersers: a comparative analysis of adaptation and constraints in plant-animal interactions. *American Naturalist,* 145, 163–191.

Koike, S., & Masaki, T. (2008). Frugivory of Carnivora in central and Southern parts of Japan analyzed by literature search. *Journal of Japanese Forest Socience,* 90, 27–36. (In Japanese with English abstract)

Koike, S., Kasai, S., Yamazaki, K., & Furubayashi, K. (2008a). Fruit phenology of *Prunus jamasakura* and the feeding habit of Asiatic black bear as a seed disperser. *Ecological Research,* 23, 385-392.

Koike, S., Morimoto, H., Goto, Y., Kozakai, C., & Yamazaki, K. (2008b). Frugivory of carnivores and seed dispersal of fleshy fruits in cool-temperate deciduous forests. *Journal of Forest Research,* 13, 215-222.

Koike, S. (2009). Fruiting phenology and its effect on fruit feeding behavior of Asiatic black bears. *Mammal Study,* 34, 47-52.

Koike, S. (2010) Long-term trends in food habits of the Asiatic black bear in the Misaka mountains, Japan. *Mammalian Biology,* 75, 17-28.

Koike, S., Masaki, T., Nemoto, Y., Kozakai, C., Yamazaki, K., Kasai, S., Nakajima, A., & Kaji, K. (2011). Estimate of the seed shadow created by the Asiatic black bear (*Ursus thibetanus*) and its characteristics as a seed disperser in Japanese cool-temperate forest. *Oikos,* 120, 280-290.

Kozakai, C., Yamazaki, K., Nemoto, Y., Nakajima, A., Koike, S., Abe, S., Masaki, T., & Kaji, K. (2011). Effect of mast production on home range use of Japanese black bears. *Journal of Wildlife Management,* 75, 867-875.

Levey, D., Silva, W., & Galetti, M. (eds) (2001). *Frugivores and seed dispersal: ecological, evolutionary, and conservation issues.* CAB International, UK, Wallingford.

Malo, JE., Jimenez, B., & Suarez, F. (2000). Herbivore dunging and endozoochorous seed deposition in a Mediterranean dehesa. *Journal of Range Management,* 53, 322–328.

Mizui, N. (1993). Ecological studies on reproduction in deciduous broadleaved tree species. *Bulletin of Hokkaido Forestry Research Institute,* 30, 1–67. (In Japanese)

Mogi, T., Katsuyama, T., Ota, K., Sakio, H.,Takahashi, H.,Ishii, H., Kigawa, S., & Nakagawa, S. (2000). *Flowers bloom in woody plants 1.* Yamatokeikoku-sya, Tokyo, Japan. (In Japanese)

Myers, JA., Vellend, M., Gardescu, S., & Marks, PL. (2004). Seed dispersal by white-tailed deer: implications for long-distance dispersal, invasion, and migration of plants in eastern North America. *Oecologia,* 139, 35–44.

Nakashima, Y., Inoue, E., Inoue-Murayama, M., Rafiah, J., & Sukor, A. (2010). Functional uniqueness of a small carnivore as seed dispersal agents: a case study of the common palm civets in the Tabin Wildlife Reserve, Sabah, Malaysia. *Oecologia,* 164, 721-730.

Noma, N., & Yumoto, T. (1997). Fruiting phenology of animal-dispersed plants in response to winter migration of frugivores in a warm temperate forest on Yakushima Island, Japan. *Ecological Research,* 12, 119–129.

Oba, H. (1989). *Prunus.* In: *Wild Flowers of Japan – Woody Plants,* Satake, Y., Hara, H., Watari, S., & Tominari, T. (eds), pp. 186–198, Heibonsya, Tokyo. (In Japanese)

O'Brien, TG., Kinnaird, MF., & Wibisono, HT. (2003). Crouching tigers, hidden prey: Sumatran tiger and prey populations in a tropical forest landscape. *Animal Conservation,* 6, 131-139.

Ohdachi, SD., Ishibashi, Y., Iwasa, MA., & Saitoh, T. (eds) (2009) *The Wild Mammals of Japan.* Shoukadoh, Kyoto, Japan.

Otani, T. (2002). Seed dispersal by Japanese marten, *Martes melampus*, in the subalpine shrubland of northern Japan. *Ecological Research*, 17, 29-38.

Otani, T. (2003). Seed dispersal and predation of fleshy-fruited plants by Japanese macaques in the cool temperate zone of northern Japan. *Mammal Study*, 28, 153-156.

Otani, T. (2004). Effects of macaque ingestion on seed destruction and germination of a fleshy-fruited tree, *Eurya emarginata*. *Ecological Research*, 19, 495-501.

Schupp, EW. (1993). Quantity, quality and the effectiveness of seed dispersal by animals. *Vegetatio*, 107/108, 15–29.

Stapanian, MA. (1982) Evolution of fruiting strategies among fleshy-fruited plant species of eastern Kansas. *Ecology*, 63, 1422–1431.

Stiles, EW. (1980). Patterns of fruit presentation and seed dispersal in bird-disseminated woody plants in the eastern deciduous forest. American Naturalist, 116, 670–688.

Takeuchi, M., Iijima, K., & Tamura, M. (2005). *Organisms around the cherry*.Tokyo Agriculture University Press, Tokyo, Japan. (In Japanese)

Tokyo Environmental Office. (2000) *The environment in Tokyo*, 2000. Tokyo Metropolitan Government, Tokyo. (In Japanese)

Tokyo Environmental Office. (2010). *The Reddata book in Tokyo*, 2010. Tokyo Metropolitan Government, Tokyo. (In Japanese)

Thompson, JN., & Willson, MF. (1979). Evolution of temperate fruit/bird interactions: phenological strategies. *Evolution*, 33, 973–982.

Traveset, A., & Willson, MF. (1997). Effect of birds and bears on seed germination of fleshy-fruited plants in temperate rainforests of southeast Alaska. *Oikos*, 80, 89–95.

Tsuji, Y. (2010). Regional, temporal, and inter-individual variation in the feeding ecology of Japanese macaques. In: *Japanese Macaques*, Nakagawa, N., Nakamichi, M., & Sugiura, H. (eds.), pp. 95-123, Springer, Tokyo.

Tsuji, Y. (2011). Seed dispersal by Japanese macaques *(Macaca fuscata)* in western Tokyo, Japan: a preliminary report. *Mammal Study*, 36, 165-168.

Tsuji, Y., Sato, K., & Sato, Y. (2011a). The role of Japanese macaques *(Macaca fuscata)* as endozoochorous seed dispersers on Kinkazan Island, northern Japan. *Mammilian Biolology*, 76, 525-533.

Tsuji, Y., Wada, K., & Watanabe, K. (2011b). Woody Plants Eaten by Wild Japanese Macaques. *Primate Research*, 27, 27-49. (In Japanese)

Willson, MF. (1993). Mammals as seed-dispersal mutualists in North America. *Oikos*, 67, 159–176.

Yagihashi, T. (2001). Effect of bird ingestion on seed germination of arboreal plants in Hokkaido, Japan. *Research Bulletin of the Hokkaido University Forest*, 58, 37–59. (In Japanese with English abstract)

Yoshikawa, T., Isagi, Y., & Kikuzawa, K. (2009). Relationships between bird-dispersed plants and avian fruit consumers with different feeding strategies in Japan. *Ecological Research*, 24, 1301-1311.

Yumoto, T., Noma, N., & Maruhashi, T. (1998). Cheek-pouch dispersal of seeds by Japanese monkeys (*Macaca fuscata yakui*) on Yakushima Island, Japan. *Primates*, 39, 325–338.

Zhou, YB., Zhang, JS., Slade, E., Zhang, LB., Palomares, F., Chen, J., Wang, XM., & Zhang, SY. (2008). Dietary shifts in relation to fruit availability among masked palm civets *(Paguma larvata)* in central China. *Journal of Mammalogy*, 89, 435–447.

Part 2

Plant Seasonality

5

Life Cycle in Natural Populations of *Canna indica* L. from Argentina

María de las Mercedes Ciciarelli
Laboratorio de Estudios de Anatomía Vegetal Evolutiva y Sistemática (LEAVES), Facultad de Ciencias Naturales y Museo, Universidad Nacional de La Plata, Argentina

1. Introduction

Several species of genus *Canna* L. (*Cannaceae-Zingiberales*) from Argentina, such as *C. ascendens* Ciciarelli, *C. coccinea* Mill., *C. glauca* L., *C. indica* L., *C. fuchsina* Ciciarelli and *C.variegatifolia* Ciciarelli were observed to grow mainly by vegetative multiplication of underground rhizomes (Ciciarelli, 1989, 1995, 2007, Ciciarelli & Rolleri 2008, 2010). For a long time, rhizomes have been considered as the most important propagation mode of many weeds to ensure plant growth and genetic uniformity. Some species, such as *C. indica*, are particularly invasive and even become weeds. This occurs both in the wild and urban areas of several tropical or subtropical American and Asian countries, precisely because of the ease of vegetative propagation through its rhizomes (Maas Van de Kamer & Maas, 2010; Tanaka 2001). Indeed, extensive populations are formed mostly by vegetative multiplication of rhizomes in Province of Buenos Aires, Argentina (Ciciarelli & Rolleri, 2008). Field work carried out during several seasons (2003-2011) demonstrated that, after removal of plant populations by human action, seeds germinate easily in the substrate assisted by external factors such as sun radiation, dew and rainfall distributed throughout the year.

This work aims at contributing with knowledge of the life cycle of some herbaceous geophytes such as species of the genus *Canna*, and with studies of the reproductive mechanisms (sexual or asexual) prevailing throughout several seasons. The present research also includes morphological and histochemical studies of rhizomes and seeds to understand underlying growth factors such as energy, water availability, and resistance to cold. Finally, germination of *C. indica* seeds under natural and in vitro conditions is studied to make a contribution to the understanding of this process in species bearing very hard seedcoats.

Although all species of the genus *Canna* mentioned above were studied in their phenology and life cycle, this analysis will focus mainly in *C. indica*: its wide distribution, resistance to eradication by human action, ability to reach a rapid coverage, capacity to cover new fields with dense colonies, to produce many flowers, fruits, and to have seeds with high and quick germination, make this species an interesting example of the characteristics of a typical weed.

Canna indica is a plant of tropical or subtropical origin in South America, distributed over a vast area, reaching southern regions as Province of Buenos Aires (PBA) in Argentina (35° south latitude). In this area, climate is temperate, soils are rich in humus, and the short

Fig. 1. (a) *Canna indica* colony (in May, autumn). (b) Detail of leaves and flowers in the same colony.

periods of winter frost may constitute a threat for populations. Winter frost may begin in May, and continue over June, July and August causing decay in the aerial parts of the plant until September-October, when they begin to recover. (Fig. 1a, b)

1.1 Description of the species

Canna indica L. Sp.Pl.1, 1753, is a mesophyte colonial terrestrial plant 2.5m height. Rhizomes are sympodial underground, with chestnut- brown cataphylls and abundant adaxial and abaxial roots. The aerial shoots are 1-3 cm diameter, with 7 - 11 sheathing leaves, patent to reflexed the inferior, ascending the superiors, and sometimes convolutes. The phyllotaxis is distichuous to spirally, and foliar sheaths are light green. Leaves are glabrous, green yellowish, adaxially fasciate, and abaxially greyish green and dull. The contour of leaves is narrow ovate with a length/ width ratio of 2:1. The third leaf is 19-37 (55) cm in length x 10-18 (30) cm wide. The apices are acute, acuminate, twisted and mostly dry at the tops, with cuneate decurrent asymmetric bases. Venation is eucamptodrome. Primary vein and foliar margins are colourless, with light green foliar sheaths and bracts. Inflorescences are politelicae, intermediate, of 40- 41 cm in length carrying 1-2 paracladia, each one bearing 1-4 nodes, each node carrying 2 flowers. The principal florescence carries 6-8 nodes and the basal internode is of 16-17 cm long. Flowers with 11-12 pieces are 7 cm long, concolours red or yellow or bicolorous, bright light red at the base and yellow in the third basal part. The floral tube is 3-4 cm long. The ovarium is inferior and green. Pollen grains are spherical or subspherical 38-61-(63) μm, equinatae, spinules conical of 3 x 3 μm. Capsules are spherical, subespherical, green of 2-3 x 2-4 cm. Capsules have 20-28 seeds, ovoid in shape, dark chestnut-brown or black in colour, and of 0,5-0,7 cm diameter.

1.2 Materials and methods

Populations were observed *in situ* to study the life cycle and phenology. Field trips were carried out over the years 2003 to 2010, from September (spring in the Southern hemisphere) to June (winter), to observe flowering and pollination processes, fruit and seed production, autumn decay of green parts and the initial emergence of new plants in each colony. Specimens were also collected and cultivated; original plants were grown from rhizomes in an experimental plot and similar observations were conducted. Rhizome branching was studied in the field, carefully digging areas of 2 m^2, previously removing green vegetative parts. The rhizomes can be studied throughout the year, but the better period is late winter, before the emergence of aerial stems. Parts of rhizomes were planted again and successive visits during the phenological study showed a rapid recovery. The terms used in describing rhizome branching were taken from Font Quer (1982). For germination *in vitro,* the seeds collected were briefly immersed in boiling water for three times then were placed in a petri dish and covered with wet cotton.

For studies of the internal morphology of rhizomes, fresh material was fixed in FAA (Johanssen, 1940). Permanent and non permanent samples of longitudinal and transverse sections of adult rhizomes were prepared. Permanent samples were coloured by safranine-fast green 2% in methyl cellosolve and mounted in Canada balsam. Characterization tests were carried out to detect the presence of mucilage (Bismarck Brown) tannins (Ferrous sulphate), and starch (iodine) (Johanssen, 1940). For MEB observations, samples of parenchymatous tissue were scrapped and placed on a double face adhesive tape, metalized with gold-palladium and examined with a JEOL JSM T-1100, from the Electronic Microscopy Service of the Facultad de Ciencias Naturales y Museo de La Plata.

1.3 Key results

Germination *in vitro* took about 18 days to produce a new independent plant. Immediately after contacting the boiling water, the seed imbibition lid was expelled and the coat began to soften. At the 3rd day the white cotyledon appeared and pushed the radicle through the micropilar region. At the 7 th day, the white cotyledon, 2.5 cm long, was observed with the cotyledon node, a short hipocotyle and the tiny radicle. Inside the seed, part of the cotyledon remains in contact with the endosperm with haustorial functions, such as solubilizing, translocating and storing nutrients through the cotyledon towards the growing point.

Fig. 2. Seedlings of *C. indica*. CS, cotyledon sheath; CR, collar root.

At the 9th day the positive geotropism of the cotyledon makes it to curve forming the *cotyledon knee* and directing the primary root towards the earth. The hypocotile is short and bears many endogenous collar roots and a short radicle at the top.

At the 18th day, the resulting axis is 6 cm long, with the growing apex being protected by the cotyledon sheath. The first green leaf appears at the top end of the sheath (before the first leaves reaches daylight, the plantlet had been nourished from the seed). The photosynthetic process starts as well as the independent life of the plant. At this time, several white ramifications are observed in the radicle, as described by Tillich (1959). Seedlings are composed of three primary leaves, the cotyledon with the haustorial part and the tubular cotyledon sheath. The hipocotyle, with endogenously shoot-borne roots, and the collar with several endogenous roots. The primary root has exogenous, short root hairs (Fig. 2).

Germination in the field was observed in a small wetland of the area under study. An adult colony of *Canna indica* completely cut down by human action, had been associated with other herbaceous plants as *Typha latifolia, Iris pseudoachorus, Verbena intermedia* and *Acmella decumbens*. After the complete removal of plants a black and humid soil was exposed, and seedlings of *Canna indica* took about two weeks to appear again. Seedlings were carefully removed from the soil to confirm that they had grown from seeds (Fig. 3). No other species grew there until that moment.

Fig. 3. Seedlings of *C. indica*. Seed (0.6-0.8 cm in diameter) is indicated in a frame.

1.4 Life cycle in C. *indica*

The life cycle of *C. indica* and other species of *Canna* is of about 9 months. It begins in the first week of September when buds from underground rhizomes start to produce the first green leaves, while aerial foliage of the previous year is completely dry.

Fig. 4. (a) Seeds of *C. indica* in the soil (one is framed).
(b) unburied parts of rhizomes, one dried, the other growing.

The weather conditions in La Plata, in Province of Buenos Aires during the first ten days of September are the following: rainfall, 54 mm; mean average temperature, 13.7°C (with average minimum of 9,2°C and average maximum temperature is of 18,2°C). The average relative humidity in this month is of 79%, with 11.8 hours of solar radiation (Servicio Meteorológico Nacional, 2011; (National Weather Forecast Service)). At the second week of September plants have 2 or three leaves and continue growing along 45 days to reach their adult length (2,5 m) and 7 to 11 leaves. Thus vegetative grow from underground rhizomes takes place from the beginning of September to the end of October. Blooming begins in late October- early November and continues until the end of May in late Autumn. During summer, and particularly at the end of February, colonies reach their maximum size and plants exhibit abundant fruits and seeds. In *C. indica* one infrutescence carries about 5-6 capsules, each bearing 4-7 seeds, so 20-28 seeds are produced in one infrutescence of the plant. Considering the whole colony, the average number of seeds in *C. indica* is relevant and greater than in other wild species of *Canna* growing in the Province of Buenos Aires. Infrutescences of other species of *Canna* produce the following average number of seeds: *C. ascendens*, 1-3; C. *coccinea*, 6-8; *C. fuchsina* 6-10; *C. glauca* 6-9 and *C. variegatifolia*, 3-6.

During the 6 months of Summer and Autumn seeds are abundantly produced and fall to the ground remaining there or else germinating if conditions are adequate. In 2011 the first frost (May 6 th) produced a slight damage in some flowers. On May 26th, after a second frost, some upper leaves, tops of inflorescences and flowers were dried. On June 2nd the vegetative part of plants was completely dried though some flowers still remained fresh. On the 26 th of June the last flowers were completely frozen. During Winter dried plants and fallen seeds remain on the ground. (Fig. 4a).

It was observed that over the last three years, the life cycle of *C. indica* has been extended one additional month as well as the size of colonies in the field.

2. Rhizomes

2.1 External morphology

Rhizomes may be monopodial or sympodial, stoloniferous or tuberous according to the species. Concerning *C. indica*, the most vigorous species growing in this area, rhizomes are sympodial with Y-shaped axes (Ciciarelli & Rolleri, 2008) and abundant roots growing both adaxially and abaxially from the nodes. Monopodial or stoloniferous rhizomes are less common than the other types and are observed in *C. glauca* and *C. variegatifolia*, while sympodial types are observed in the other species. Tuberous rhizomes only appear in *C. coccinea*, and in spite of their diverse external morphology, all species of *Canna* store abundant starch granules.

In *C. indica*, the rhizome is sympodial, and the plagiotropic axis after producing 5-6 nodes, curves its direction upwards to form the aerial plant. From the base of the last node an axillary bud restarts propagation by repeating the pattern. These "active" nodes can also produce up to three aerial branches, so three new plants grow very close to each other. From the base of those active nodes, abaxial and adaxial roots are abundant. By repeating this growth mecanism, one row of plants can be formed. Besides, the main axis can also divide into two other branches forming a "Y" branching mode, so other 2 rows of plants are produced parallel to the previous row. "Y" branching propagation of the rhizome increases not only the number of plants but also the underground connection among them, occupying a wider soil surface. This pattern is usually disordered and crossover of branches coming from different axes may occur. This branching system is similar to that of *Alpinia speciosa* (Bell, 1979). (Fig. 4b, 5a).

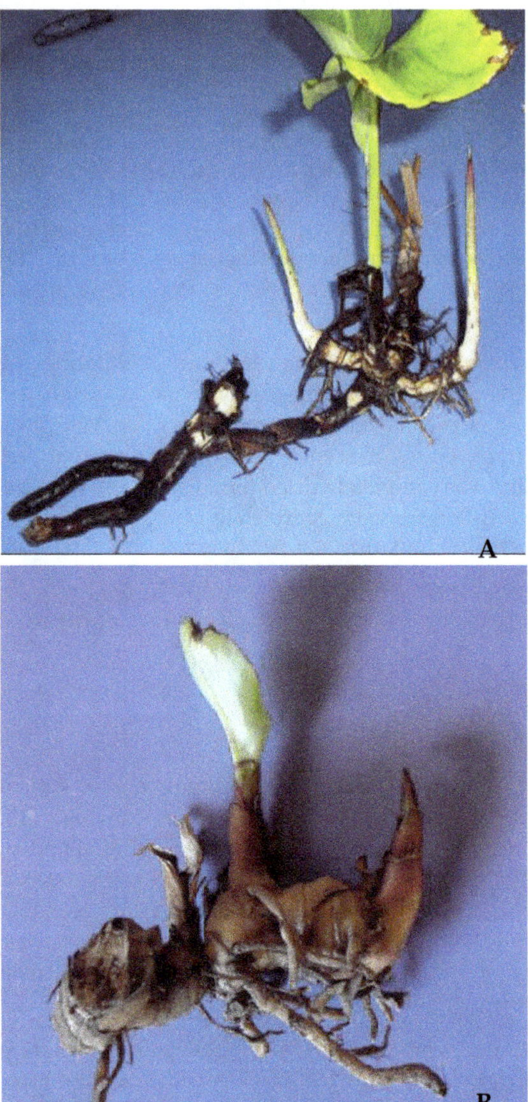

Fig. 5. (a) Simpodial rhizome of *C.indica*. "Y" ramification, and 3 aerial branches.
(b) Early development of simpodial rhizome in *C. fuchsina*

2.2 Internal morphology

The rhizome of *Canna*, as seen in transverse section, has the following organization: epidermis, hypodermis, cortex, endodermis, pericycle, vascular plexus and central cylinder. In all species the epidermis cell walls are scarcely cutinized. Beneath the epidermis there is a three layered hypodermis, which exhibits cells with subpolygonal outline and thickened walls. The cortex is a relatively thin zone placed between the hypodermis and the endodermis. It is mainly composed by a parenchymatous tissue, which store abundant starch granules and tanniferous contents. It also reveals an outer ring of fibrous strands, many internally spread fibrovascular strands as well as foliar and root traces. In the palustrine species of *C. glauca,* schyzogenous lacunae are also observed in the cortex. One endodermal layer may be present as a continuous ring in *C. ascendens, C. coccinea, C. glauca, C. indica* and *C. variegatifolia* or as several arcs separated by parenchymatous tissue in *C. coccinea* and *C. compacta*. In the former species a well defined cortex is observed and most endodermal cells present Casparian bands. In *C. compacta* the thickening of the cell walls is complete or in "O". A pericycle is always evident and presents one layer of cells. It surrounds a zone composed of close and anastomosing vascular strands with many xylem vessels. These are disposed between parenchymatous cells, some with tanniferous contents. The presence of abundant xylem elements in this zone of the rhizome is related to the production of root and cataphylls traces. According to Tomlinson (1962) this zone is called "vascular plexus", and is quoted for rhizomes of families belonging to the Order Scitaminae.

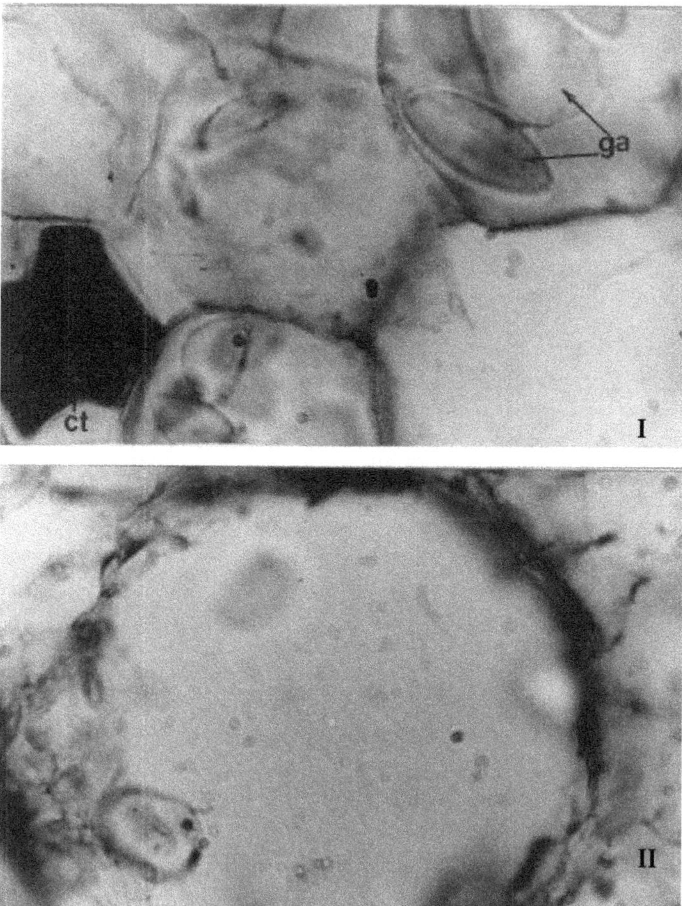

Fig. 6. (a) Central cylinder in *C. glauca*. I. starch granules (ga) and tanniferous cells (ct). II. Mucilage duct. Black bars represent 100 μm

The central cylinder presents a parenchymatuous tissue specialized in storage of ovoidal and excentric starch granules, and other substances as tannins, and prismatic crystals of calcium oxalate (fig. 6a, I). Fibro-vascular and collateral bundles are irregularly scattered in the cylinder. In *C. glauca,* each one bears 1-2 metaxylem elements and phloem, adaxially and abaxially surrounded by fibres. In the other species the bundles are bigger in size and have more metaxylem elements, until 10 as in *C. coccinea.* Mucilage ducts, are regularly distributed in the central cylinder, they have a circular outline in *C. glauca* (fig. 6b, II) and ellipsoidal in the other species.

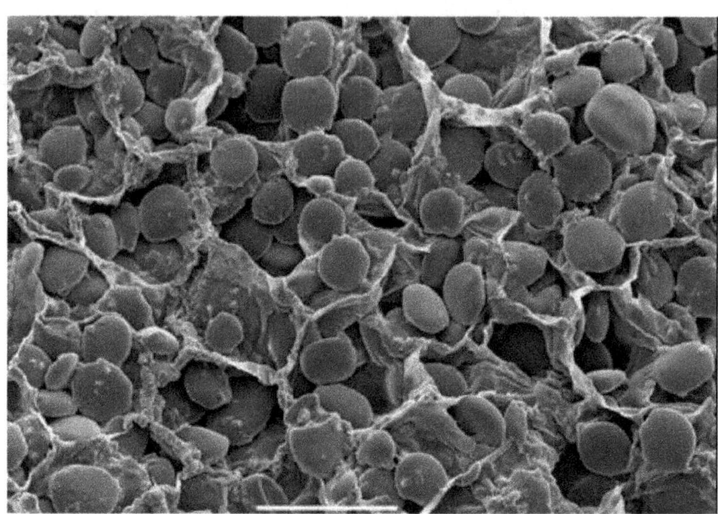

Fig. 6. (b) Scanning electron micrograph of starch granules and calcium oxalate crystals in *Canna sp.* The white bar at the bottom represents 100 µm

In *C. glauca,* the central cylinder bears an aerenchyma with large cells, surrounding the lacunae. Many cells of this tissue also contain starch granules. Aerenchyma is not present in other species, and there are other differences in shape and size of mucilage ducts. C. *coccinea* and C. *compacta* have elliptical mucilage ducts, not circular as in the other species.

3. Seeds

Seeds are regularly produced in all species and strewn out of the capsule when moved by the wind or touched by passing animals. The production of seeds is different in each species of *Canna.* Thus, *C. indica* presents 20-28 seeds, *C. coccinea* 6-8 seeds, *C. variegatifolia* 3-6, *C. ascendens* 1-3, *C. glauca* 6-12, and *C. fuchsina* 6-10. Seeds initially white and when mature, black or black, with chestnut brown spots are protected with a smooth coat. In *C. indica, C. ascendens* and *C. variegatifolia,* seeds are spheroidal, black, small (0.6 to 0.8 cm). In turn, seeds are spheroidal black and larger (1- 1.2 cm) in *C. coccinea* and *C. compacta,* while they are ellipsoidal and black, spotted of chestnut brown, 0,5- 0,9 cm in size, for *C. glauca.*

Seed microstructure consists of a massive chalaza, surrounded by an extremely hard, completely impermeable seedcoat. No germination is possible under normal conditions of temperature and humidity, because the seed cannot absorb water, therefore, it does not swell, remaining hard. The embryo is linear and bears one large cotyledon and a tiny plumule and rootlet.

According to Grootjen & Bouman (1988), the seed coat is mostly formed by chalaza tissue and in a lower proportion of integumentary and raphal tissue.

The chalazal part of the seed coat consists of four layers: the exotesta which is formed by the epidermis, and the three-layered endotesta.

The epidermis is composed by a palisade layer of long and narrow cells with very thickened walls called "Malphigian cells" : in them, Graven et al. (1996) recognise two zones, the lower hydrophilic and the upper hydrophobic. The first zone is mainly composed by cellulose, while the upper one contains silica, callose, and lignin. In *Cannaceae* the epidermis becomes the seed exotesta.

The endotesta comprises three layers of chalazal origin: the subepidermal, the vascularized as well as the "tanniferous". In the latter layer, polyphenols are abundant cell contents, which are considered to impart a secondary dormancy after shedding of the seed imbibition lid (Graven et al. 1997).

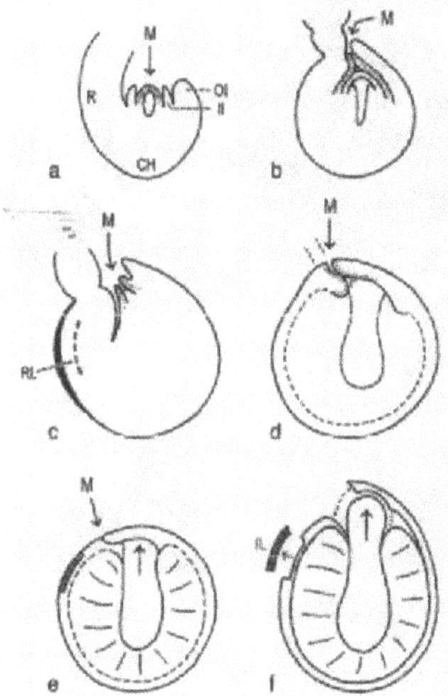

Fig. 7. Details of developing ovule and seed (CH = chalazal tissue, II=inner integument; IL = imbibition lid; M= micropyle; OI = outer integument; R= raphal tissue; RL = rupture layer.) a. Longitudinal section of developing ovule; b.Mature ovule; c.Developing seed showing rupture layer; d.Developing seed with vestigial micropylar collar; e. Mature seed, showing imbibition lid; F.Germinating seed after shedding of imbibition lid (After Grootjen & Bouman 1988).

Integumentary tissue: the area around the micropyle and the micropyle itself (which represents only about 1/15th part of the seedcoat surface) is formed by the inner integument (Fig.7, a, II). The integumentary seed coat is similarly composed of 4 layers: epidermis or palisade layer, subepidermal, vascularized, and tanniferous layer. The subepidermal layer grows inwards around the apex of the embryonal root, to form a small micropylar collar (Fig.7 d).

Raphal tissue: the raphal part of the seed coat form an 'imbibition lid' of 1.5-3 mm diameter. At the beginning of the imbibition process, the only visible part of the seed coat is an elevated circular area devoid of stomata close to the hilar-micropylar slit. As a consecuence of hydration the more or less circular boundaries surrounding the lid area loosen and the lid is expelled (Fig. 7c, RL). The lid itself is composed of only the two outer layers of the seed coat, the exotesta, and the lowermost subepidermal layer of endotesta which both constitutes the rupture layer. It is suggested that during imbibition, a difference in pressure between the hydrophylic and hydrophobic parts of the Malpighian cells is developed in the exotesta. This effect induces shedding of the lid along the rupture layer (Fig. 7 e, f). Graven et al.(1997) have found that the mechanism activating the opening of the imbibition lid is an elevated incubation temperature of 50 °C during 24 hours in wet surroundings.

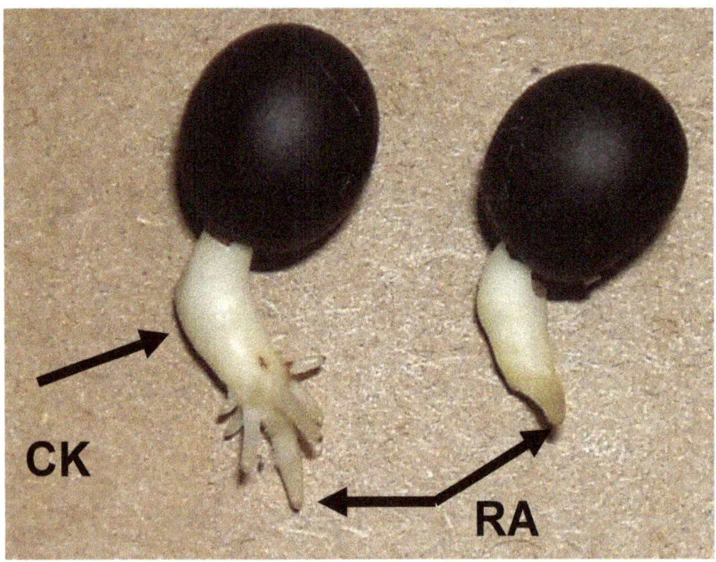

Fig. 8. (a) *C. indica* seeds after 7 days of germination. CK= cotyledon knee; RA= radicle.

Ciciarelli (2011) observed that germination, as well as the production of leafs from rhizome in the field, start in the beginning of September. Weather conditions at that time are an average temperature is of 13, 7° C, and average relative humidity of 79 % and 54 mm average of rainfall monthly. Seeds germinate under natural conditions when placed in a moist, airy, free and humiferous soil. A daily sun radiation of 11,8 h and maximum temperatures up to 18,2 °C produces the necessary temperature raise that allows the shedding of the lid. Minimum average of temperature registered for the month of 9,2 °C, does not avoid germination once initiated. This last observation coincides with the opinion of Cooke (2001). Following this, hydration and translocation of water inside the seed are possible, especially through the vascularized layer of the inner integumentary in direction to the endosperm, where enzymatic reactions triggers germination, later allowing the emergence of the cotyledon-radicle from the seed (Fig. 8a).

In *Cannaceae* the main mechanical properties of hardness and also the lid shedding mecanism through the rupture layer are provided mainly by the exotesta. The secondary dormancy is related to the tanniferous layer of the endotesta and the hydration and translocation of water is through the vascularized layer of endotesta.

In horticulture, and *in vitro* conditions the imbibition lid is removed deliberately to make germination possible.

This extremely hard coat and its indirect hydration mechanism allowed seeds of *Canna compacta* from the North West of Argentina to protect germ during 600 years under dry conditions (Sivori et al. 1968). After being brought to the laboratory one of the seeds germinated by scarification of the seed coat, growing a plant that bloomed and produced fruits and seeds.

Fig. 8. (b) *C. indica* seedlings after 18 days of germination *in vitro*. The imbibition lid (framed) is shown.

4. Discussion and conclusions

The external morphology of the simpodial rhizome shows a very invasive "Y"-branching way of propagation similar to those of other colonial species as *Alpinia speciosa (Zingiberaceae)*, as well as other species of *Canna*. In *Cannaceae, C. indica* produces the largest colonies in the Buenos Aires region.

The most important features of the internal morphology of rhizomes are the presence of parenchyma with abundant starch granules, as well as mucilage ducts, calcium oxalate crystals, and tanniferous cells. It is suggested that tannins may provide impermeability in plants. In *Cannaceae* these substances are stored in cells of rhizomes, stems and leaves, and may provide cold resistance in rhizomes, and,according to Graven et al. (1996), a secondary dormancy in seeds. Mucilages have always been related to the storage and availability of water in plants. In *Cannaceae* (Ciciarelli,1989) these substances are visible in circular or ellipsoidal ducts of longitudinal and transverse sections of rhizomes, stems and inflorescences supposedly related to maintain internal moisture in plants as well as the functions mentioned above. Abundant starch granules provide the nutritious source of energy for cell division and growth.

The presence of starch in species of *Canna* has been very important as a source of food for many cultures of the Peruvian Andes (Gade, 1966), in other regions of South America and Mexico. Rhizomes are cultivated and eaten in China, Vietnam and Japan, among other asian countries. (Tanaka, 2001, 2004, 2006).

Concerning the life cycle, if plants are not removed, the colonies will increase very much in size, due to asexual reproduction. The substrate is fully occupied by the growing colony and there are no room for seeds to grow from the soil, and, therefore, no seedlings are observed under adult plants. However, if part or the whole colony is cut down, and the substrate becomes exposed, fallen seeds from the progenitors are able to naturally grow. Weather conditions at the beginning of September in La Plata (PBA) are adequate for the growth of leaves from underground rhizomes as well as the germination of seeds in the field. In the first ten days, warm temperatures were from 14-18 °C, high average relative humidity of 80% and rainfall of 54 mm. The humiferous, airy and fertile soil has to be cleared from other plants to allow seeds to germinate. Sun radiation of 11,8 h daily and warm temperatures with a water intake from rain and dew triggers the germination mechanism described above and posibilitates plant growth. Seedlings with two leaves arising from seeds are visible in one or two weeks in this region (Fig. 3). Germination " in vitro" took about 18 days to produce a seedling with 2 leaves (Fig 8 B).

Seedlings were not observed in the surroundings of old colonies in PBA, but seeds remain in the soil under their progenitors. No germination was observed under adult plants, but in the boundaries, where the soil was free from plants, seeds can naturally grow.

This situation is similar to that of forests of *Sequoia* sp in North America where a large colony of trees were produced from the neck of the roots (Jepson, 1923). When an old tree dies, buds from roots produce trees scattered in a circle surrounding the original stum. Sequoias are known to live up to 1300 years (Jepson, 1923). The author estimates that in such a forest the 80% of adult trees were produced by asexual reproduction while the remaining 20 % by seed germination. In the central part of the forest, reproduction is almost exclusively by buds, while seedlings settle down in open and peripheral zones (Grant, 1989).

Some traits of *Canna indica* colonies as the remarkable height of plants, profuse seeds production, rapid growth, abundant descendants in adult colonies are, from an ecological point of view, typical of an r-strategist. Descendants can be quickly produced from seedlings in 1 to 3 weeks. If plants are young when frosts kill them, small rhizomes may survive underground until the following spring.

In *Canna,* seeds may represent a long term genetic reservoir that fell naturally from dried and opened fruits and remain in the ground or elsewhere. If the progenitors die, or the colony is destroyed by biotic or abiotic factors, or even remain in a very dried place, seeds are able to produce a new generation of plants with different genotypes from that of the original colony. This hypothesis has been confirmed in *C. compacta,* a species from the north west of Argentina, in which hard seedcoats are known to protect germ for centuries under dry conditions (Sivori et al.1968). However, when brought into the laboratory one of the seeds germinated, producing a plant that bloomed and produced fruits and seeds. Seedcoat hardness, extreme longevity, a water absorption-regulating mechanism consisting of an impermeable palisade layer and an imbibition lid are seed features that favor the insertion of the genotype in the environment.

The genetic variability provided by sexual reproduction (seeds) may be useful to colonize a new environment, or cleared soil after cutting part of the original population.

Rhizomes may represent a short term genetic reservoir, safer and faster, it remains underground, and in spring, axillary buds start growing and continue the production of plants until June, resulting in a uniclonal colony. However, at the end of June (winter) seeds that fell from capsules remain in the ground, and if conditions are adequate as those described above, new genetically different clones can grow from seedlings and enrich the genetic reservoir of the colonies, and most importantly, of the species. Cross pollination between different colonies may occur, and provide an explanation for such a variability in several morphological traits, particularly flowers in this species. The variation in number, pattern of colour of staminodes and labella were characters utilized by authors to describe several varieties under the name of *C. indica;* for instance, Tanaka (2001) have described four varieties in Asia.

The phenological studies of colonies in the field may contribute to improve knowledge on the life cycle of the plant, to evaluate the possibilities of cross- pollination between different colonies as well as the production of new viable genotypes from seeds. Phenological studies may also help elucidate taxonomic problems in the *Cannaceae* familiy which may even become prototype for other families as well.

Possibly as a consequence of global warming, in the last three years colonies were observed to extend their survival period in one month compared with previous years.

This longer life cycle of *C. indica* as a result of warmer seasons contributes to enlarge colonies in the field.

5. Acknowledgements

This work was supported by the Facultad de Ciencias Naturales y Museo, Universidad Nacional de La Plata, and was held in the LEAVES, led by Dr. Cristina H. Rolleri, to whom the author wishes to express her thanks for many suggestions and advice.

6. References

Bell A. D. 1979. The hexagonal branching pattern of rhizomes of Alpinia speciosa (Zingiberaceae). Annals of Botany 43 (2): 209-223.

Ciciarelli M. M. 1989. Las *Cannaceae* Link. Argentina. Tesis Doctoral. Facultad de Ciencias Naturales y Museo de L Plata, Argentina. 246 pags.

Ciciarelli M. M. 1995 *Canna variegatifolia* Ciciarelli (*Cannaceae- Zingiberales*), una nueva especie para Argentina. Rev. Museo La Plata, Bot. 14 (103): 333-341

Ciciarelli M.M. 2007. *Canna ascendens* Ciciarelli (Cannaceae) una nueva especie de la provincia de Buenos Aires y comentarios de otras especies argentinas de este género. *Darwiniana* 45: 188-200. ISSN 0011-6793. Publicaciones del Instituto de Botánica Darwinion. San Isidro. Argentina.

Ciciarelli M. M. & Rolleri, C.H. 2008. Morfología, taxonomía y caracterización de siete especies neotropicales del género *Canna* (Cannaceae- Zingiberales). *Botanica Complutensis* 32: 157-184.

Ciciarelli M. M., Rolleri, C. H., & M. C. Gonzalez Dubox. 2010. *Canna fuchsina* (Cannaceae- Zingiberales), una especie nueva para la Argentina y sus relaciones con otras especies del género. *Botanica Complutensis* 34: 49-55. ISSN 0214-4565. Publicaciones de la Universidad Complutense de Madrid, España. Indexed by SCOPUS.

Cooke, I. 2001. The Gardener's Guide to Growing *Canna*. Timber Press. ISBN 978-0-88192-513-5

Font Quer , P. 1965. Diccionario de Botánica. Ed. Labor, S.A., Barcelona.

Gade, D. W. 1966. Achira, the edible Canna, its cultivation and use in the Peruvias n Andes. Economic Botany 20 (4): 407-415.

Grant, V. 1989. Especiación Vegetal. Noriega Eds. Mexico.

Graven, P.,C. G. de Koster, J.J. Boon & F. Bouman. 1997. Functional aspects of mature seed coat of the Cannaceae. Pl. Syst. Evol. 205: 223–240.

Grootjen, C.J. & F. Bouman. 1988. Seed structure in Cannaceae: Taxonomic and ecological implications. Ann. Bot. (Oxford) 61: 363–371, f. 1–4.

Jepson, W. L. 1923 .The trees of California . Associated Students Store. Berkeley. Calif.

Johanssen, W. A. 1940. Plant Michrotechnique. New York.

Maas van de Kamer, H., Maas P.J.M. 2008. The Cannaceae of the World. *Blumea*.53: 247-318 (72)

Tanaka, N. 2001. Taxonomic revision of the family Cannaceae in the New World and Asia. *Makinoa* 1: 1- 74.

Tanaka, N. 2004. The utilization of edible Canna plants in southeastern Asia and southern China. Economic Botany 58: 112-114.

Tanaka, N., N. Inouch, and T. Koyama. 2006. Edible canna and its starch: an underexploited starch-producing plant resource. Food & Food Ingredients Journal of Japan 211: 319-325.

Tillich, H. J. 1959. Seedlings and systematics in Monocotyledons. In: P.J. Rudall, P.J. Cribb, D.F.Cutler & C.J. Humphries, Monocotyledons: systematics and evolution: 303–352. Royal Botanic Gardens, Kew.

Tomlinson P.B. 1962. Phylogeny of the *Scitaminae*: morphological and anatomical considerations. Evolution 16: 192-213.

Sivori, E., Nakayama, F. & Cigliano, E. 1968. Germination of "*achira*" seed (*Canna* sp.) approximately 550 years old. *Nature* 219 (5.160): 1269-1270.

6

Phenological Observation in the Czech Republic – History and Present

Lenka Hajkova et al.*
Czech Hydrometeorological Institute Prague,
Charles University Prague,
Czech Republic

1. Introduction

Periodicity in the life of plants and animals is considered to be an indirect indicator for the periodicity in the climate. This is because plants and animals continuously respond to changing climatic influences. It is at the same time a supporting science for biogeography, ecology and phytocenology. Phenology is a branch of science which deals with the study of periodically recurring natural life cycles in the course of time, called phenological phases of plants and animals, as they relate to environmental conditions, particularly climate and weather. The base of the word phenology emanates from the Greek word fainó, which means "I reveal" (Krška, 2006). Dating from ancient times, the observation of the surrounding environment has been one of the basic features of everyday life, specific expressions in the acquired pieces of knowledge arise from weather proverbs e.g. "September brews the wine, in October we drink it". The weather proverbs had of course great importance even for peasants and other professions. Phenological observations, which are according to the kind of observed organisms divided into phytophenological and zoophenological. In climatology the basic time period is a year together with its months but the growth and subsequent development of plants and animals do not follow our calendar. Natural life cycles of organisms are determined, simply said, by internal (genetic) and external (climatological) conditions. And we divide the year into phenological periods according to the response of nature to the real course of the weather (Rožnovský & Havlíček, 1999). Phenology season onset in relation to synoptic situation studied Hájková & Kožnarová & Sulovská (2011). Agroclimatological classification is based on relation of plants to selected climatological characteristics as well as agroclimatological indicator of temperature, agroclimatological indicator of irrigation and agroclimatological indicator of hibernation (Žalud, 2010).

In this chapter we briefly describe history of phenological observations in the Czech Republic and phenological observations at present including concisely description of

* Jiri Nekovar[1], Dasa Richterova[3], Vera Koznarova[3], Sona Sulovska[4],
Ales Vavra[4], Alena Vondrakova[4] and Vit Vozenilek[4]
[1]*Czech Hydrometeorological Institute Prague, Czech Republic*
[2]*Charles University Prague, Czech Republic*
[3]*Czech University of Life Sciences Prague, Czech Republic*
[4]*Palacky University Olomouc, Czech Republic*

current methodology used at CHMI (Czech Hydrometeorological Institute) phenological stations. The Czech Hydrometeorological Institute cooperates with other European countries within the phenology research, very short description is provided. From all observed species we have chosen the Common Hazel as one of the most important pollen allergen. Phenological results in the Common Hazel were evaluated in the Czech Republic within the period 1991-2010 including climatological characteristics.

2. History of phenological observation in the Czech Republic

Phenological observations have a long tradition in Czech Lands. We can divide it into two parts - phenological observations executed in the last centuries and observation of phenological stages at present.

2.1 History of phenological observations in 18th and 19th century in Czech countries

The first Czech meteorologists J. Stepling, A. Strnad and M. A. David, whose activities are known from the second half of the 18th century and David's from the 19th century as well, devoted themselves to studying the influence of weather on the life of plants and animals. A. Strnad attached his remarks to regular measures, which he carried out at the Prague observatory from the 1st January, 1775 up to nearly the end of his life (23rd September, 1799). A number of these phenological observations is also attached to his paper "Meteorologischer Beytrag auf das Jahr 1792". A longer article, containing an economic survey of the year 1791, was published by Strnad in Mannheim Eefemeridas with the heading „Conditio anni generalis" (Seydl, 1954).

The Mannheim or Falc meteorological society (Societas meteorologica Palatina), was assigned on the 15th September 1780 as a "meteorological class" to the Academy of Sciences, which has been in existence in Mannheim since 1763, and worked till 1799. The society welcomed, apart from meteorological measurement, also notes on phenology and nosology, especially the reports on the blossoming and ripening of fruits of the most important cultural plants, on the results of haymaking, harvest and fruit harvest, both from the point of view of the quantity and quality, and the data on the diseases of plants and on damage caused by insects. And the reports on arrivals and departures of migrating birds, e.g. storks, swallows, nightingales and cuckoos were welcomed. The first phenological calendar in our literature was published by Med. Dr. Tadeáš Haenke in his longer paper „Blumenkalendar fűr Noumen in Jahre 1786". The author carried out in the years 1784 and 1785 a detailed observation on the earlier and later beginning of spring, on its course and the changes of plants during this time (Seydl, 1954).

The principles for regular and methodologically unified phenological observations in a station net were laid by the Swedish botanist Carl von Linné. He established the network of 18 stations in Sweden in the years 1752–1755. Professor ing. Dr. Octavianus Farský (Farský, 1942) said, that phenology of botany cultivated from the time of Linné has not only purely scientific (biological, phyto - a zoographical, ecological and climatological), but first of all practical importance. Regular phenological observations in the Czech countries were first introduced by the Patriotic – economic company, the successor of K. k. Ackerbau-Gesellschaft, based in 1769 as an order of the Empress Maria Teresia in Prague for the enhancement of agriculture. The following phenological elements were observed: the development of buds into leaves, the beginning of blossoming, the end of blossoming and maturation of seeds. Further, some animals e.g. bats, hamsters, badgers, snakes and lizards, frogs, which do not leave our countries and hibernate in winter, mainly their awakening in spring and the beginning of hibernation were observed. The Prague lawyer Karl Fritsch as a significant part of his work in the field of phenology. His first work on phenology was devoted to the influence of weather on vegetation. In the paper „Elemente zu einer Untersuchung űber den Einfluss der Witterung auf die Vegetation"Fritsch explained the link between the yearly amount of warmth and moisture to the most important phases of the development of a plant, he presents eight charts of meteorological data (e.g. gradual total of positive values of the air temperatures, the differences in gradual total of precipitation) (Seydl, 1954).

2.2 The first phenological observations in the Czech Republic in the 20th century

The state phenological service was organized in Moravia by the Department for soil science and agricultural meteorology of the Regional research office of agriculture in Brno. Prof. Ing. Dr. Václav Novák (1888–1967) set up in 1923 one of the first national phenological services in the world. The observation net was soon so extensive (with 650 observers involved in its activities), that it was unsustainable in the long term. The organization of phenological services was so sophisticated, several challenges in newspapers and professional press were published in order to acquire other observation sites, national schools and public corporations were asked for cooperation. The results of these observations were gradually processed in a long-term average of phenological phases, the so-called phenography. Coming out of these observations, phenological yearbooks were published, with map enclosures for the years 1923 and 1924 - and thus the principles of the beginning of the Czech phenography were laid. The material was processed together with professor Novák and professional officer ing. Josef Šimek. Phenological yearbooks were put together in the years 1927–1937 (Novák & Šimek, 1926).

In the year 1931 prof. ing. Dr. Bohuslav Polanský (1901–1983) jointed in by organizing phenological observations in the Moravian - Silesian country. In the year 1942 docent ing. Dr. Octavianus Farský, later professor at the School of agriculture in Brno, tried to introduce hunting - phenological observations in the area of the Bohemia and Moravia protectorate. He collectively described his hunting - phenological observations in the journal „Stráž myslivosti", no. 27, volume XIX (Farský, 1942).

As O. Farský stated, already in the last century the „Fysiokratická společnost", based on the impulse of the author, teacher and naturalist Dr Karel Slavomil Amerling (1807–1884), started to practise phenology and gained merit in the development of our applied natural sciences and progress in agriculture and forestry. In hunting - phenological observations, the following species of our commercial game were observed: feathered: *Scolopax rusticola L., Tetrao tetrix L., Tetrao urogallus L., Perdix perdix L., Phasianus colchicus L.* and furry: *Lepus europaeus Pall., Sus scrofa L., Ovis musimon Pall., Dama dama L., Cervus elaphus L., Capreolus capreolus L.* (Farský, 1942).

The phenological phases follow one after the other in a certain stable order; the first phenological calendar was published by the above mentioned T. Haenke. Arising from these long-term phenological observations, we can create the so called "Calendars of nature" for a certain place and its surrounding. We can also add a border data of the beginnings of these phases (the earliest and the latest, their amplitudes, phenoanomaly, the curve of phenodynamics) (Brablec, 1952).

Vyplněný záznamní arch zašlete koncem roku (v listopadu neb v prosinci) na adresu:
St. výzkumné ústavy pro agroped. a bioklimatologii v Praze II., Karlovo nám. 3.

Čtěte především návod na poslední straně!
Pište inkoustem!
Znamenejte jen spolehlivé údaje!

052431

Nadmořská výška: 290 m

Fenologická pozorování

(o význačných zjevech v životě rostlinstva.)

Rok: 1924.

Místo pozorovací (obec): Střednice u Mělníka. Pošta: Vysoká u Mělníka.

Okres soudní: Mělník. Župa: Mladá Boleslav.

Pozorovatel (jméno, zaměstnání): Vladislav Nový, dělník.

Druh půdy (jílovitá, hlinitá, písčitá a pod.) dle praktického označení:

Převládá rovina či svah a na kterou stranu světovou obrácený?

a) u kultur polních: hlinitá a písčitá. Převládá rovina.

b) u kultur lučních: ~ Na jedné straně naléhá

c) u kultur lesních: písčitá. se táhlá pahorkatina, obrácená k jihu.

Upozornění: Jména zvláště důležitých rostlin jsou význačněji tištěna. Fáze zejména důležité jsou orámovány.

Všeobecné úkony polní.

Vyplniti vždy, i když nejsou konána na polních kulturách pozorování jednotlivá

Druh práce	Datum	Poznámka
Počátek jarních prací polních	1. dubna	Setí započato 3. dubna.
Počátek senoseče	5. června	Rozumí se set. jetelovin, luk ...
Počátek žní žitných (kosení žita)	17. července	—

Polní kultury.

(Pozorujte hlavně odrůdy v kraji převládající! Jména odrůd pokud možno vždy uveďte!)

Pojmenování	Den setby (sázení)	Počátek vzcházení	Většina vymetala	V plném květu	Počátek seče (sklizně)	Vysvětlivky	Poznámky zpravodajovy
Brambory rané	29. dub.	20. květ.	—	1. června	28. června	Uveďte případné poškození mrazy.	Rohlíčky.
Brambory pozdní	1. května	26. květ.	—	15. června	15. září		—
Cukrovka	11. dub.	30. dub.	—	—	22. září		Originál, Dobrovice!
Chmel	8. dub.*	22. dub.	13. květ.	6. června	21. srpna		* Rozumí se vytahování sadby. + Rozumí se zavádění na drát.
Ječmen jarní	6. dub.	20. dub.	20. června	30. června	28. června	Uveďte, zda se jedná o ječmeny nízl či o vyplímené	Hanácký. Žlutý se nesměs. s ječmeny lehčejšími.
Ječmen ozimý *)	—	—	—	—	10. července		—
Kukuřice	—	—	—	—	—	Udati, je-li drobnozrnná či velkozrnná a zdali dozrála	—
Oves	15. dub.	1. května	1. června	8. června	10. srpna		—
Pšenice ozimá *)	22. září	30. září	8. června	14. června	25. července		Přesnice dobrovická. Setba a vzcházení v r. 1923.
Žito ozimé *)	14. září	25. září	19. květ.	1. června	18. července	Důležité jest záznam o vymrznutí	Žito petkuské.

Fig. 1. Report of phenological observations at station in Střednice, year 1924.

The operative of the Hydrometeorological institute Josef Brablec processed and published the calendar of nature for Plumlov in Moravia during the period 1926–1947 (Brablec, 1952), for Hodonín in Moravia during the period 1924–1949 and for Střednice at Mělník in Bohemia during the period 1923–1949 (Brablec, 1953). At the above mentioned station Střednice at Mělník, Vladimír Nový, a farmer, carried out some meteorological observations and noticed not only prescribed species of plant and field crops (Fig. 1), but he also recorded data about weeds and other plants and his records have been up to the present the only long-term observation rows at our disposal.

In 1939, all meteorological services in the area of protectorate Bohemia and Moravia were brought together and the Central meteorological institute for Bohemia and Moravia was established, from the year 1940, phenological observations were overtaken by the Czech meteorological service with the whole net (about 1 000 places) and the archive of data since 1923. From that time up to the present, the phenology makes up a part of the meteorological service, included in 1954 in the Hydrometeorological institute (Miháliková, 1983; Krška & Vlasák, 2008). Phenological observations were conducted according to the principles included in the Handbook for phenological observers from the year 1956 (Pifflová et al., 1956). It was determined for the observers of the institutes for general phenology, which served mainly the needs of agriculture production (Fig. 2).

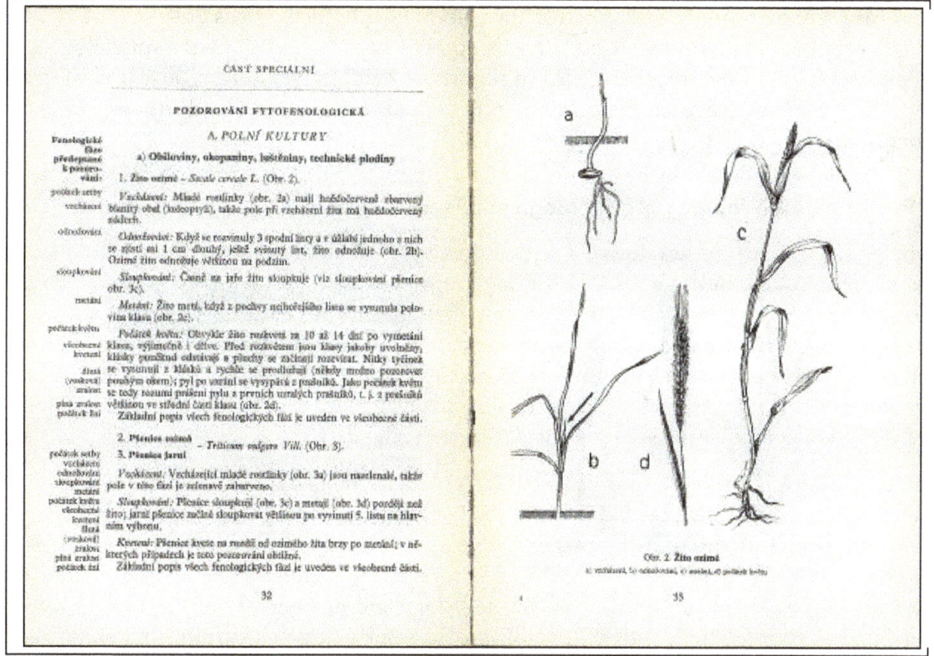

ČÁST SPECIÁLNÍ

POZOROVÁNÍ FYTOFENOLOGICKÁ

A. POLNÍ KULTURY

a) Obiloviny, okopaniny, luštěniny, technické plodiny

1. Žito ozimé – *Secale cereale* L. (Obr. 2).

Vzcházení: Mladé rostlinky (obr. 2a) mají hnědočerveně zbarvený blanitý obal (koleoptyli), takže pole při vzcházení žita má hnědočervený nádech.

Odnožování: Když se rozvinuly 3 spodní listy a v úžlabí jednoho z nich se zjistí asi 1 cm dlouhý, ještě svinutý list, žito odnožuje (obr. 2b). Ozimé žito odnožuje většinou na podzim.

Sloupkování: Časně na jaře žito sloupkuje (viz sloupkování pšenice obr. 3c).

Metání: Žito metá, když z pochvy nejhořejšího listu se vysunula polovina klasu (obr. 2c).

Počátek květu: Obvykle žito rozkvete za 10 až 14 dní po vymetání klasu, výjimečně i dříve. Před rozkvětem jsou klasy jakoby uvolněny, klásky poněkud odstávají a pluchy se začínají rozevírat. Nitky tyčinek se vysunují z klásků a rychle se prodlužují (někdy možno pozorovat prostým okem); pyl po uzrání se vysypává z prašníků. Jako počátek květu se tedy rozumí prášení pylu z prvních uzralých prašníků, t. j. z prašníků většinou ve střední části klasu (obr. 2d).

Základní popis všech fenologických fází je uveden ve všeobecné části.

2. Pšenice ozimá
3. Pšenice jarní – *Triticum vulgare* Vill. (Obr. 3).

Vzcházení: Vzcházející mladé rostlinky (obr. 3a) jsou nazelenalé, takže pole v této fázi je zelenavě zabarveno.

Sloupkování: Pšenice sloupkuje (obr. 3c) a metá (obr. 3d) později než žito; jarní pšenice začíná sloupkovat většinou po vyvinutí 5. listu na hlavním výhonu.

Kvetení: Pšenice kvete na rozdíl od ozimého žita brzy po metání; v některých případech je toto pozorování obtížné.

Základní popis všech fenologických fází je uveden ve všeobecné části.

32

Fig. 2. Handbook for phenological observers (1956) - demonstration.

A constituent part of the handbook was also zoophenological observations. The following phenomena were observed e.g. the date of arrival, mass arrival, the first singing, herding and departure of thrushes, martins, quails, cuckoos, swifts, larks, starlings and swallows. With bee melliferous, the first flight and the first congeries of pollen were observed.

3. Phenological observation in the CHMI at present

A significant change in phenological observations came in the year 1983; observation sites were divided into stations observing field crops and fruit trees. Separate instructions for both types of stations were issued for observers. The transformation was finished in the year 1987 by issuing methodological instructions for the activities of phenological stations observing forest plants (from January 1st 1987). From January 1st 2005, Phenological atlas (Coufal et al., 2004) became an aid for observers in the net of the CHMI.

In the year 2004 a trial run of the database Oracle Phenodata started, for acquiring and storing phenological data (application pod Clidata in environment Oracle) and since 1st Jan. 2005 phenological data have been stored in this database up to the present. Older data were transferred to this database from environment Excel, where they had been stored till that time.

The present phenological net in the area of the Czech Republic is consisting of the three types of stations (forest plants, fruit trees, field crops). Voluntary observers monitor the beginnings of phenophases according to the methodological instructions of the CHMI (art. 2, 3 and 10 – in the year 2009 new, updated instructions for observers were issued), data are recorded in the phenological notepad and then transferred to current reports. The current reports are sent by post or in the electronic version to the branch offices of the CHMI, where the data are checked and transferred to application Oracle Phenodata by the phenologists of the branch offices of the institute. Current reports of field crops are sent once a week from 10th to 49th week of the year, reports on fruit trees 10 times a year and reports on forest plants 14 times a year.

3.1 Present methodology for phenological observations

Phenological observations are carried out at three types of stations – at field, fruit and forest stations (Fig. 3). The stations are selected with regard to that fact that their position should represent the respective area, it should not defy the local conditions in any way (e.g. clenched valley, frost basin, slopes with geographical polarity, places with atypical soil conditions), and at the same time it has to be close to a climatological station. The Czech phenological network consists of 79 field phenological stations, 34 fruit trees phenological stations and 49 wild plants phenological stations at present.

With field stations, such areas are chosen, which are in the given land register and used as arable land. Within these, selected parts of the overgrowth of the plant were then observed, where the plant was sown in the respective year, and the species of it are recorded. **With fruit stations**, such sections of the soil with cultivated fruit trees were chosen, where the plants are homogeneous in terms of the terrain, slope and its orientation, spacing etc. and at the same time they are uniformly cultivated, i.e. the same pruning, nutrition, cultivation and protection. In these areas 5 to 15 individuals are selected, which have identical biological features (the same species, variety, age, rootstock and shape) such individuals are then monitored until they perish or their health state is considered sufficiently representative. **In the case of forest station**, where plants are found in the surroundings, not standardized by the human, the rules are as follows: the locality has to be sufficiently far from any built-up area, outside the reach of urbanization, industrial, building, mining or agricultural activities;

it should contain just one type of botanical community and one type of macro relief, it should be rich in varieties, which means preferably mixed forests to cultural pine groves, or forest edges to their denser and poorer centres, and the plant should have optimal living conditions. In these localities such areas are chosen, which are homogenous from the ecological point of view. From these approximately five healthy individuals are chosen, in the case of shrubs 20-40 years old, in the case of trees 40-60 years old. Within each area, the location, elevation, exposition and slope, soil characteristics, irrigation or moisture conditions, in the case of forest stations also subsoil, botanical community, irradiation and the grade of legal protection are registered.

Fig. 3. Map of CHMI phenological station at present (situation in 2010).

The observer then carries out observations each lasting two days during the vegetation period, and outside the vegetation period once or twice a week. The vegetation period is defined at field stations from March to November, at fruit stations from April to October and at forest stations from March to October. The results of the observations are to be reported during the vegetation period at field stations every week, at fruit stations in the given terms from April to May each fortnight, from June to mid July once every three weeks, till the beginning of October once every six weeks and then at the beginning of December, at forest stations they should be handed in from March to June once a fortnight, from July to December always at the end of a month. Besides the data about the observed phenophases, the observer has to provide information concerning possible damage, both the damage caused by weather conditions, insects and fungi and other pathogenic agents, in the case of field crops also concerning the length of plants and the number of leaves and agro technological interventions carried out. If any of the station conditions in the observed area significantly change, (e.g. a change in cultivation, new building site nearby, the dehydration of the area), the observation there has to terminate.

At field station, 19 crops are observed: Winter and Spring Wheat (*Triticum vulgare*) - Fig. 4, Winter and Spring Barley (*Hordeum vulgare*), Winter Rye (*Secale cereale*), Oat (*Avena sativa*), Sugar Beet and Fodder Beet (*Beta vulgaris*), Potato (*Solanum tuberosum*), Mays (*Zea mays*), Broad Bean (*Vicia faba*), Field Pea (*Pisum sativum*), Green Bean (*Phaseolus vulgaris*), Cultivated Flax (*Linum usitatissumum*, Winter Rape (*Brasica napus*), Oppium Poppy (*Papaver somniferum*), Lucerne (*Medicago sativa*), Trifolium (*Trifolium pratense*) and Hops (*Humulus lupulus*). With these crops, the following phenophases are observed: 1) sowing, 2) emergence, 3) leaf bud burst, 4) first leaves, 5) tillering, 6) the beginning of leaf sheath elongation, 7) the beginning of stem elongation, 8) first node, 9) second node,10) swelling of the sheath of the last leaf, booting, 11) heading, 12) closing rows of the growth, 13) total closing of the growth, 14) beginning of the side shoots growth, 15) beginning of decortications, 16) inflorescence emergence, 17) beginning of flowering, 18) beginning of male flowers flowering, 19) beginning of female flower flowering, 20) full flowering, 21) end of flowering, 22) cone development, 23) green ripeness, 24) milky ripeness, 25) milky wax ripeness, 26) yellow ripeness, 27) full ripeness, 28) harvest ripeness, 29) dead tops, 30) harvest.

Fig. 4. Winter Wheat (*Triticum vulgare*) - full ripeness (Photo: Lenka Hájková).

At **fruit tree stations** the following 15 plants are observed: Apple (*Malus domestica*) - Fig. 5, European Pear (*Pyrus communis*), Prunus (*Prunus domestica & P. insititia*), Wild Cherry (*Cerasus avium*), Sour Cherry (*Cerasus vulgaris*), Apricot (*Armeniaca vulgaris*), Peach (*Persica vulgaris*), Red Current (*Ribes rubrum*), Black Current (*Ribes nigrum*), Gooseberry (*Grossularia uva-crispa*), Common Walnut (*Juglans regia*), Common Hazel (*Corylus avellana*), and Vine (*Vitis vinifera*). The following phenophases in the respective plants are observed: 1) beginning of blooding, 2) leaf bud burst, 3) flower bud burst, 4) mixed bud burst, 5) first leaves, 6) inflorescence emergence, 7) beginning of flowering, 8) beginning of flowering (male flowers), 9) beginning of flowering (female flowers), 10) full flowering, 11) beginning of petal fall, 12) end of flowering, 13) bud creation, 14) end of shoots elongation, 15) bunches begin to hang, 16) softening of berries, 17) harvest ripeness, 18) harvest, 19) end of leaf fall.

Fig. 5. Apple - variety Champion (*Malus domestica)* - flowering (Photo: Lenka Hájková).

Fig. 6. Blackthorn (*Prunus spinosa*) - flowering (Photo: Lenka Hájková).

At wild plants stations are 24 species of wood plants and 21 species of herbs observed. Among the wood plants are: Norway Spruce (*Picea excelsa*), European Larch (*Larix decidua*), Scotch Pine (*Pinus sylvestris*, Scrub Pine (*Pinus mugo subsp. Mughus*), Wild Cherry (*Cerasus avium*), Blackthorn (*Prunus spinosa*) - Fig. 6, Rowan *(Sorbus aucuparia)*, Midland Hawthorn *(Crataegus oxyacantha)*, Robinia *(Robinia pseudoacacia)*, European Hornbeam *(Carpinus* *betulus)*, Common Hazel *(Corylus avellana)*, Silver Birch *(Betula pendula)*, Common Alder *(Alnus glutinosa)*, Grey Alder *(Alnus incana)*, European Beech *(Fagus sylvatica)*, Pedunculate Oak *(Quercus robur)*, Goat Willow *(Salix caprea)*, Harewood *(Acer pseudoplatanus)*, Norway Maple *(Acer platanoides)*, Lime Tree *(Tilia cordata)*, European Dogwood *(Cornus sanguinea)*, Cornelian Cherry *(Cornus mas)*, Black Elder *(Sambucus nigra)* and Racemic Elder *(Sambucus racemosa)*.

Observed herbs are as follows: Marsh Marigold *(Caltha palustris)*, Wood Anemone *(Anemone nemorosa)* - Fig. 7, Hepatica *(Hepatica nobilis)*, Meadow Buttercup *(Ranunculus acris)*, Wild Strawberry *(Fragaria vesca)*, Trefoil *(Trifolium repens)*, Perforate St. John's-wort *(Hypericum perforatum)*, Rosebay Willowherb *(Chamerion angustifolium)*, Heath *(Calluna vulgaris)*, Bilberry *(Vaccinium myrtillus)*, White Dead Nettle *(Lamium album)*, Ox-eye Daisy *(Chrysanthemum leucanthemum)*, Coltsfoot *(Tussilago farfara)*, Butterbur (*Petasites hybridus)*, White Butterbur *(Petasites albus)*, Colchium *(Colchicum autumnale)*, Lily of the Valley *(Convallaria majalis)*, Common Snowdrop *(Galanthus nivalis)*, Cocks Foot *(Dactylis glomerata)*, Meadow Foxtail *(Alopecurus pratensis)* and Reed *(Phragmites australis)*. The following phenophases are observed in these plants: 1) bud burst, 2) first leaves, 3) fully leaved, 4) inflorescence emergence, 5) beginning of flowering, 6) end of flowering, 7) bud creation, 8) first fruit visible, 9) St. John's sprouts, 10) pulpy sprouts begin to lignifying, 11) leaves colouring, 12) leaves fall, 13) fruit ripeness, 14) haymaking and second haymaking.

Fig. 7. Wood Anemone (*Anemone nemorosa*) - flowering (Photo: Lenka Hájková).

4. The Czech phenology and Europe

The Czech Hydrometeorological Institute is involved in the international cooperation in phenological observations in the project called **IPG - International Phenological Garden** since the end of 1990s. Since the half of the 20th century, the network of international phenological gardens has been working, supervised by the Humboldt University in Berlin.

The network contains more than 70 stations, in the Czech Republic there are 3 stations: at the observatory in Doksany (50°28' N, 14°10' E, 158 m asl, IPG-No. 85), in the arboretums of forestry schools in Kostelec nad Černými lesy (50°00' N, 14°52' E, 345 m asl, IPG-No. 86) and Křtiny at Brno (49°19' N, 16°45' E, 475 m asl, IPG-No. 87). The administrator of IPG delivered the observed trees from the nursery for trees in Ahrensburg at Hamburg for free, the biological material is thus genetically identical. The removal and planting is ensured by the operator. The observation is not paid, after the end of each calendar year, a report is sent to the administrator, who processes the results, prepares a yearbook and sends it for free to all the stations of the network (Nekovář et al., 2008). The main idea of the IPG was to standardise the conditions for observation - to use unified regulations for observations realized in similar surroundings only for professionals and thus on clone plants coming from one nursery. The aim of the project IPG is to study the presupposed impacts of changes in the climate on forest plants.

Fig. 8. 3-D model of phenological station in Doksany.

The International Phenological Garden in Doksany was set up in the year 2000 on the premises of the CHMI observatory, thanks to the Humboldt University in Berlin, which provided forest plants for free. The IPG is situated about 50 km of Prague, in the Ústí region. Its elevation is 158 m above sea level, its average yearly air temperature is 8.5 ° C and the average amount of precipitation is 450 mm. The IPG in Doksany was established as a first in the framework of the entire Czech Republic. In total, 9 phenological phases in 19 plants are observed. The wood plants have been placed in an area of the observatory in Doksany in such a manner, that they will either impede each other in the future, or influence the phenological measuring. The IPG is situated in a representative place, which is not influenced by local microclimatic conditions. The situation of the IPG is apparent in Fig. 8, which has been created by means of a 3D model, showing the garden in spring, summer, autumn and winter. Phenological observations in forest growth can be very demanding especially in difficult to access terrain. In recent years, an effort has been gathering strength, to use automatic monitoring with the help of phenological cameras and satellites. A greater objectification and an increase in the precision of phenological observations is thus to be expected.

Fig. 9. Foliage within the year - Common Hazel (*Corylus avellana*) (Photo: Martin Možný).

In September 2006, in the CHMI, a piloting project of an all-year-long monitoring of phenological phases by a digital camera was commenced, within the framework of the phenological garden (IPG) in Doksany. Figure 9 shows an example of the phenological development taken in automatic regime. A digital camera Canon PowerShot S3 IS was used, it was positioned in a closed container with a sight, its interior heated by the heating foil and it was attached to a swivel stand.

In Europe, the international cooperation in phenological observation and the analyses of data by the use of programme COST is greatly intensifying (**European Cooperation in Scientific and Technological Research**). The CHMI was a member of COST 725 (**Establishing a European Phenological Data Platform for Climatological Applications)** and COST ES603 (Assessment of production, release, distribution, health impact of allergenic pollen in Europe).

In the year 2010, the project PEP 725 (Pan European Phenology DB) was commenced. It was another step in the continuation and enlargement of the European phenological database for scientific purposes, education and research. Project PEP 725 is proposed to run until 2014 with the presumption of prolongation at a later date.

Further cooperation was initiated by the activities of the **Joint Research Centre (JRC)** of the European Committee in the town of Ispra in Italy. **The European Agrophenological Net (EAgPN)** was established, which will provide "country truth" for calibration of satellite snaps and for calculation of the quantity and quality of the harvest of cereals, corn, rape and sunflower. Since the end of the 90s, the University of Waningen initiated the **European Phenological Net (EPN)**, which is the creation of purpose-built websites, informing about national nets, general number of stations, plants and phases (Nekovář & Rožnovský, 2006).

5. "A case study - phenology observations of Common Hazel (*Corylus avellana*) in Czechia"

5.1 Climatological characteristics of the Czech Republic during the period 1991–2010

The country's natural and environment is characterized by a moderate, humid climate and four alternating seasons. Its vegetation is determined by the merging of the Hercynian and Carpathian forest areas and the warm Pannonian steppe. The overall character of the landscape reflects vertical variation in the georelief.

The climate is generally favourable and has rather maritime character. Despite the small surface area of the country, the climate is highly varied. The elongated shape of the territory results in a slight increase in continentality as one moves east (Tolasz et al, 2007).

The maps (Fig. 10–12) introduce the average values of the selected climatic elements (air temperature, atmospheric precipitation, duration of sunshine) in the Czech Republic during the period from 1991–2010 in the warm half of the year, i.e. from April to September.

Air temperature, sunshine and precipitation are closely connected with growth of plants. The average values of air temperature ranges between 10 and 16 °C (the maximum values are in the South Moravia and Polabská Lowland), the sunshine duration oscillates between 1,050 and 1,350 hours (in the part of south and North Moravia show the maximum values) and the precipitation totals are between 300 and 650 mm (the highest values are in the mountain areas of Krkonoše, Jeseníky, Beskydy and Šumava Mts.).

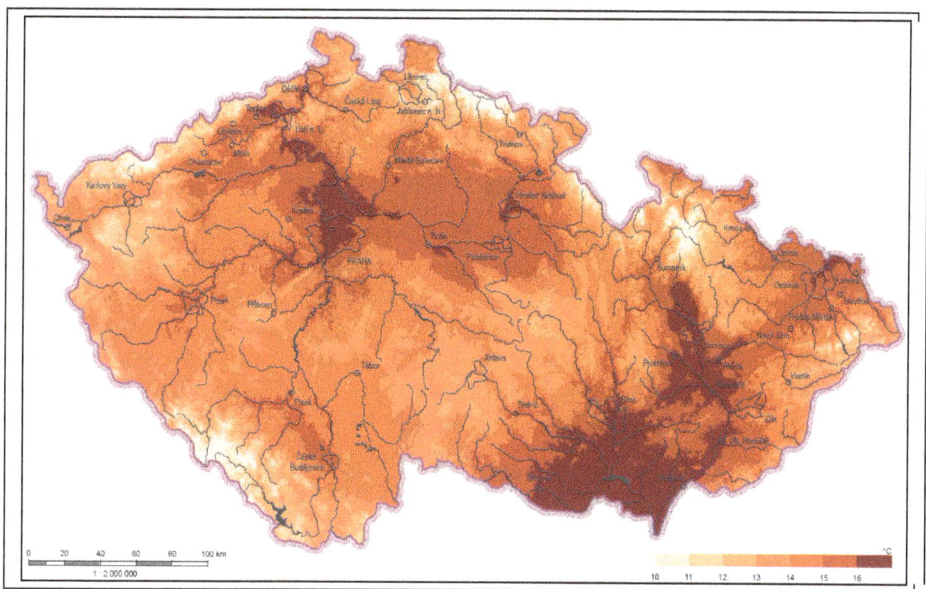

Fig. 10. Average air temperature in warm half-year (April–September).

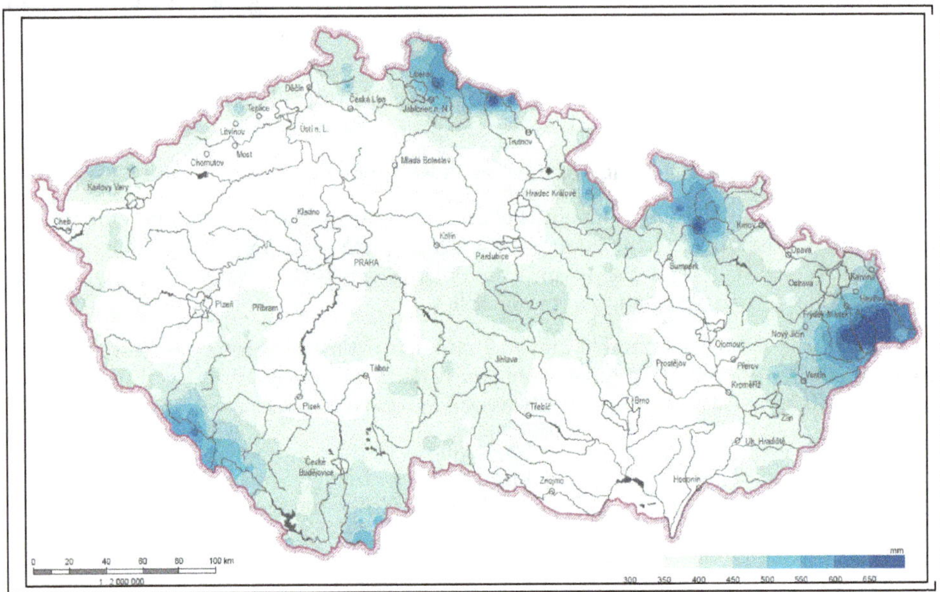

Fig. 11. Average precipitation total in warm half-year (April–September).

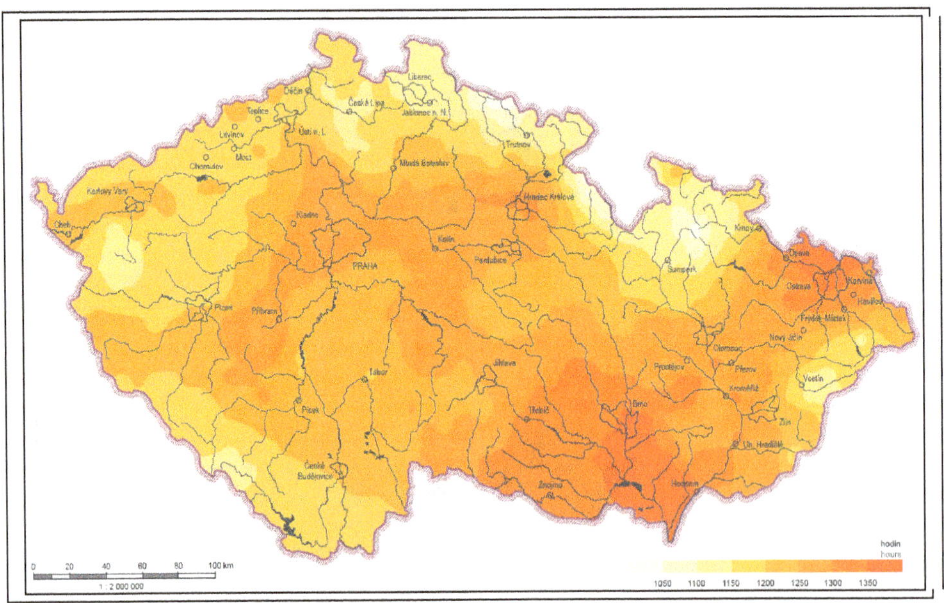

Fig. 12. Average total of sunshine duration in warm half-year (April–September).

5.2 Botanical description and occurrence of the Common Hazel in the Czech Republic

Common Hazel - *Corylus avellana* is a medium sized spherical shrub 2 to 8 m high, with straight branches with diameter of 30 cm. The bark is smooth, brown-gray. The young twigs are densely hairy. The buds are sessile, glabrous, oval, and bluntly spiky. They have simple, rounded leaves with double-serrated margins, shortly petiolated, 7 to 12 cm long, on the face sparsely and densely hairy on the back. The roots are mainly spreaded on the surface. The male flowers are arranged in pendant catkins, 10 cm long. The new unblooming catkins occur on the shrubs in mid of summer. Small female flowers are sessile to twig, unobtrusive, similar to buds. They grow usually in pairs, and are completely enclosed in support scales, which protrude from red stigmas. The plant is monoecious. The Hazel flowers as early as 8-10 years, however, it becomes, that pistil's flowers are damaged by frost, and it does not produce fruit annually. The fruit is a nut surrounded by an involucres (husk) which partly to fully encloses the nut, it called „punčoška". Shell (pericarp) is cinnamon brown. The Hazel can be found as undergrowth in light forest, in forest edges, as a part of shrubs in the surrounding of water, in the clear-cut areas and bounds, it is often planted. It reaches ages of 60–80 years. Taxonomy classification is mentioned in Table 1.

Order	*Fagales*
Family	*Corylaceae*
Genus	*Corylus*
Species	*Corylus avellana* L.

Table 1. Taxonomy subsumption.

The Hazel has no special moisture requirements and grows even in the areas of poor rainfall. It is modest in demands on soil, hates peat and muddy soil. It is resistant to climate, it is a thermophilic tree species in our country that searches often thermally favourable position with the southern exposure.

It expanded in the Termophyticum and the Mesophyticum scattered to abundantly (Hejný & Slavík et al., 2003), rarely in the Oreophyticum. The Hazel is very abundant at some localities, particularly in the upper hill country belt to submontane belt of southwest Bohemia and part of south Bohemia (here is rich in shrub forest floor, but it also creates extensive separate bushes formation). In the lowland belt to the submontane belt, exceptionally above the cirques (max.: Hrubý Jeseník Mts., Velká kotlina, 1,310 m asl; Krkonoše Mts., Schlusterova zahrádka, 1,100 m asl).

In forestry is considered for weed tree species, are grown mainly large-bed cultivars for the nuts. Nuts have a high content of protein, fats and vitamins B. Pressed oil is used in parfumery and painting. Hazel rods are used in basket making. Hazel is the first grazing for bees due to early flowering period. Infusions of the leaves are used in folk medicine (skin rashes and intestinal illnesses treatment). Dried leaves were used during the war as a tobacco substitutes.

5.3 The Common Hazel as an allergen

Pollen grains of Hazel are triangular, flat, about a size of 20 up to 27 μm in diameter with three porous apertures, smooth and unsticky (Fig. 13). The production of pollen is plentiful, if species does not grow in the shrub floor and is overshadowed (Rybníček et al., 1997).

Fig. 13. Common Hazel (*Corylus avellana*) - pollen (Photo: Lenka Hájková).

Fig. 14. Common Hazel (*Corylus avellana*) - flowering (Photo: Jana Škvareninová)

Pollen is transmitted by wind very well. Pollen grains of the Hazel contain at least allergen Cor a 1 (Špičák, Panzner et al., 2004). The Hazel belongs among moderately significant to very important pollen allergens. Primarily it reacts across with pollen botanically related species (birch - hornbeam - alder), (beech - oak).

The flowering of the Common Hazel (Fig. 14) lasts on average 22–23 days in the selected elevation zones, standard deviation ranges between 2.3 and 5.3 days. Common Hazel begins to flower on average between 24th February and 24th March. The probability of the beginning of flowering between 12th March and 21st March is 19 %, between 2nd and 11th March 16 % and between 21st and 29th February 16 %. The earliest beginning of flowering was recorded on 12th January 1994 (Lednice), the latest onset was registered on 12th April 2006 (Měděnec).

5.4 Analysis of phenological data in the Common Hazel

Within the framework of phenological observations and according to the methodology of the CHMI, the following phenophases of the Common Hazel are observed: the bud burst (10 %), the first leaves (10, 50, 100 %), the fully leaved (100 %), the inflorescence emergence (10 %), the beginning of flowering (10, 50, 100 %), the end of flowering (100 %), the bud creation (10 %), the first fruit visible (10 %), the herb sprouts begin to lignifying (10 %), the leaves colouring (10, 100 %), the leaves fall (10, 100 %), the fully ripe (10, 50, 100 %). The size of the crop has also been recorded. The Common Hazel is in the network of CHMI phenological stations (wild plants) observed at stations at elevations from 155 m (Doksany) to 830 m (Měděnec). The observed areas are located at stations on plains or on moderate slopes (slope up to 10°), at very bright stands. Typical moisture conditions are mainly hygromezophytic and mezophytic.

The phenological data of the selected plants were evaluated in the environment Excel, but especially a space analysis in the environment of geographical information systems was carried out. For the depiction of maps, the method Clidata-DEM was used with a horizontal differentiation of 500 m and of regressive semi-diameter 40 km. This method is based on local linear regression between the measured value (average data of the onsets of the selected phenophases in the period 1991–2010) and a digital model relief. For each station, regressive coefficients from the nearest stations by means of the method of the smallest squares were calculated, which were later consequently interpolated in the space distribution, and by means of map algebra, and a straight line equation, a space distribution of the given phenophase was acquired. The maps are processed from the observed data from the phenological stations; in the area above the boundaries of the present occurrence the map expresses potentially possible values. Termopluviograms were constructed according Kožnarová et al. (1997), some results of termophenopluviograms have already been published with Apple, Pear, Apricot, Morello and Walnut (Kožnarová & Sulovská & Hájková, 2011).

In place of a phenological observation station, which climatological stations lack in most cases, so-called technical rows for necessary methodological elements for geographical coordinates of phenological stations were assessed for the calculation of complementing meteorological characteristics. The methodology for the calculation was as follows: before the individual calculation of technical rows, the entrance data from climatological stations of the CHMI were carefully checked from the point of view of their quality using the software ProClimDB (Štěpánek, 2009). The methodology of the error detection in calculation combines several procedures e.g. comparison of differences between neighbouring stations, comparison with the expected value calculated by means of geostatistical methods and so on, the result of this combination is that it is easy to automate. After the error correction, the rows were homogenised by the use of several statistical tests, different reference rows, iteration testing-assessment-correction of nonhomogenities. The correction of nonhomogenities was carried out directly in the daily data (more e.g. in Štěpánek et.al, 2009). For each station, the missing values from the period 1991–2010 were further completed by means of geostatistical methods. And with the use of these homogenised and completed station rows, new technical rows in selected points were finally calculated (in relation to the position of the original stations). The calculation of technical rows itself arises from the methodology IDW, whereby the applied data of the neighbouring stations are at first standardised for elevation of the point at which we calculate a new row (Štěpánek et.al, 2011) and then a new value by means of weighted average is calculated. The adjustment of parameters for the calculation differs for each meteorological element. The results of the onsets of phenological phases of plant species of the corresponding phenological station in the given year were associated with technical rows for geographical coordinates of phenological stations in the period 1991–2010 for further processing. And such prepared data were used for calculating meteorological characteristics according to the following procedure: pentad air temperature was calculated as the average air temperature for five consecutive days before the day of the onset of the selected phenophase in the given year. The sum of the air temperatures, the total of sunshine duration and the number of days with precipitation total of at least 1 mm during the selected phenophase intervals were calculated during the entire mentioned interval (so they include the day of the phenophase onset). The accumulated temperatures are processed as sums of average daily air temperatures in the given phenophase interval (°D e.g. DD), the duration of sunshine as a total of daily amounts of sunshine in the phenophase interval (hour) and the number of days with precipitation total of 1 mm or more comes out of daily precipitation totals in the selected days of the phenophase interval corresponding to the selected criterion.

5.5 Temporal and spatial variation in Common Hazel phenology

Temporal and spatial variation of phenophase onsets in Common Hazel is very different in given year. Fig. 15 and 16 show annual deviations of phenophase onsets (the bud burst, the inflorescence emergence, the beginning and the end of flowering, the leaves fall) of Common Hazel from the average within the twenty year period 1991–2010.

The years 2007, 2008, 1998, 2002, 1994 and 1995 had earlier onset with hazel, the years 2006, 1996, 2005 and 2010 were recorded later phenological phases onsets. The highest negative deviations, i.e. an earlier phenophase onsets, were as follows: the sprouting -9 days (2007), the inflorescence emergence -28 days (2007), the beginning of flowering -28 days (2007), the end of flowering -23 days (2007) and the leaves fall -7 days (1994). The highest positive deviations, i. e. later phenophase onsets, were with the sprouting +12 days (2006), the inlorescence emergence +28 days (2006), the beginning of flowering +29 days (1996), the end of flowering +25 days (2006) and the leaves fall +11 days (2001). The differences in the deviations with the inflorescence emergence, beginning and end of flowering are almost one month.

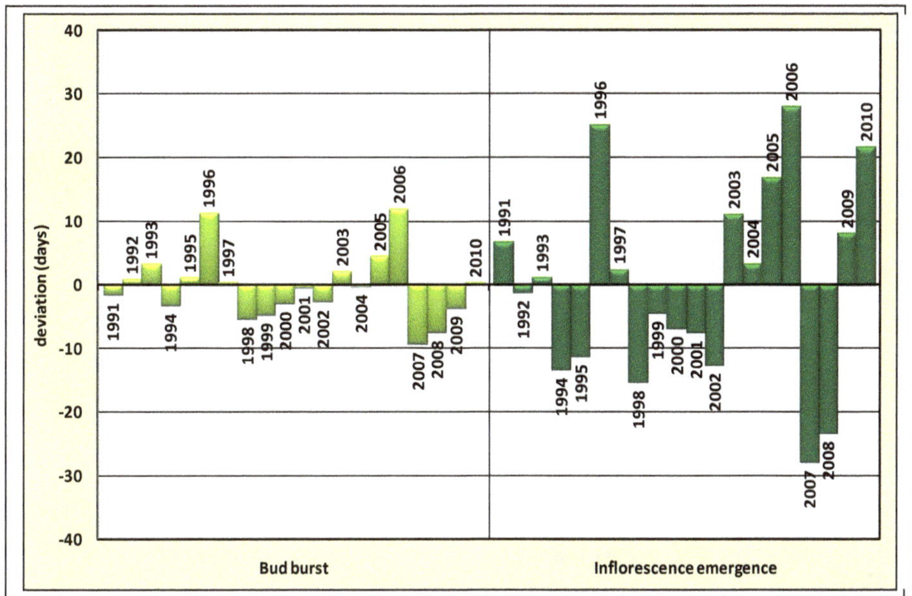

Fig. 15. Annual deviations of phenophase onsets (bud burst, inflorescence emergence) of Common Hazel (*Corylus avellana*) from the long term average 1991-2010.

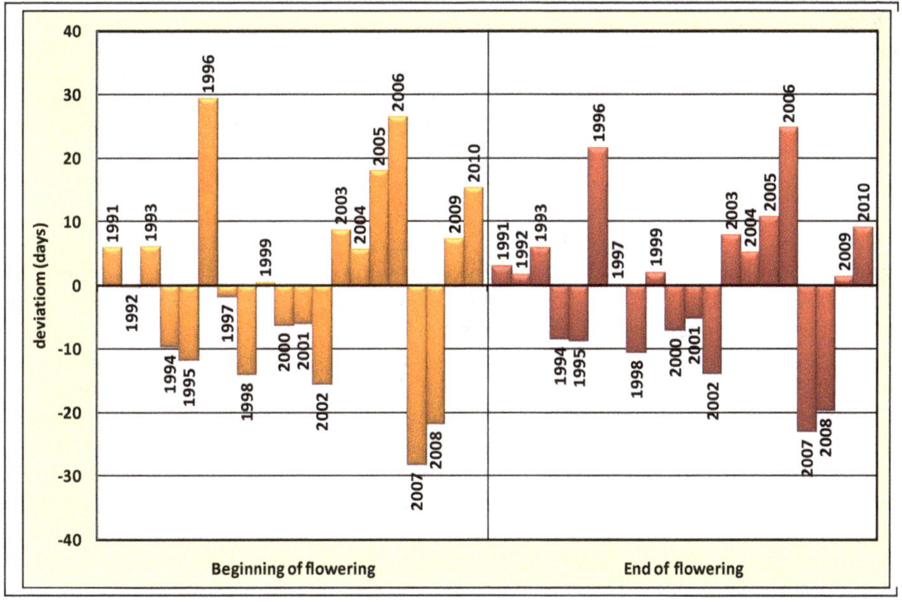

Fig. 16. Annual deviations of phenophase onsets (beginning of flowering, end of flowering) of Common Hazel (*Corylus avellana*) from the long term average 1991-2010.

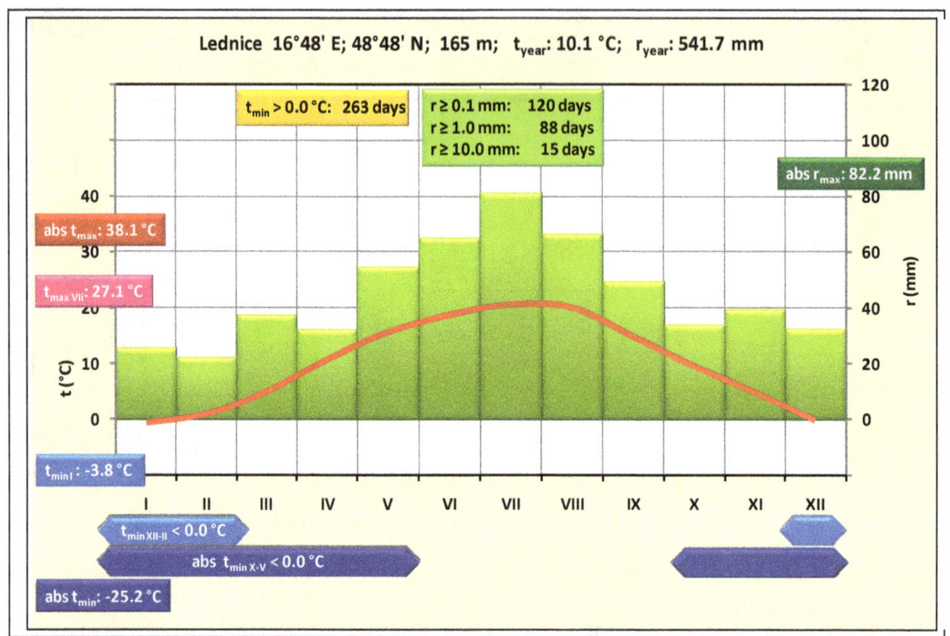

t = average monthly air temperature (°C); r = average monthly total precipitation (mm); t_{year} = average year air temperature (°C); r_{year} = average year total precipitation (mm); abs t_{max} = absolute maximum of air temperature; $t_{max\ VII}$ = average monthly maximum of air temperature of the warmest month; $t_{min\ I}$ = average monthly minimum of air temperature of the coldest month; abs t_{min} = absolute minimum of air temperature; $t_{min\ XII\text{-}II} < 0.0$ °C = months with average minimum of air temperature < 0.0 °C; abs $t_{min\ IX\text{-}VI} < 0.0$ °C = months with absolute minimum of air temperature < 0.0 °C; $t_{min} > 0.0$ °C = average count of days with air temperature > 0.0 °C; abs r_{max} = absolute maximum of daily total precipitation; $r \geq 0.1$ mm= average count of days with total precipitation ≥ 0.1 mm; $r \geq 1.0$ mm = average count of days with total precipitation ≥ 1.0 mm; $r \geq 10.0$ mm = average count of days with total precipitation ≥ 10.0 mm.

Fig. 17. Climagram of Lednice.

For detailed analyse of phenological phases were used two stations Lednice (165 m asl) and Přibyslav (533 m asl). The average duration of growing season in Lednice station is 231 days, in Přibyslav station it is 223 days. The length of growing season or more precisely the growth of plants and phenological phases onsets are closely connected with climate, mainly temperature influences the course of phenological stages. Figures 17 and 18 present climatological conditions at stations in Lednice and Přibyslav by means of climagram. The climagram provides overall climatological description of specified station. In the climagram are described the course of main climatological elements (air temperature - average, minimum, maximum; precipitation total; number of days with precipitation total ≥ 0.1; 1.0; 10.0 mm). These climatological characteristics were evaluated in the same period as phenological phase's onsets (1991-2010) to show the climatological conditions of the Common Hazel in the selected stations. The stations are situated at different conditions - Lednice station is situated in Dyje lowland at elevation of 165 m and Přibyslav station is situated in „Českomoravská vysočina" at elevation of 530 m. The air temperature has the principal influence on phenological phase onset. The onsets of phenological phases in

Přibyslav station are delayed - it is correlated with the course of average monthly air temperature. The average year air temperature in Přibyslav is 7.4 °C; in Lednice it is 10.1 °C, also the occurrence of absolute minimum temperature < 0 °C is higher at station in Přibyslav within months IX-XII and I-VI; in Lednice station is the occurrence of days with absolute minimum temperature < 0 °C minor just during months X-XII and I-V. On other side the number of days with t_{min} > 0.0 °C is higher at station in Lednice (263 days). The absolute minimum temperature is lower in Přibyslav station (-26.4 °C). Even though the precipitation total has not so important influence on phenological stages onsets, the climagram includes it to present the whole climatological conditions. The annual precipitation total is naturally higher at station in Přibyslav (693.8 mm) in comparison with Lednice station (541.7 mm). Number of days with precipitation total above 0.1mm; 1.0 mm and 10.0 mm is also higher in Přibyslav station.

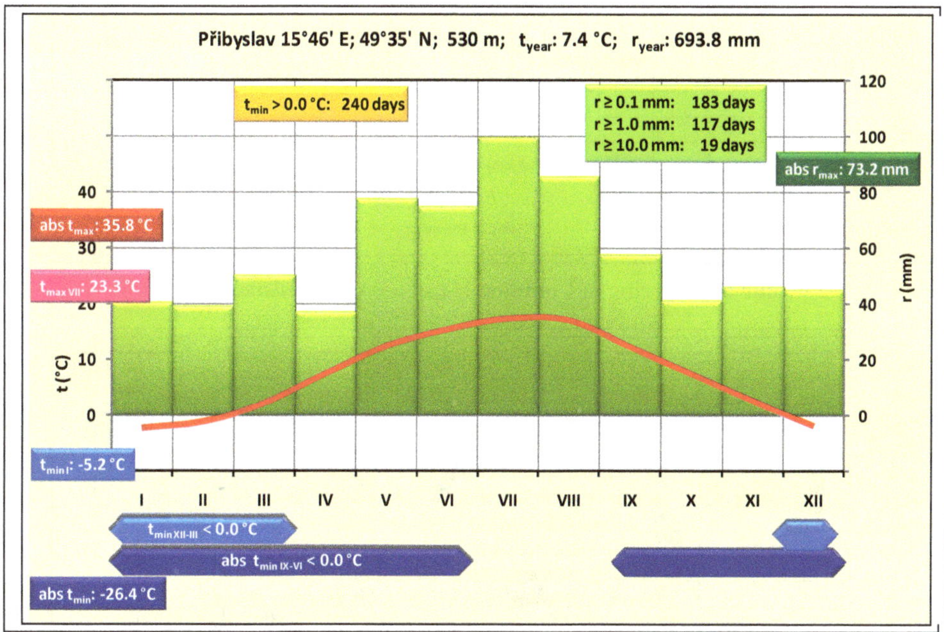

Fig. 18. Climagram of Přibyslav.

Duration of the flowering at stations in Lednice and Přibyslav illustrates Fig 19. Each year is very variable and the time of flowering is different in Lednice and Přibyslav station. The average time of flowering at station in Lednice is 23 days; at station in Přibyslav is 19 days. The shortest duration of flowering was in station at Lednice 6 days (2010), in Přibyslav station it was 7 days (1994). On the other hand the longest duration of flowering was recorded in the year 1997 (45 days) in Lednice station; in Přibyslav station the longest duration of flowering was in the year 2008 (37 days). The extreme duration of flowering was not recorded in the same year at both stations. The average onset of the beginning of flowering within the whole Czech Republic starts on 6th March, the average onset of the end of flowering begins on 28th March. The average duration of flowering is 22 days.

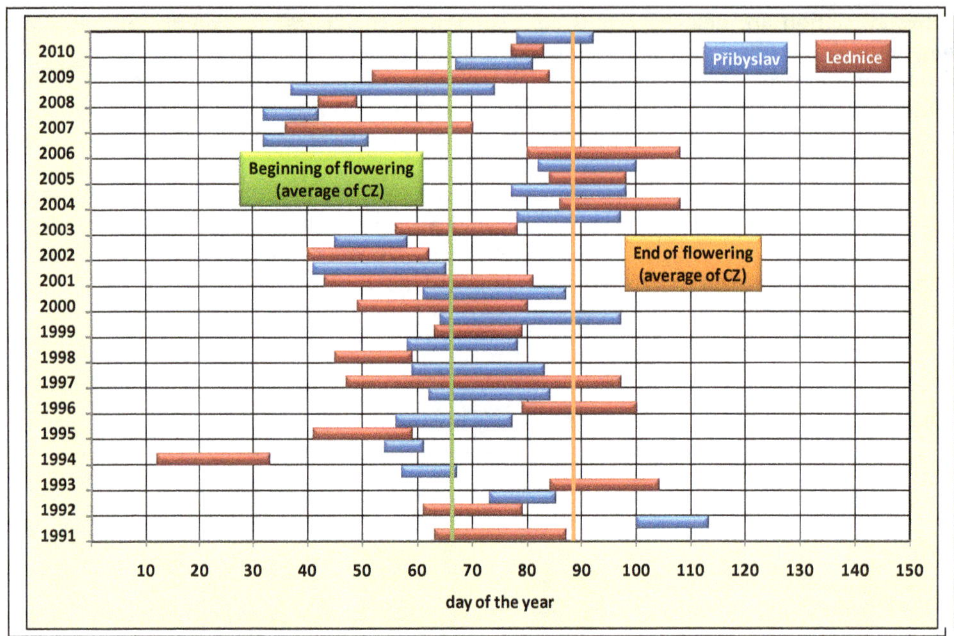

Fig. 19. Average time of flowering of Common Hazel (*Corylus avellana*).

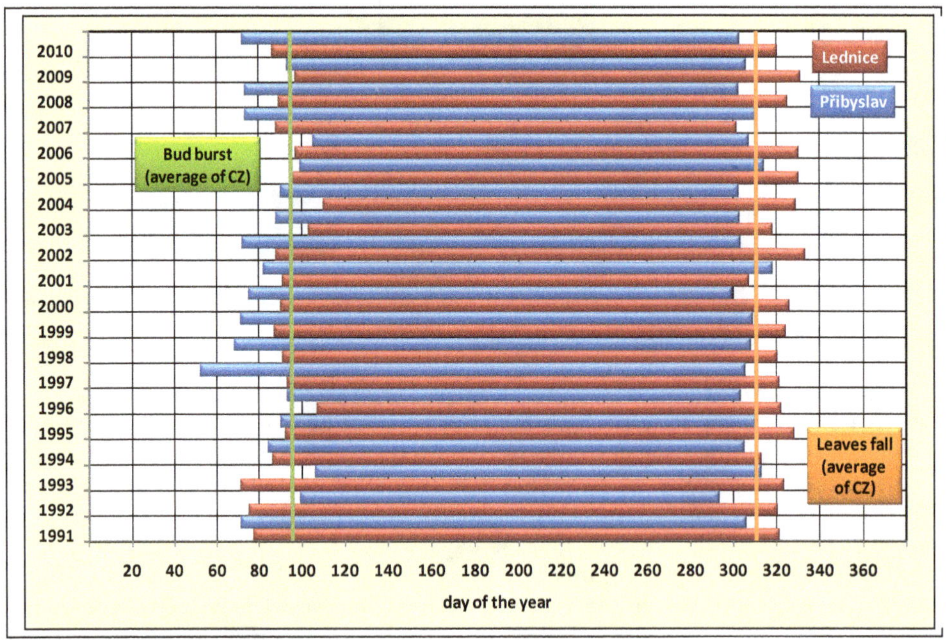

Fig. 20. Average time of growing season (bud burst–leaves fall) of Common Hazel (*Corylus avellana*).

The duration of growing season (bud burst-leaves fall) at both stations demonstrates Fig. 20. The average duration of growing season at station in Lednice is 231 days, at station in Přibyslav 223 days. The average onset of the bud burst of the Common Hazel within the whole Czech Republic begins on 7th April; the average onset of the leaves fall is on 5th November. The shortest duration of growing season was 213 days (2007) and the longest duration of growing season was 252 days (1993) in Lednice station. At station in Přibyslav was the shortest duration of vegetation season was recorded in the year 1992 with 194 days, the longest duration of vegetation season was recorded in the year 1997 with 253 days.

The phenotermopluviogram on Fig. 21 represents relations between air temperature and precipitation total in period of flowering on location Lednice and Přibyslav. The longest period of Common Hazel on location Lednice was 50 days (1997), with sum of temperature ($\sum t$) 255.7 DD and precipitation total ($\sum r$) 43.4 mm, on location Přibyslav 37 days (2008), $\sum t$ = 75.5 DD and $\sum r$ = 49.6 mm. The shortest interval of flowering was in location Lednice 6 days (2010); $\sum t$ = 56.7 DD and $\sum r$ = 0 mm, and in Přibyslav 7 days (1994), $\sum t$ = 26.5 DD and $\sum r$ = 9.8 mm. Long-term mean of Lednice $\sum t$ = 119.1 DD; $\sum r$ = 25.0 mm; duration of flowering = 23 days. Long-term mean of Přibyslav $\sum t$ = 35.1 DD; $\sum r$ = 30.1 mm; duration of flowering = 19 days.

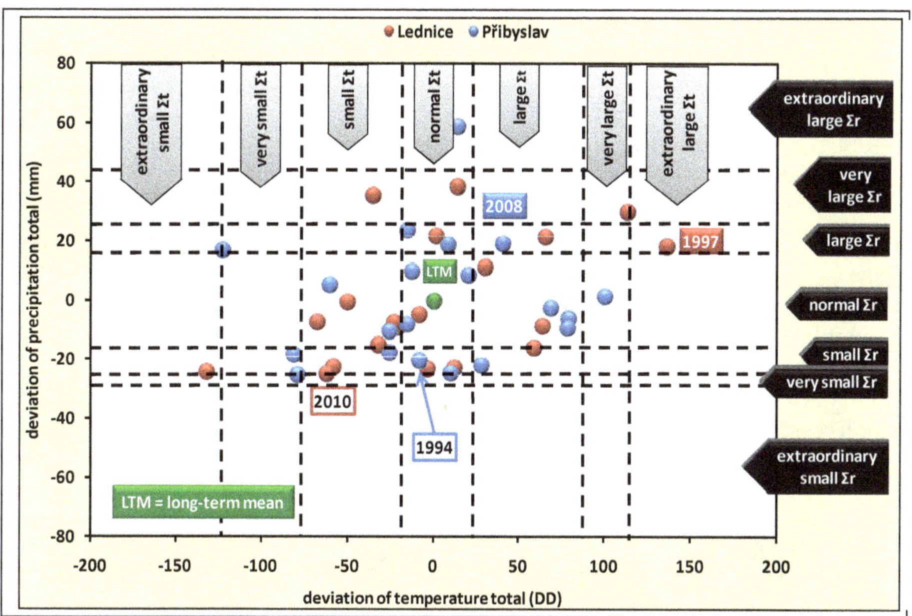

Fig. 21. Phenotermopluviogram of flowering of Common Hazel (*Corylus avellana*).

In Tables 2-5 some statistic results (average, median, 25% quartile, 75 % quartile, standard deviation and variation coefficient) of selected phenophase are enunciated. The variation coefficient presents the balance of time interval, it means a ratio of standard deviation and average. In the Common Hazel were chosen subsequent phenophase stages: the bud burst

(BBCH 07), the beginning of flowering (BBCH 61), the end of flowering (BBCH 69) and the leaves fall (BBCH 97). From the results is apparent the dependence on elevation. Table 6 presents the pentad air temperature to phenophase onset at elevation zones.

Statistical characteristic	Elevation (m)				
	≤ 200	201-400	401-600	601-800	≥ 801
Average	4. IV.	8. IV.	14. IV.	20. IV.	27. IV.
Median	31. III.	2. IV.	9. IV.	19. IV.	25. IV.
25 % quartile	27. III.	27. III.	1. IV.	13. IV.	20. IV.
75 % quartile	4. IV.	9. IV.	16. IV.	27. IV.	29. IV.
Standard deviation	8.9	6.5	8.9	10.3	6.3
Variation coeff. (s_x %)	9.47	6.63	8.56	9.36	5.38

Table 2. Statistical characteristics of bud burst (BBCH 07).

Statistical characteristic	Elevation (m)				
	≤ 200	201-400	401-600	601-800	≥ 801
Average	24. II.	28. II.	7. III.	15. III.	24. III.
Median	23. II.	1. III.	9. III.	22. III.	22. III.
25 % quartile	12. II.	13. II.	21. II.	6. III.	13. III.
75 % quartile	9. III.	15. III.	21. III.	31. III.	3. IV.
Standard deviation	11.2	10.0	11.2	8.8	8.7
Variation coeff. (s_x %)	20.36	16.95	16.96	11.89	10.48

Table 3. Statistical characteristics of beginning of flowering 10 % (BBCH 61).

Statistical characteristic	Elevation (m)				
	≤ 200	201-400	401-600	601-800	≥ 801
Average	18. III.	23. III.	30. III.	7. IV.	15. IV.
Median	22. III.	22. III.	2. IV.	5. IV.	20. IV.
25 % quartile	7. III.	10. IV.	19. III.	17. III.	18. IV.
75 % quartile	30. III.	4. IV.	13. IV.	17. IV.	25. IV.
Standard deviation	10.6	6.2	9.2	10.7	11.4
Variation coeff. (s_x %)	13.77	7.56	10.34	11.03	10.86

Table 4. Statistical characteristics of end of flowering (BBCH 69).

Statistical characteristic	Elevation (m)				
	≤ 200	201-400	401-600	601-800	≥ 801
Average	15. XI.	12. XI.	8. XI.	4. XI.	29. X.
Median	17. XI.	4. XI.	5. XI.	31. X.	18. XI.
25 % quartile	10. XI.	29. X.	30. X.	27. X.	16. XI.
75 % quartile	24. XI.	10. XI.	12. XI.	6. XI.	25. XI.
Standard deviation	14.0	11.1	10.2	10.5	11.6
Variation coeff. (s_x %)	4.39	3.51	3.27	3.41	3.84

Table 5. Statistical characteristics of leaves fall (BBCH 97).

Elevation (m)	Bud burst (BBCH 07)	Beginning of flowering (BBCH 61)	End of flowering (BBCH 69)	Leaves fall (BBCH 97)
≤ 200	5.8	3.1	5.2	6.3
201 až 400	5.9	3.0	4.9	5.9
401 až 600	6.0	2.6	4.7	4.5
601 až 800	6.6	2.8	4.6	3.5
≥ 801	6.8	2.8	4.8	1.5

Table 6. Pentad air temperature (°C) to phenophase onset.

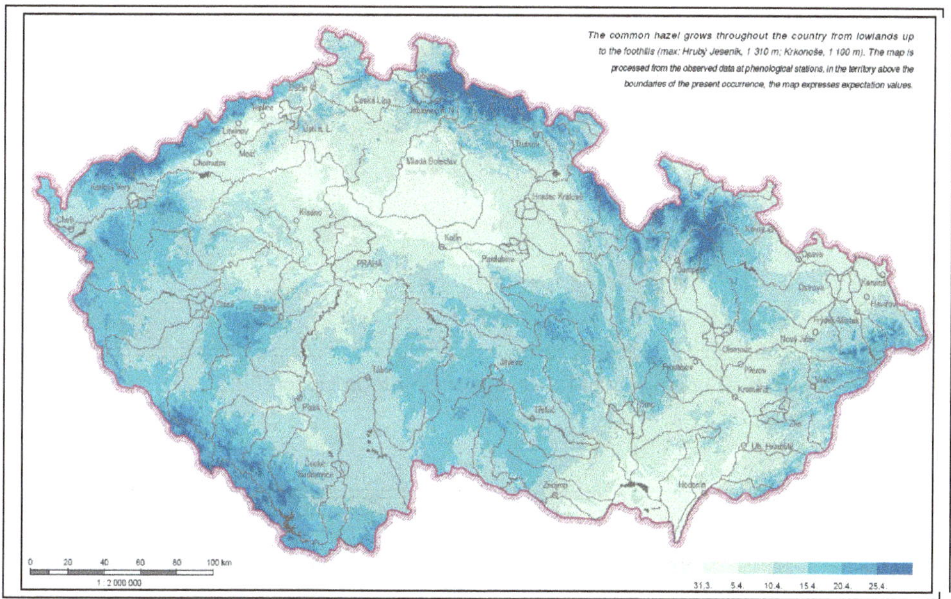

Fig. 22. The average date of bud burst of Common Hazel (BBCH 07).

The average date of selected phenological phase's onsets within the period 1991–2010 show figures 22–25. The onset of the bud burst of the Common Hazel begins between 31st March (lowlands) and 25th April (mountain areas), the beginning of flowering starts between 24th February (south Moravia, Polabská lowland) and 26th March (Krušné hory Mts., Šumava Mts., Krkonoše Mts., Jeseníky Mts., Beskydy Mts.), the end of flowering occurs between 16th March and 20th April and the leaves fall comes between 2nd November (the highest mountain elevations) and 14th November (Polabská Lowland, South Moravia).

The Common Hazel reveals a great variability of different phenophase onset in the observed period 1991 to 2010. The phenological phases were most accelerated in the year 2007; on the contrary the onsets were most delayed in the year 2006.

The vertical phenological gradient (it means gradient of average date of phenological stage onset) for selected phenological stages are subsequent: for the bud burst 2 days/100 m of elevation, for the beginning of flowering is 3 days/100 m of elevation, for the end of flowering is 3 days/100 m of elevation and for the leaves fall 2 days/100 m of elevation.

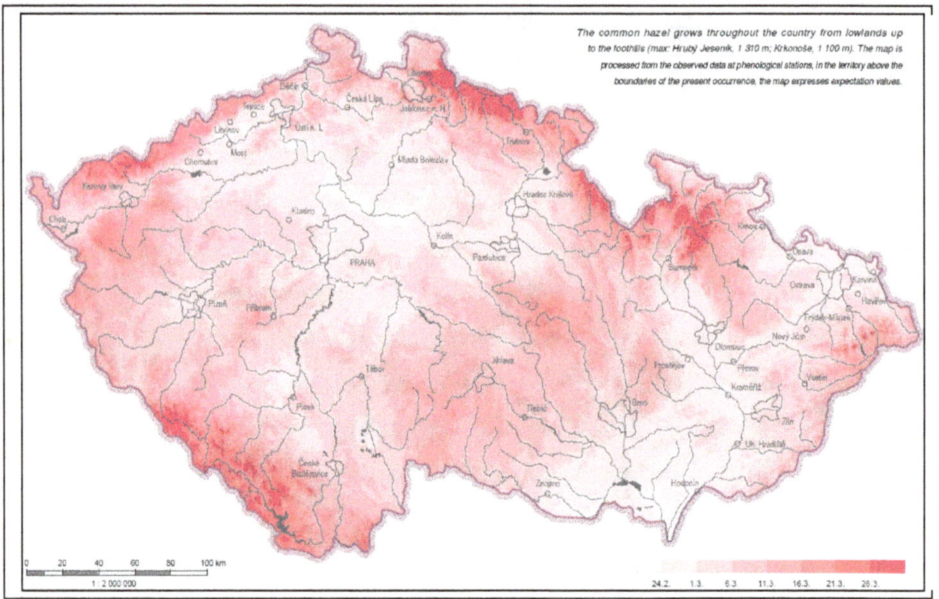

Fig. 23. Average date of beginning of flowering of Common Hazel (BBCH 61).

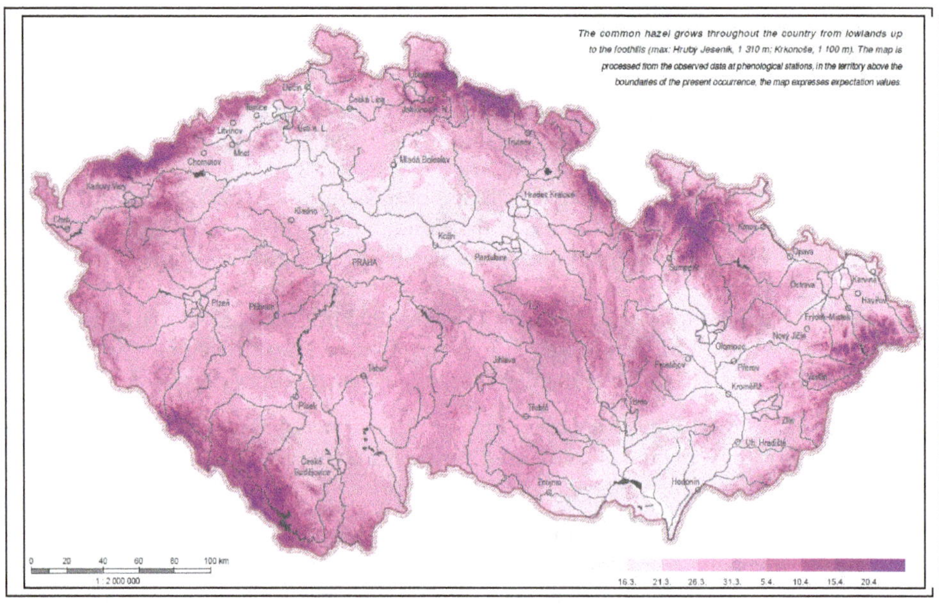

Fig. 24. Average date of end of flowering of Common Hazel (BBCH 69).

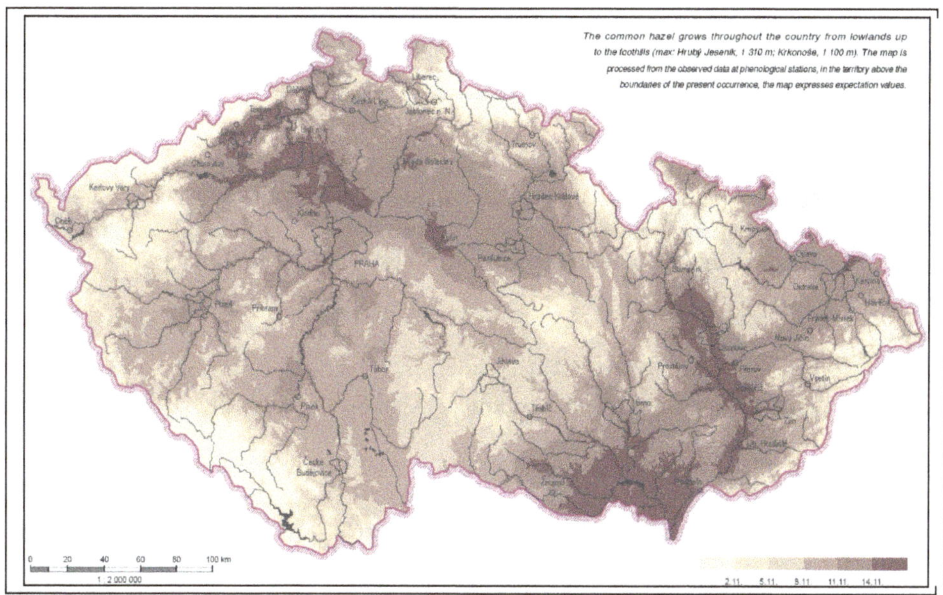

Fig. 25. Average date of leaves fall of Common Hazel (BBCH 97).

The Common Hazel begins to sprout on average between 4th and 27th April, the beginning of flowering comes between 24th February and 24th March, the end of flowering starts between 18th March and 15th April and the leaves fall is between 29th October and 15th November at elevation zones. The period between the bud burst and leaves fall takes from 185 and 225 days on average with the sum of air temperatures from 2,277 to 3,261 °D, the duration of sunshine 1,286–1,366 hours, precipitation total from 430 to 460 mm and 63.5 till 67.8 days with precipitation total of at least 1 mm.

6. Summary

Phenological observations in the Czech Republic have a long tradition; the beginning of phenological research began already in the 18th century. The Hydrometeorological Institute assumed existing phenological data in 1954 and phenological network within the Hydrometeorological Institute has been developed according the own methodology (Pifflová et al., 1956). There were some changes in the methodology of phenological observation till present, nowadays is the phenological network divided into three types of stations (field crops, fruit trees, forest plants). Data are stored in database FENODATA including historical records. The Czech Republic cooperates with other European countries within the phenology research (e.g. IPG, COST 725, COST ES603 and PEP725). From all observed species in the CHMI phenological network was processed detailed evaluation of phenological results in the Common Hazel within the period 1991-2010. The Common Hazel is very important pollen allergen; the time of flowering appears between 24th February and 24th March and lasts 22–23 days on average. The vegetation period takes 185–225 days (bud burst-leaves fall) on average. The phenophase onsets are closely connected with climate, detailed analysis is shown in climagrams and phenotermopluviogram.

7. Acknowledgement

The authors gratefully thank to the Czech Ministry of Education, Youth and Sports, research project OC 09029 (Phenology atlas of Czechia) and research project 6046070901 (Sustainable agriculture, quality of agricultural products, sustainable use of natural and landscape resource) for funding this research. This work was also supported by the grant SVV-2011-263 202.

8. Reference

Brablec, J. (1952). Two studies about phenology. Meteorological Bulletin 5, č. 3, pp. 74 - 75

Brablec, J. (1953). Phenology calendar of Střednice by Mělník, Meteorological Bulletin 6, č. 3, pp. 78-79

Coufal, L. et al (2004). Phenology atlas. 1. edition. Praha: CHMI. 264 p. ISBN 80-86690-21-0

Farský, O. (1942). Gamekeeper-phenology observations. Special issue of the journal „Stráž myslivosti", year XIX, č. 27

Hájková, L., Kožnarová, V. & Sulovská, S. (2011). Phenology season onset in relation to synoptic situations in the Czech Republic within 1991–2010. *Proceedings of International konference on current knowledge of climate change impacts on agriculture and forestry in Europe*. pp. 129–130, Topolčianky, Slovakia, 3rd–6th May 2011

Hejný, S. & Slavík, B. et al. (2003). Flora CR 2. Praha, Academia, ISBN 80-200-1089-0, 540 p.

Kožnarová, V.; Klabzuba, J. & Bureš, R. (1997). The Use of termopluviogram to evaluate agrometeorological year, season and month. Pametnik Pulawski 110, Pulawy, ISSN 0552-9778, pp. 71 - 78

Kožnarová, V. & Klabzuba, J. (2010). The conventional and modern approaches in evaluation of weather and climate in biology. Current possibilities of physiology and agricultural research (selected chapters). Crop research institute, v. v. i. , Praha 2010. ISBN: 978-80-7427-023-9. p. 4 - 33

Kožnarová, V.; Sulovská, S. & Hájková, L. (2011). Temporal variability of phenophase onsets of fruit trees observed in the CHMI phenological network during the period 1991–2010 in relation to synoptic situations. In: *Plants in variable climate*. Lednice. 12 p. (in press)

Krška, K. (2006). Phenology as a science, method and tool. In: *Phenology response to climate change* Brno. 37 p. + CD. ISBN 80-86690-35-0

Krška, K. & Vlasák, V. (2008). History and present of hydrometeorological service in South Moravia. Praha 2008, ISBN 978-80-86690-52-0

Meier, U. et al. (1997). Growth stages of Mono- and Dicotyledonous plants. BBCH Monograph. Blackwell Wissenschafts-Verlag, Berlin - Wien, ISBN 3-8263-3152-4, 622 pp.

Menzel, A. et al. (2006). European phenological response to climate change matches the warming pattern. Global Change Biology, 12, pp. 1969–1976

Methodology instruction No. 2 (2009). Methodology instruction for phenological stations - field crops. Praha: Czech Hydrometeorological Institute. 84 p.

Methodology instruction No. 3 (2009). Methodology instruction for phenological stations - fruit trees. Praha: Czech Hydrometeorological Institute. 90 p.

Methodology instruction No.10 (2009). Methodology instruction for phenological stations - wild plants. Praha: Czech Hydrometeorological Institute. 74 p.

Miháliková, I. (1983). 60 years of phenological network. Meteorological Bulletin 36, č. 6, pp. 187-188

Pifflová, L. et al. (1956). Methodology instruction for phenological observers. Hydrometeorological Institute, Praha, D-571503, 152 p.

Nekovář, J. et al. (2008). The history and current status of plant phenology in Europe. COST Office & METLA Finland, 182 pp. ISBN 978-951-40-2091-9

Nekovář, J. & Rožnovský, J. (2006). Phenological service of Czech Hydrometeorological Institute, Proceedings of international science seminar Brno, Phenology response of climate variability, 22. 3. 2006, ISBN 80-86690-35-0, p. 19

Novák, V. & Šimek, J. (1926). Phenological observations in Moravia and Silesia in the years 1923 and 1924. Reports of agricultural research institutes. No. 16. Edition of Ministry of Agriculture in Prague 1926

Rožnovský, J. & Havlíček, V. (1999). Bioclimatology. Mendel university in Brno, reprint 1999, first edition 1998, ISBN 80-7157-291-8

Rybníček, O. et al (1997). Miniatlas of pollen allergens, Municipal health station Brno 1997, 1. edition, 46 p.

Seydl, O. (1954a). Mannheim society of meteorology (1780–1790). Meteorological Bulletin 7, No. 1, pp. 4–11

Seydl, O. (1954b). Czech historical phenological studies. Meteorological Bulletin 7, No. 6, pp. 153–164

Sobíšek, B. et al. (1993). Meteorological dictionary explanatory and terminological. Ministry of the Environment CR, Praha, ISBN 80-85368-45-5, p. 87

Škvareninová, J. et al. (2009). Plant's phenology within the changing conditions of environment. 2009. Publisher of Technical University in Zvolen. 103 p. ISBN 978-80-228-2059-2

Špičák, V. et al. (2004). Allergology. Praha: Galén. ISBN 80-246-0846-4, 348 p.

Štěpánek, P. (2009). ProClimDB - software for processing climatological datasets. CHMI, Regional office Brno. Available from: <http://www.climahom.eu/ProcData.html>

Štěpánek, P. & Zahradníček, P. & Skalák, P. (2009). Data quality control and homogenization of the air temperature and precipitation series in the Czech Republic in the period 1961-2007. Advance Science Research., 2009, 3, s. 23–26.

Štěpánek, P., Zahradníček, P. & Huth, R. (2011). Interpolation techniques used for data quality control and calculation of technical series: an example of Central European daily time series. Idöjárás, 2011 (in press)

Tolasz, R. et al. (2007). Climate atlas of Czechia. CHMI. Praha. ISBN 978-80—86690-26-1. 255 p.

Žalud, Z. Bioclimatology [online] (2010). Brno: > Mendel University in Brno, 2010 p. 137. Available from: <http://web2.mendelu.cz/af_217_multitext/ke_stazeni/bioklimatologie/bioklimatologie_texty.pd>

7

Phenology of Woody Species Along the Climatic Gradient in West Tropical Africa

Josiane Seghieri[1], Frederic C. Do[2],
Jean-Louis Devineau[3] and Anne Fournier[4]
[1]IRD-UMR HydroSciences Montpellier, Université Montpellier 2,
[2]IRD-UMR Eco&Sols, Campus SupAgro-INRA,
[3]CNRS-Museum National d'Histoire Naturelle, Département Hommes, Natures, Sociétés, Eco-Anthropologie et Ethnobiologie,
[4]IRD-UMR PALOC, Paris,
France

1. Introduction

West tropical Africa (WTA) is known to be one of the most exposed regions in the world to climate change. The seasonality of rainfall and of atmospheric conditions, including a dry season lasting for several months in most of the area, strongly influences vegetation activity (Bourlière & Hadley, 1983; Breman & Kessler, 1995). Indeed, this region was already affected by severe and prolonged droughts in the 1970s-1980s. These events represent one of the major climate variations of the 20th century recorded at the global scale (Giorgi, 2002; Neelin et al., 2006; Redelsperger et al., 2006). The average annual rainfall deficit varied by ± 20% in higher rainfall zones and by 50% in lower rainfall zones. However, while these droughts were relatively uniform over most of WTA, the 1990–2007 period was characterised by a more complex pattern including large spatial variability (Lebel & Ali, 2009). Current projections of global change predict higher temperatures and lower rainfall, although opinions on temperatures are contradictory (Mearns et al., 2001; Haarsma et al., 2005; IPCC, 2007; Funk & Brown, 2009). More frequent extreme events are also expected in some parts of Africa (Hély et al., 2006; Frappart et al., 2009; Lebel & Ali, 2009).

In WTA, the current vulnerability of agro- and eco-systems to climate change is dramatically amplified by the rapidly growing population (Thornton et al., 2008; Heubes et al., 2011). In the next half century, the fastest growing populations will be mainly located in sub-Saharan Africa (Garrity et al., 2010). Current projections predict a doubling of the population (Funk & Brown, 2009). To feed these increasing numbers of people, between 1990 and 2006 the area of cultivated land already grew by more than 10% annually, to the detriment of the natural woody cover. However, woody cover is still a crucial component of the agro-sylvo-pastoral ecosystems in WTA.

The phenology of woody plants could be thus a significant indicator of the vulnerability and ability of existing agro-sylvo-pastoral ecosystems to respond to local and global changes, through their relationships with regional climate, soil distribution, and land use (Chuine & Beaubien, 2001; Scheiter & Higgins, 2009). On the other hand, changes in the woody cover may also interact with the effect of global change on ecosystems through changes in leaf phenology (variations in the leaf area index). The relationships between environmental conditions and woody plant phenology are still poorly known, notably in seasonally dry tropics (Eamus & Prior, 2001). Studies on this topic and long-time series of phenological data are particularly rare in WTA. Deciduous and evergreen species are recognized as being the two extremes of a continuum of survival strategies which optimise annual carbon gain and water use under environmental constraints such as seasonal or prolonged droughts (Givnish, 2002; Scheiter & Higgins, 2009). Knowledge of their distribution along the WTA bioclimate gradient and of the factors that control their phenophases is thus of great value for large scale modelling of carbon and water cycles, and for forecasting land surface responses to climate change (Harrison et al., 2010). This chapter thus has three objectives: (1) to present a state of the art of the knowledge on woody species phenology along the WTA bioclimatic gradient, (2) to examine available information on the environmental causes of phenological variation, and on phenophase predictors and triggers, (3) to make recommendations for future phenological studies to help identify the possible consequences of global changes in WTA.

The first phenological studies in the humid part of the gradient were conducted in the 1960s and mainly concerned leaf shedding and increments of the stand basal area, especially in forests. Studies in the driest part of the gradient were mainly classical qualitative characterisation of leafing, flowering and fruiting at the population scale. However, the most recent studies conducted in Sahelian and Sudanian bioclimates included quantitative data analyses and modelling and made a thorough study of the most probable determining factors of phenology. Given their importance, these approaches are presented in this chapter.

Most studies were conducted along a transect from Guinean to Sahelian areas, and focused on the description of specific phenological strategies. The aim of most of these studies was to determine the distribution of the availability of woody forage over the year (Le Houerou, 1980; Hiernaux et al., 1994; Breman & Kessler, 1995; Ickowicz et al., 2007; Sanon et al., 2007). The aim of more theoretical studies was to understand the functioning of the savanna by characterising the general phenological patterns of their components at the community scale. Most authors concentrated on the adaptive strategies which enable species both to survive in a particular local environment and to coexist with other species (Fournier 1991; Seghieri et al. 1995; De Bie et al., 1998; Devineau 1999, Mahamane et al., 2007). They described species-specific annual phenology. We reviewed these studies to identify the distribution of phenological strategies along the WTA bioclimatic gradient.

2. Methods used

2.1 Definitions of bioclimates and phenological strategies

In WTA, the strong gradient of climate and surface conditions from the sub-humid area in the south to the arid area in the north, and its possible impacts at global scale have attracted international research programs on climate and global changes including the Hydrological and the Atmospheric Pilot Experiment (HAPEX-Sahel, Goutorbe et al., 1994), the SAvannas on the Long Term (SALT, core IGBP program, Koch et al., 1995), and more recently, the African Monsoon Multidisciplinary Analysis (AMMA, Redelsperger et al., 2006). WTA supports a wide range of natural vegetation along its south-north bioclimatic gradient:

moist forests, mosaics of forests and wooded savannas, dry deciduous forests, bush-grass savannas, and arid steppes (grass bush and thickets) just before the Sahara desert begins. The area is characterised by two rainy seasons which progressively decrease to become one summer rainy season which, in turn, becomes progressively shorter. In parallel, the average annual rainfall decreases, and the variability of rainfall distribution increases. The part of the transect used for the analysis of the phenological strategies along the bioclimate gradient stretches from the Guinean to the north-Sahelian bioclimates (table 1).

Vegetation zones	Annual rainfall (mm)	AETFAT, 1959	Aubreville, 1949	Knapp, 1973	Peyre de Fabrègues, 1980	White, 1983	Saadou, 1990	Adjanohoun et al., 1989	Le Houérou, 1989
Sahara	0-200	Subdesert steppe and desert	Saharian	Vegetation of the southern Sahara	Steppe vegetation: sparse and contracted	Sahara regional transition zone	Southern-Saharian contracted steppe		Contracted and diffuse steppes
Northern Sahel	200-400	Woody steppe with abundant *Acacia* and *Commiphora*	Sahelo-Saharian	Very dry, semi-desert, thorny-bush savanna	Open shrubby steppe	Semi-desert grassland and shrubland	Thorny and shrubby steppe		Sahel *sensu stricto*, Mimosaceae savanna with annual grass layer
Southern Sahel	400-600		Sahelo-Sundanian (woody savannas)	Moderate dry savanna	Shrubby and woody steppe	Sahelian *Acacia* woody grassland and deciduos bushland	*Combretum* thickets, steppes, savannas		Sudano-Sahelian Combretaceae savanna with annual grass layer
Northern Sudanian zone	600-900	Relatively dry type of woodlands, savannas, and steppes		Dry savanna	Shrubby and woody savanna	Undifferentiated Sundanian woodland	Low dry forest, open forest, savannas, riparian forests	Dry forests with dominant Combretaceae and enclosed thorny steppes	Sahelo-Sudanian Combretaceae savanna and woodland with perennial grasses
Southern Sudanian zone	900-1100					Sudanian woodland with abundant *Isoberlinia*		Savanna with enclosed miombo (Brachystegia genus tree)	Sudanian *sensu stricto* mixed savanna and woodland, perennial grass layer
Northern Guinea zone	1100-1200	Woodlands and savannas with abundant *Isoberlinia doka* and *I. dalzielii*	Sudano-Guinea (woody savannas)	Northern moist savanna				North Guineo-Sudanian zone Miombo	Sudano-Guinean zone
Southern Guinea zone	1100-1200	Relatively moist types of woodlands and savannas						South Guineo-Sudanian zone Dense dry forest	
Guinea-Congolia zone	1100-1400	Forest-savanna mosaic	Guinea forest (old dense litoral forests)	Southern moist savanna		Mosaic of lowland rain forest and secondary grassland		Semi-deciduous forest and impoverished deciduous forest	
Coastal zone	900-1400		Togo and Benin coasts	Coastal type of derived savanna					

Table 1. Definition of WTA bioclimates according to several authors , and areas concerned in the present work (boxes surrounded by dotted lines).

In the studies we reviewed, which mainly focused on savannas and steppes, we identified four groups:

1. Deciduous species with a short leafing period whose canopies are bare for more than two months in the year;
2. Deciduous species with a long leafing period. These include species for which no case of canopy evergreeness has yet been reported, despite the shortness of the leafless

period, which rarely lasts more than two months. The flush of new leaves rapidly follows leaf shedding.

3. Semi-evergreen species. Unfortunately the authors used different criteria when assigning this strategy to species, which makes them difficult to compare. The group includes species which shed their leaves over a short period, two months at the most according to Breman & Kessler (1995) (which corresponds to our group 2), but also species in which a smaller proportion of trees in the sample population shed their leaves than those which remained evergreen (Devineau, 1999; Seghieri et al., submitted), and finally species which lose more than 50% of their canopy (Eamus & Prior, 2001; Do et al., 2005). We found it more convenient to classify as "semi-evergreen" species referred to in the literature as evergreen (if not riparian) in at least one location, and otherwise as "long leafing". In this way, we show that these species can actually be both evergreen and deciduous, depending on environmental conditions.
4. Evergreen species. This group comprises species for which no leafless period of the canopy has been reported in the literature at any time during the year and irrespective of where their phenology was studied, except due to damage to the canopy by fire or locust attacks, for example. In these species, old leaves were still on the tree when the new leaf flush occurred.

Before presenting the results, we summarise the most widely used methods of field observation at the tree scale and at the twig scale, then the more common and the more recent data analysis. The discussion focuses on the explanation of the group distribution along the bioclimate gradient and on the role of leafing in the maintenance of the tree water balance.

2.2 Inventory of field observation methods

In WTA, most field observations at local scale were qualitative and concerned the individual tree. However, some authors quantified leaf development, and the number and density of leaves at the twig scale. These observations were generally combined with monitoring of environmental variables (rainfall, soil water availability, radiative energy, evapotranspiration, temperature, air humidity, day length, etc.).

2.2.1 At the tree scale

At the wettest end of the gradient, phenological studies mainly concerned leaf shedding patterns and variations in trunk size. Periodic observations identified variations in leaf shedding, and band dendrometers were used to measure variations in trunk girths. Annual variations in cambial activity (Liming, 1957) and periods of water stress undergone by trees (Borchert, 1999) can be deduced from the latter. An important study was carried out in the semi-deciduous forest at Lamto (Guinean bioclimate) by Devineau (1984, 1991). Band dendrometers were installed on all trees with a trunk diameter of more than 20 cm at breast height. Leaf shedding and trunk girths were monitored monthly over a period of ten years (1972-1981).

However, most reports on dry forests, savannas, and steppes in WTA were based on periodic observations of successive phenophases of the canopy: leafing, flowering and fruiting. A few individuals to a few dozen trees were usually sampled per species. Phenophases were recorded at different stages, which varied in number depending on author. They corresponded to the percentage of the canopy covered by buds, mature, or senescent organs. By far the most commonly used method in WTA was developed by Grouzis & Sicot (1980). The stages of each phenophase were recorded using the following classification:

1. Organs are absent.
2. Beginning of the phase, buds are swelling, expanded leaves and open flowers are rare, fruits are in the early setting stage.
3. Early development, both buds and open organs are present, 10-50% of the canopy is covered with mature or ripe organs, fruits are at all stages up to maturity.
4. Peak development, most leaves and flowers are open, more than 50% of the canopy is covered with mature or ripe organs, fruits are mature.
5. Decline, leaves and flowers are open and dry, 10-50% of the canopy is covered with senescent organs, fruits are ripe and dissemination is beginning (fruits starting to fall).
6. End, over 50% of the canopy is covered with senescent, dry, and falling organs, shedding flower elements, fruits falling or in open pods.

More often, to establish phenograms, the stages were combined as follows:

0, 1 and 5 combined as "non occurring",
2, 3 and 4 combined as "occurring".

Considering a tree as leafing, flowering or fruiting from stage 2 to 4 enabled exclusion of miscarrying primary states and long periods during which woody plants still displayed sparse organs at the end of a phenophase (especially fruiting, Grouzis & Sicot, 1980; Seghieri et al., 1995; Seghieri & Simier, 2002). But to express the magnitude of canopy fullness, Do et al. (2005) preferred to use the midrange percentage values of the stages to estimate the percentage of branches bearing mature green leaves: stage 0 = 0%; stages 1 and 5 = 5%; stages 2 and 4 = 30%, stage 3 = 75 %.

2.2.2 At the twig scale

Although small-scale approaches at the twig scale are time and labour consuming, they provide significant details about phenological processes such as the number of leaves, leaf area, biomass, and life span, and facilitate quantitative analysis. First, Hiernaux et al. (1994) randomly selected one calibrated twig, whose basal diameter was 1 cm, from 6-20 shrubs per species. The twig was representative of the mean phenological stage and leaf density of the canopy. Total leaf biomass and leaf area were estimated for each twig. Samples were harvested every two weeks, once a month during the rainy season, and less frequently during the dry season. Later, Goudiaby (2003) combined classical phenological observations on leafing at the canopy scale on 40 individual trees with detailed observations on 12 tagged twigs on three trees. Following tree architectural concepts (Hallé et al., 1978), the terminal branches or twigs were divided into growth units of successive age according to their length and basal diameter. The appearance and development of buds, leaves, stem ramification and growth, average leaf area and average leaf life span were quantified.

2.3 Summary of data analyses

2.3.1 Characterisation of phenology and its variations

Analyses of time series from dendrometer data revealed trends and periodic components of trunk growth for each indivual tree. Multifactorial analysis made it possible to ordinate trees according to the seasonal distribution of their trunk growth (Devineau 1984, 1991). Cumulative trunk growth provided the stand basal area increments.

Over time, analytical approaches based on observations of the canopy stages (0-5) of each phenophase (leafing, flowering and fruiting) evolved from qualitative to more quantitative analyses. In the first studies, individual phenograms (figure 1) showed inter-individual variability, and enabled the calculation of the mean duration of each phenophase or each stage. Phenological spectra (figure 1) were the basic tools used to describe phenological patterns of woody species. They revealed the same characteristics as phenograms but at the population scale. In this way, variations in the magnitude of stages or phenophases were quantified and linked to the seasonal course of environmental variables (Grouzis & Sicot, 1980). Until recently, the influence of environmental factors was mainly assessed using this type of analysis (Poupon, 1979; Menaut, 1983; Seghieri et al., 1995; De Bie et al., 1998).

An original analysis based on the same observations consisted of comparing species plasticity using a synchrony index (Devineau, 1999). For each species sampled, the index was the ratio of the mean individual duration of a given phenophase to its overall duration at the scale of the sample population. The higher the ratio, the greater the coincidence between individuals.

To describe between-species variability, Seghieri & Simier (2002) used centred principal components analysis (PCA). This graphical representation was enhanced by plotting the centres of gravity for each site and the link between shrubs and sites, representing respectively between- and within-site variability. In addition, to obtain an unbiased typology of fallows, these authors performed hierarchical cluster analysis using the average-linkage method originally developed by Sokal & Michener (1958).

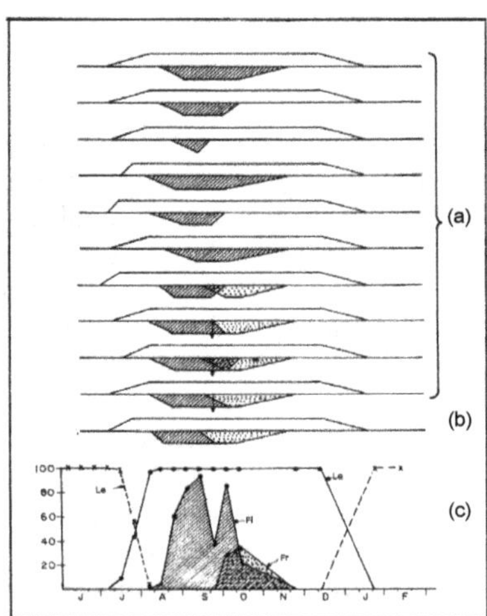

Fig. 1. Examples of phenograms of ten woody individuals among a sample population (a), average phenogram of the same sample population (N=35 individuals) (b), phenological spectrum of the same sample population over one phenological cycle, i.e., percentage of the trees leafing (Le, solid line) and without leaf (Le, dotted line), flowering (Fl, hatch patterns), and fruiting (Fr, stippling pattern) (c), after Grouzis & Sicot (1980).

2.3.2 Quantification of the relationships between phenophases and environmental factors

Recently, attempts have been made to quantify the relationships between ordinal phenological observations and quantitative environmental predictors in WTA (Do et al., 2005; Seghieri et al., 2009, Seghieri et al., submitted). These studies tackled difficulties inherent in analysing time series dealing with temporal auto- and inter-correlations among variables. These authors quantified the level of statistical significance of the correlation between the occurrence of phenophases and variations in environmental variables using logistic regression models. Do et al. (2005) focused on the predictions of interannual variations in leaf flush and leaf shedding peaks respectively, while Seghieri et al. (2009) and Seghieri et al. (submitted) analysed the global occurrence of the phenophases in the sample population. The models linked the cumulative log of the odds of an occurrence of a given phenophase to quantitative (environmental) predictors in linear form (Agresti, 1996).

Generally, different time lags of environmental variables were included as predictors to account for the inertia of the vegetation response. The coefficients of the models were computed using maximum likelihood estimation. The global quality of each model was assessed by the Akaike information criterion (AIC), which compares the fit of a given model including predictors (constrained model) with the fit of a model excluding predictors (unconstrained model). McFadden's Pseudo R^2 indicated the proportion of the phenophase variance explained by the model (Seghieri et al., 2009). The level of significance of the model coefficients (validity of the regression for each variable) was assessed by the Wald test. The c-statistic provides a measure of the discriminative power of the logistic equation, i.e., a comparison of the predictive power of environmental variables (Do et al., 2005). It measures the degree to which probabilities agree with actual outcomes, and is a quick and simple indicator of the accuracy of a predictive model.

However, to obtain relevant results from logistic regression models certain conditions have to be fulfilled. Indeed, the multicolinearities between environmental variables (especially climate variables) are rarely taken into account, which may lead to bias in the results. To avoid skewed results, Seghieri et al. (submitted) ran logistic regression models with uncorrelated pseudo-predictors generated from the principal axis of a PCA applied to original environmental predictors. However, the direct meaning of the coefficients obtained in the models is then lost. To identify the most significant predictors, these authors mapped the PCA pseudo-variables back to the original environmental predictors, i.e. they fitted a multivariate linear regression model that links pseudo-observations in the PCA space to observations in the environmental space.

The main results synthetized from the literature are displayed in the following part. They focus on the characteristics of each phenological group and their distribution along the bioclimate gradient, and then on environmental and human factors that have been reported as determining.

3. Review of main results

3.1 Phenological strategies: characteristics and distribution

In the literature we reviewed, the phenophases (leafing, flowering, fruiting) of 125 species were described along a transect from the Guinean to the Sahelian bioclimate (table 1).

Riparian species were excluded from our analysis since in the three studies which mention this particular category of tree, they were all classified as evergreen (Breman & Kessler, 1995; De Bie et al., 1998; Mahamane et al., 2007).

Fifty-seven percent of the species were deciduous, including short-leafing and long-leafing types. Twenty-six percent were evergreen, and 17% were semi-evergreen species (according to our definition).

Short-leafing deciduous species: 89% started leafing at the end of the dry season or at the beginning of the rainy season (figure 2). Forty-nine percent shed their leaves at the beginning of the following dry season, and 33% later. Eighteen percent had a short leafing period limited to the rainy season *sensu stricto*, including *Acacia seyal, Adenium obesum, Commiphora africana, Euphorbia balsamifera*. Most often, they flowered at the same time as leafing during the rainy season, and 55% started flowering at the end of the dry season or at the beginning of the rainy season (figure 3). However, 25% completed reproduction entirely during the dry season, including *Pterocarpus erinaceus* and *Ziziphus mucronata*. Consequently, only 20% reproduced during the most favourable season after a minimum of photosynthetic activity (*Acacia adansonii, Feretia apodanthera, Acacia gerardii, Detarium microcarpum*). Several species were also reported to have either variable or several flowering periods, including *Adansonia digitata, Grewia bicolor, Grewia lasiodiscus, Grewia venusta*. Atypical phenologies were also found in this group, especially in the Sahel. This is the case of *Leptadenia pyrotechnica* which bears atrophied ephemeral leaves and is able to flower continuously, the peak of abundance being at the end of the rainy season (Seghieri et al. submitted).

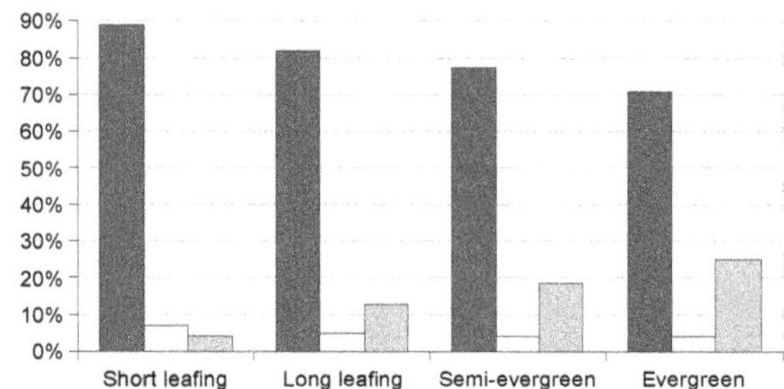

Fig. 2. Distribution of the start of leafing periods for each phenological strategy (see definition in section 2.1.) of the woody species in west tropical Africa cited in the literature. Leafing started at the end of the dry season or at the beginning of the rainy season (dark grey), leafing started in the middle of the rainy season (white), leafing started in the middle of the dry season (light grey).

Long-leafing deciduous species: 82% started leafing at the end of the dry season or at the beginning of the rainy season vs. 13% which started leafing in the middle of the dry season. Only five percent of species in this group started leafing in the middle of the rainy season (figure 2). Forty-four percent bore leaves during the dry season and shed them in the middle or at the end of the dry season. But 52% shed their leaves at the beginning of the dry season like in the previous group. This is not surprising since 37% of the species in this group were observed in the Sudanian bioclimate vs. 31% in the Sudano-Sahelian area (figure 4), despite the fact that most species recorded in the literature were monitored in the Sudano-Sahelian area (50% *vs.* 25% in the Sudanian area). Forty-five percent of species belonging to this group started flowering while leafing at the end of the dry season or at the beginning of the rainy season, while 30% completed reproduction entirely during the dry season (figure 3). The contribution of this group to flora were higher in Sudanian and Guinean bioclimates - including in the mosaic landscape of savannas and forests in the Lamto region (Menaut, 1971; Menaut & César, 1979) - than in drier bioclimates (figure 5). One species with reverse phenology was observed in this group: *Faidherbia (Acacia) albida.* This species bore leaves during the dry season and was leafless during the rainy season (Hiernaux et al., 1994; Depommier, 1998; Roupsard et al., 1999).

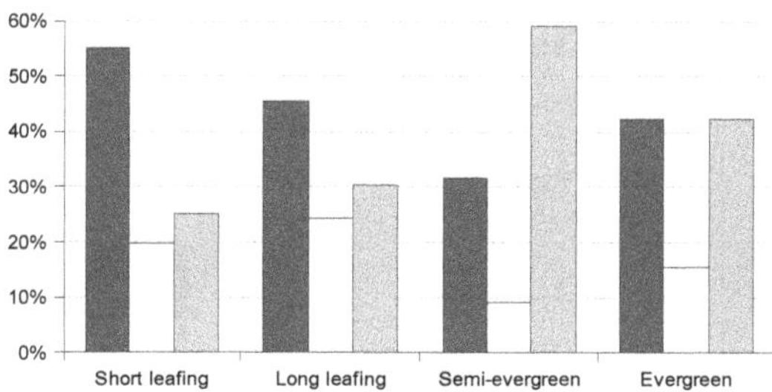

Fig. 3. Distribution of reproductive periods according to each phenological strategy (see definition in section 2.1.) of the woody species in WTA reported in the literature: flowering started at the end of the dry season or at the beginning of the rainy season (dark grey), in the middle of the rainy season (white), reproduction completed during the dry season (light grey).

Semi-evergreen species: like in the previous group, most (78%) of the species replaced their leaves at the end of the dry season or at the beginning of the rainy season (figure 2). Only 18% replaced their leaves in the middle of the dry season. However, 42% shed their leaves in the middle or at the end of the dry season. Twenty-nine percent shed their leaves during the rainy season and 29% at the beginning of the dry season. Unlike in the two previous groups, in this group, 59% reproduced entirely during the dry season, and started flowering at the end of the rainy season, at the beginning or in the middle of the dry season (figure 3). Only *Guiera senegalensis* was observed by Devineau (1999) to flower at any time in the Sudanian bioclimate.

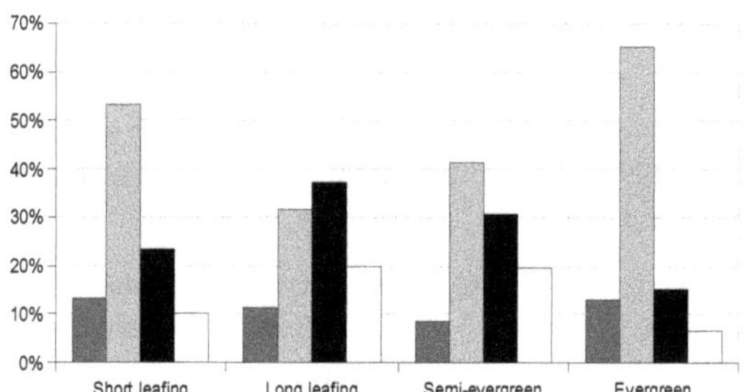

Fig. 4. Proportions of the bioclimates investigated along the WTA latitudinal gradient in the literature as a function of each phenological strategy (definition in section 2.1.), from the driest to the wettest bioclimate: Sahelian (dark grey), Sahelo-Sudanian (light grey), Sudanian (black), Guinean (white).

Evergreen species: these species were mainly observed in Sahelo-Sudanian bioclimate (figure 4), in which they contribute to a greater part of the flora (figure 5). Their contribution was also high in the Sahel. They replaced their old leaves at any time (*Balanites aegyptiaca, Cadaba farinosa, Maytenus senegalensis*) or at a specific period, mostly (72%) at the end of the dry season or at the beginning of the rainy season (figure 2), and 25% in the middle of the dry season (*Balanites aegyptiaca, Maytenus senegalensis, Nauclea latifolia, Piliostigma reticulatum*). Their reproductive period mostly occurred at a fixed time of the year: 42% started flowering at the end of the dry season or at the beginning of the rainy season, and 42% completed reproduction during the dry season (figure 3). Several flowering periods were observed in *Balanites aegyptiaca,* but generally only one was followed by fruiting (Seghieri et al., 1995; Seghieri et al., submitted), with peak flowering during the dry season.

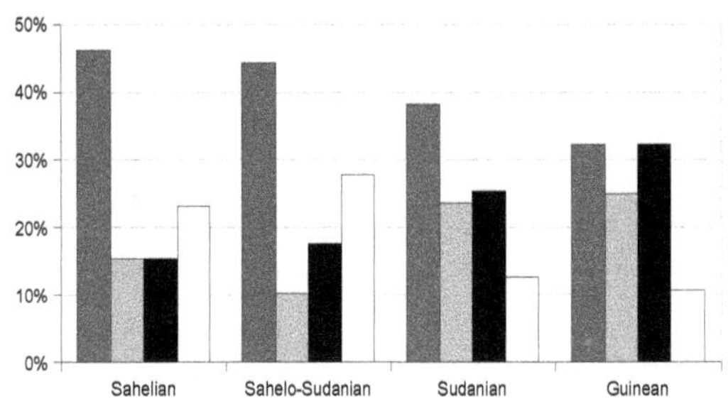

Fig. 5. Contribution of each phenological strategy cited in the literature (see definition in section 2.1.) to the woody flora investigated in the successive bioclimates along the WTA latitudinal gradient: short leafing deciduous species (dark grey), long leafing deciduous species (light grey), semi-evergreen species (black), evergreen species (white).

There was a clear decrease in the proportion of short-leafing deciduous woody plants from the driest to the wettest bioclimate, along with an increase in the proportion of semi-evergreens (figure 5).

3.2 Factors controlling phenology and its variations

3.2.1 Environmental factors

The total rainfall amount and the length of the rainy season were most often suggested to explain global variations in woody phenology along the south-north gradient of increasing climatic aridity (Fournier, 1991). This explanation is well illustrated along the whole WTA bioclimatic gradient from rain forests to Sahelian steppes (table 1).

At the wettest extremity of the gradient, studies were performed in a natural tropical rain forest in southern Ivory Coast (Bernhard, 1970; Bernhard-Reversat et al., 1972), which has a typical equatorial climate. At the community level, a distinct unimodal seasonal rhythm of leaf shedding was observed during the main dry season, although leaf shedding continued throughout the year (litter-fall, figure 6a). In a tropical dry semi-deciduous forest in the Guinean bioclimate (Hall & Swaine, 1981), both basal area increment and leaf shedding were found to be more seasonal (figure 6b; Devineau, 1976).

Further north in the dry forests and savannas, basal area increment and leaf shedding were found to be more closely linked to climate seasonality. The period with no (or reduced) cambial activity, which lasted one month in the semi-deciduous forest (figure

6b), lasted almost three months in the *Anogeissus leiocarpa* dry forest (figure 7a) and for five to eight months in the North Sudanian savanna (figure 7b). However, in both the dry forest and the savanna, where woody plants are not active throughout the year, the proportion of leafing trees varied between communities but was never nil (Devineau 1982, 1999).

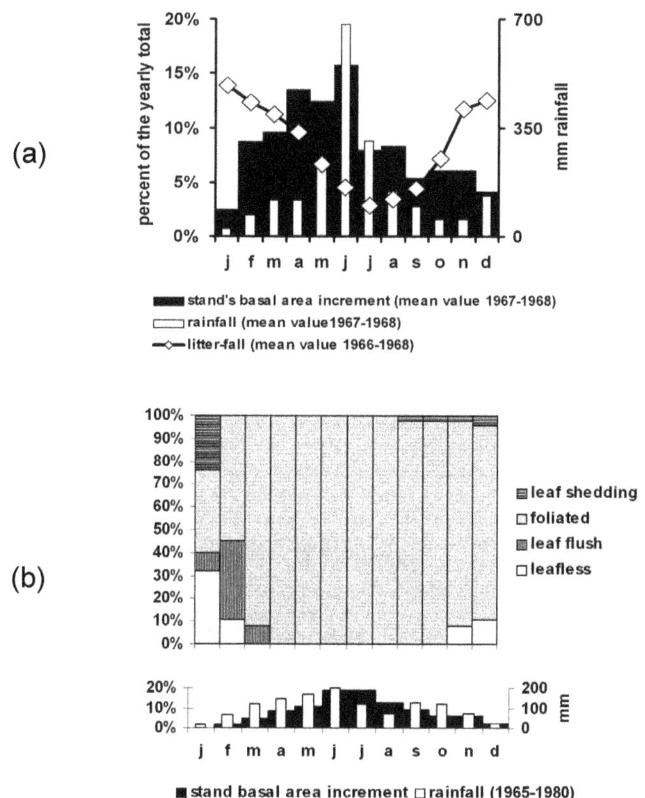

Fig. 6. Leaf phenology and stand basal area increments as a function of rainfall distribution in a typical equatorial bioclimate, the Banco National Park (5.39N -4.05W) with four seasons and 2 100 mm of mean annual rainfall (a) compiled from Bernhard-Reversat et al. (1972), and in a tropical dry semi-deciduous forest, at the Lamto research station (6.22° N, -5.02°W) with four seasons and 1 280 mm of mean annual rainfall (b) compiled from Devineau (1984, 1991).

In the most arid open environments along the bioclimate gradient, phenological variations gave way to a number of different strategies in response to water stress and its variations. Poupon (1979) found significant between-year variability in the beginning and in the length of the leafing period in a Sahelian woody community. Significant variations in the duration of the leafing period were also observed between years (up to two months over 5.5 years) and between individuals (25% of 39 individuals on average) in the semi-evergreen species *A. tortilis* ssp. *raddiana* in the northern part of the Sahel (Do et al., 2005).

Nevertheless, locally, a number of other factors were also reported to determine woody phenology. In the natural tropical rain forest in southern Ivory Coast (2 100 mm of annual rainfall), at the community level, leaf shedding was significant throughout the year and appeared to be correlated with the number of hours of sunshine, while the role of day length was questioned. In spite of this clear pattern, Bernhard (1970) and Bernhard-Reversat et al. (1972) reported wide diversity and asynchrony in leaf phenology and in cambial activity between species and between individuals. In these moist forests, a large part of the within and between species variability of growth patterns was mainly related to biotic interactions, resulting in competitive or niche partitioning strategies. For instance, Devineau (1984, 1991) showed that the phenology of cambial growth in a moist semi-evergreen forest (1 280 mm of annual rainfall) varied according to the position of the tree in the vertical strata, its biological type, and its stage of maturity. In dominant trees, cambial growth peaked during

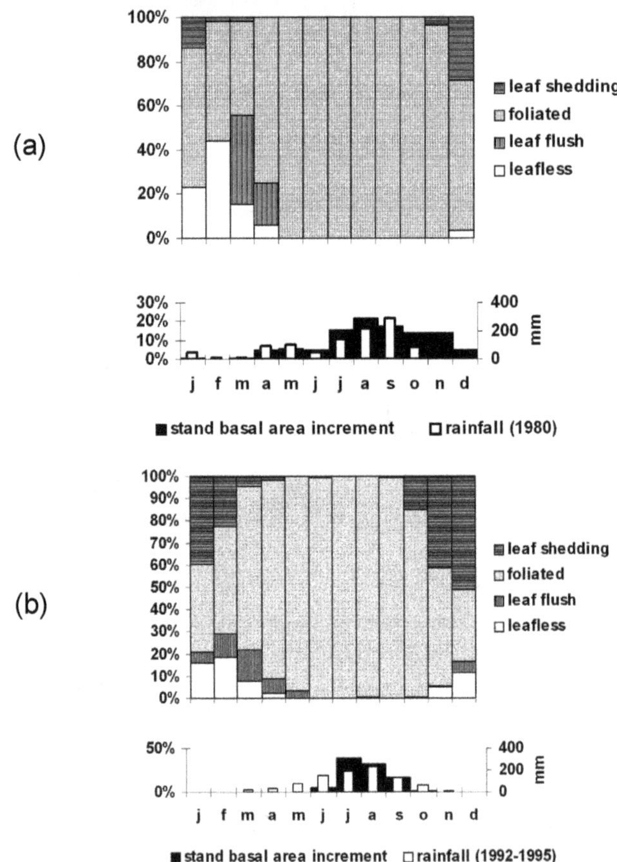

Fig. 7. Leaf phenology and stand basal area increments as a function of rainfall distribution in an *Anogeissus leiocarpa* dry forest, Ouango Fitini. (9.58N , 4.02W) with 1080 mm of mean annual rainfall (a) compiled from Devineau (1984, 1982), and in a north Sudanian savanna (Bondoukuy 11.85°N, -3.73 W) with 900 mm of mean annual rainfall (b) compiled from Devineau (1997, 1999).

the rainy season (June, July), whereas in the understory strata, significant cambial activity occurred during the dry season (December-May). Understory strata took advantage of leaf shedding by taller trees to access more light, and were consequently more active during this period. In this semi-evergreen forest, cambial activity peaked at the beginning of the main rainy season, but did not appear to be linked to rainfall distribution. However, the same author also reported that stand basal area growth rates varied according to rainfall distribution and to the evaporation rate, especially their extreme values, as well as to the water deficit in the preceding months (Devineau, 1989). Even if leaf shedding was most abundant during the main dry season, it also occurred during the rainy season. A large part of the dominant trees were deciduous whereas understory woody species were evergreen.

Menaut (1983) reviewed the most frequently cited determining factors of phenophase in studies conducted in African savannas. This author pointed to an increase in atmospheric vapour pressure or in relative humidity. Indeed, both occur just before the beginning of the rainy season, when most species start leafing. He also indicated the seasonal change in day length and an increase in temperature, in terms of values as well as diurnal ranges. In the north Sudanian savanna, in most species he studied at Bondoukuy, Devineau (1997) found that trunk shrinkage peaked just before the beginning of the rainy season, when the leaf flush occurred (*Entada africana, Terminalia avicennioides, T. laxiflora, Vittelaria paradoxa, Pteleopsis suberosa*). Then, at the beginning of the rainy season, trunk girth increased rapidly with the first regular heavy rainfalls and soil water supply. Growth generally did not begin before at least half the canopy was covered by new foliage (Ashton et al., 1983). As soon as the soil dried out at the end of the rainy season, trunks started to shrink again and continued throughout the dry season. In drier areas, phenological variability was also related to variations in both climatic and soil water conditions, depending on the individual and on the species. Seghieri et al. (1995), Seghieri & Galle (1999), and Seghieri & Simier (2002) confirmed the possibility of high phenological within-species variability according to soil water regimes in southern Sahel and northern Sudanian bioclimates (table 1). These authors showed that this determining factor depends not only on topography and soil texture (Seghieri et al., 1995; Seghieri & Simier, 2002), but also on the nature and distribution of soil surface crusts (Seghieri & Galle, 1999).

Based on observations at the twig scale, Goudiaby (2003) demonstrated that the annual phenological pattern observed over a year in *A. tortilis* ssp. *raddiana* in North Senegal (northern Sahel) corresponded to the overlap of successive leaf cohorts of different magnitudes, three of which appeared to be particularly high. The first leafing period started between April and May, two months before the onset of rains. The second started in July, in the middle of the rainy season, and had the longest leaf life span (6-7 months). The last occurred in December. This author deduced that in this species, the annual duration of leafing mainly depended on the life span and density of the first and the last cohorts because they are the most exposed to the risk of atmospheric and soil drought. In addition, Diouf (2003) found that the individuals with the longest annual leafing duration had a better leaf water status (less negative predawn leaf water potential) and were mainly located on the downslope of micro-topography with more water available in the upper soil layers.

Finally, whatever the bioclimate, the magnitude and duration of leafing and growth rates were found to be more closely linked to the regularity of rainfall distribution than to heavy rainfall events which do not infiltrate, or annual rainfall amount. Unfortunately, very few authors quantitatively tested the environmental factors actually most strongly related to woody phenology in WTA and evidence is consequently still very scarce.

3.2.2 Predictive power of some factors

Do et al. (2005) focused on the comparison of water environmental triggers which could explain inter-annual variation in the peaks of leaf flush and leaf shedding in *A. tortilis* ssp. *raddiana* in northern Senegal. Logistic models including upper soil water availability and rainfall as predictors did not predict variation in either of these two phenophases, but models including the daily maximum vapour pressure deficit, i.e., the driver of leaf transpiration, did.

Seghieri et al. (2009) tested a global predictive approach at the stand scale by mixing species and studying the occurrence of phases (binary response) over the course of the year. The test included data from a two-year monitoring period at a Sahelian site (370 mm mean annual rainfall) and from a three-year monitoring period at a Sudanian site (1 200 mm mean annual rainfall). Only a small proportion of phenophase variability was explained by environmental factors. Nevertheless, significant relationships were identified and provided global insights. Leafing probability was linked to an increase in cumulative rainfall at the Sahelian site but not at the Sudanian site. Conversely, 66% of leafing occurrence was highly significantly explained by an increase in day length and a decrease in air temperature at the Sudanian site but not at the Sahelian site. These results suggest that the water regime is not limiting at Sudanian sites but is at Sahelian sites at the scale of the community.

At the species scale, Seghieri et al. (submitted) found that at the same Sahelian site, temperature variations during the preceding months with different time lags were often the strongest environmental predictor of flowering in *Acacia raddiana* and *Acacia seyal*, and of leafing and flowering in *Balanites ægyptiaca* and *Acacia ehrenbergiana*. Day length, air relative humidity and/or cumulative rainfall with different time lags were the best predictors of flowering in *Combretum glutinosum, Leptadenia pyrotechnica* and *Acacia senegal.*

3.2.3 Human factors

It is now widely recognized that fires are fully integrated in savanna ecosystems and favour the resilience and pseudo-stability of the system (Wright, 1974). For this reason, savannas have been called "fire sub-climax" (Clements, 1936), despite the fact that in WTA, fires are currently mainly caused by human activities. As a result, a reduction in the number of -or an end to- fires can be observed in many Sudanian areas, where they were previously the most common. These modifications may lead to significant biological changes, particularly in phenology (Devineau et al., 2010). The main effect of fire is leaf burning. This synchronizes the natural diversity of leaf shedding patterns in a given site, but simultaneously desynchronizes leaf shedding between burned and unburned sites. Furthermore, in savannas, fires are mainly superficial. Their occurrence and intensity are mainly determined by existing biomass and the state of dehydration of the grass layer, which is the most abundant combustible. That is why Caillaut (2011) simulated current regional spatial patterns of fire in western Burkina Faso based on the simple assumption that fires occur as soon as the grass layer is flammable, i.e. has dried out. Consequently, the disturbance of leaf shedding-and-flush patterns of the woody cover depends on the phenological stage of the herbaceous layer when the fire occurs. However, no major phenological differences were observed by Menaut (1971) when he compared tree phenology between a burned savanna and a neighbouring unburned savanna. The phenological switch caused by fire disappeared after only a few weeks. Seghieri et al. (1995) reported limitation of the leafing period by fire in northern Sudanian savannas, but only in semi-evergreen or evergreen species (*Piliostogma reticulatum, Acacia hockii, Combretum glutinosum, Balanites aegyptiaca, Ziziphus mauritiana, Acacia seyal, Combretum fragrans*), irrespective of when the fire occurred (early or late in the dry season). However, when fires occur early, they may also affect long leafing species. Finally, fire appears to exacerbate the impact of dryness, but does not seriously change phenophase periods, perhaps because of the long-term selection of adapted species to this human factor.

Devineau (1999) compared the phenological plasticity of species in the north Sudanian savanna (Bondoukuy), taking into account within-species synchrony/asynchrony in leafing and flowering duration, fruit ripening and seed dispersal periods. His results pointed to land use as the discriminating factor. Along the axis of increasing plasticity, species were distributed from trees to shrubs and from species which dominate uncropped savannas to species which dominate younger fallows. In addition, Leguminosae were separated from Combretaceae, the latter generally being considered as an indicator of fallow agrosystems (White, 1983). According to Kochmer & Handel (1986) and Pau et al. (2011), the segregation of species phenology according to families suggests the possible influence of phylogenetic constraints.

More particularly, it was reported that in south Sahelian agrosystems in Niger, there was a significant reduction in biodiversity due to pressure from crop and wood harvesting to the benefit of the semi-evergreen cut-resistant Combretaceae, *Guiera senegalensis* (Delabre, 1998; Seghieri & Simier, 2002; Seghieri et al., 2005). Thanks both to this species phenological and physiological plasticity (Seghieri & Simier, 2002, Seghieri & Laloë, 2004) combined with its relative drought resistance (Seghieri & Galle, 1999), it is increasingly dominating the woody cover on sandy soils (70-90% at present), a trend that is facilitated by its significant abundance in the original flora.

Another reason for dominant woody phenological patterns in WTA to change locally is that when farmers clear land for cultivation, they do not eliminate woody species that are of economic interest to them. These species are now sparsely distributed in north-Guinean, Sudanian, and south-Sahelian agricultural parklands. Following cultivation, they become dominant in fallows. This is mainly true of long-leafing and semi-evergreen species *Vittelaria paradoxa* (shea tree), *Parkia biglobosa* (African locust bean), *Faidherbia albida* (Gao), and the short leafing *Adansonia digitata* (Baobab). Further south, evergreen Anacardiaceae plantations are becoming progressively more common along with mango (*Mangifera indica*), and cashew trees (*Anacardium occidentale*). These changes in vegetation composition may show up in remote sensing images and may alter the balance of the energy transfers between the surface and the atmosphere through modifications of the leaf area index and the vegetation functioning.

4. Overall analysis

4.1 Balance between deciduous and evergreen species

Species which start leafing at the end of the dry season or at the beginning of the rainy season were found to be dominant in WTA, whatever the phenological group. In this way, these species minimise the risk of drought during the period of most intensive photosynthetic activity in young leaves, which do not yet have the necessary adaptative traits to limit transpiration. However, this implies that the dry period just before the rainy season is critical for most species, particularly if the first regular and abundant rains are delayed, as observed by Devineau (1997) in the Bondouky savanna or by Seghieri & Galle (1999) after they have prevented the runoff from arriving on a banded thicket in southern Niger.

In dry tropical environments, deciduous species are acknowledged to "avoid drought", while evergreens are assumed to be "drought resistant", two contrasted strategies to optimize water use. But as we will see, the distinction between arido-passive and arido-active species is not really clear.

Short-leafing deciduous species were called also "drought deciduous" by Breman & Kessler (1995) because they almost always had no leaves during the dry season. These species limit water stress by reducing transpiration (leafing) during the unfavourable season, although most can survive short periods without rain. These species start leafing before regular abundant rains and are bare after the last rains of the season, often thanks to residual moisture available in the soil (Seghieri et al., 1995). However, in some Sahelian species in this group, other adaptative characters may compensate for the shortness of the leafing period. Hiernaux et al. (1994) reported thin chlorophyllous bark in *Commiphora africana* and *Euphorbia balsamifera*. An extreme case is *Leptadenia pyrotechnica*, which has flexible stems engorged with water which are permanently photosynthetically active. The stems compensate for atrophied and ephemeral leaves to such an extent that the species was classified among "evergreens" by Breman & Kessler (1995). Thanks to roots that were observed at a depth of 11.5 m, this shrub is able to tap sufficient water resources to grow and reproduce. Such stems and roots are an efficient adaptation to climatic droughts, even in desert conditions (Migahid et al., 1972; Batanouny & Abdel Wahab, 1973).

The majority of long-leafing deciduous species take advantage of the longer favourable period of the Sudanian bioclimate. The fact that the majority of species in this group shed their leaves at the beginning of the dry season and flower when leafing during the rainy season justifies the closeness of this group to the previous one. Their longer leafing period appears to be mainly linked to the longer rainy season in the Sudanian bioclimate. The reverse leafing cycle in the Sudanian-Sahelian species *Faidherbia albida* is possible thanks to its deep roots. Roupsard et al. (1999) observed roots at a depth of 7 m, where they vanished in the vicinity of a permanent water table. The fraction of annual rainfall used by these trees was estimated to be less than 5%.

Semi-evergreen species are generally assumed to have access to water resources during the dry season, e.g. internal plant reserves (root, trunk), deep soil layers, or the water table. Indeed, roots which accessed a 30 m deep water table were observed by Do et al. (2008) in *A. tortilis* ssp. *raddiana* at a north Sahelian site. These authors cited root depth as the main determining factor of the length of leafing (6.7 months) as well as the intensity of transpiration throughout most of the dry season (Do et al., 1998; Do et al., 2005; Do et al., 2008). Furthermore, some species (e.g. Combretaceae) have xeric traits or hairs on their leaves to limit transpiration, but not all (e.g. Fabaceae). Chlorophyllous bark was also observed in *Acacia seyal* by Hiernaux et al. (1994). These adaptations enable the majority of species in this group to reproduce during the dry season.

Like other arido-active plants, evergreen species are also assumed either to have access to water located deep below the surface or to mobilize reserves from the roots or the trunk. Most species in arid or semi-arid bioclimates have scleromorphic features (*Boscia* ssp., *Combretum* ssp.) to reduce water losses, but again not all.

However, information is still lacking on adaptative arido-active traits in most of the species mentioned here irrespective of the group. Insufficient detailed information is available on their phenology and physiology, as well as on the strategies they use to access water during the unfavourable season.

The dominance of deciduous species in WTA is assumed to be due to their higher competitive ability (Menaut, 1983). After Eamus & Prior (2001), analyses of the cost benefit of their maintenance and construction revealed significant differences in comparison with evergreens, not only in terms of water use. Species with short-lived leaves (e.g., < 6 months) displayed a suite of related characteristics including high N investment, high surface leaf area, a high rate of light-saturated assimilation, high relative growth rate, high maintenance cost, but low construction cost, high xylem hydraulic conductivity, but also marked sensitivity to embolism. These traits ensure high short term efficiency of carbon assimilation, and are an opportunistic strategy in terms of carbon gain. By contrast, long-lived leaves ensure long term efficiency and represent a conservative strategy in terms of carbon gain (Eamus & Prior, 2001). Evergreen species do not invest large amounts of nitrogen in their leaves. Their assimilation and relative growth rates are consequently lower. Despite the fact they do not dominate, the contribution of evergreens *sensu stricto* was found to be higher in the driest bioclimates where their carbon gain strategy must consequently be more efficient than in the others.

However, the schematic separation of deciduous and evergreen physiological traits concerns species at the extreme end of the two categories. A deciduous tree with a leaf life span of nine months probably has more in common with an evergreen tree with leaf life span of 12 months than with a deciduous tree with leaf life span of three months. This hypothesis is also supported by the following trend extracted from our review of the literature on WTA savannas and steppes (figure 2): the longer the leafing period, the lower the contribution of species which leaf at the end of the dry season or the beginning of the rainy season. The corollary, which we also stress in this chapter, is that the contribution of species that start leafing in the middle of the dry season increases with an increase in the length of the leafing period, at least in seasonally dry areas.

At the ecosystem scale, Eamus & Prior (2001) reported considerable variation in the relative proportions of different phenological strategies. These authors indicated that Indian savannas are dominated by deciduous species, whereas evergreen species dominate Llanos savannas in South America. In addition, seasonally dry forests in Costa Rica are dominated by semi-evergreen or deciduous species, whereas tropical dry forests in Venezuela are dominated by deciduous species. Australian savannas differ from other savannas in having approximately the same number of species in each of the equivalent of our four groups, but evergreens dominate in terms of canopy cover (90%). Eamus & Prior (2001) presumed that this was a consequence of the extreme dominance of the predominantly evergreen *Acacias* and *Eucalypts* families. But this pattern would also be consistent with these authors' emphasis on the relative photosynthetic return on the cost of investment. Indeed, the lower cost of replacing nutrients and sclerophyllous leaves to better survive drought would be advantageous for evergreens in these savannas, which are drier and hotter than elsewhere (Bowman & Prior, 2005). In WTA, a detailed comparison of trends in terms of cover has not yet been made. Breman & Kessler (1995) claimed that semi-evergreens are important both in Sahelian and Sudanian bioclimates, especially with regard to their canopy cover, but this literature review relativized their assertion in terms of the number of species concerned (figure 5).

4.2 Leaf phenology: a physiological adjustment to the environment

The increase in the proportion of semi-evergreen woody plants with a decrease in the severity of drought along the gradient, along with a decrease in the proportion of deciduous species, has been reported in all continents except Australia (Williams et al., 1996; Eamus & Prior, 2001). This review confirmed that leafing duration and basal area increments largely depend on water conditions. The distribution of phenological strategies in WTA reported in this chapter supports the hypothesis that the alternation of leaf and leafless conditions, which gives the characteristic seasonal physiognomy of the woody component of WTA savannas and dry forests, is directly linked to the high periodicity of the dry season. Under this hypothesis, deciduous species would dominate in climates with a clear boundary between favourable and unfavourable seasons for plant growth, and with moderately long unfavourable periods. The greater contribution of evergreens in the driest bioclimates could thus also result from reduced predictability of rainfall accompanied by a decrease in annual rainfall amount. Breman & Kessler (1995) concluded that, in terms of contribution to species diversity and canopy cover, the proportion of evergreen but also, contrary to our results, of semi-evergreen species decreases from the north to the south of the Sahelo-Sudanian gradient. Paradoxically, these authors recognize that water availability is a determining factor of leaf flush and shedding, and of flowering and fruiting. This apparent contradiction emphasises the crucial need in WTA for 1) generic definitions of phenological observations, analyses and partitioning of strategies, 2) regional monitoring networks of land surface phenology with sufficient permanent field observation sites. This would enable precise and robust characterisation of the phenology of the woody cover and of its variability along the gradient, which is indispensable for validating data acquired by remote sensing.

Furthermore, it is now widely recognized that the general phenological patterns observed at the community scale conceal a wide range of behaviours both among species and among individuals (Seghieri & Galle, 1999; Seghieri & Simier, 2002; Pau et al., 2011). Seghieri & Simier (2002) showed that even if variability between topographical locations is highly significant, within-individual variability at the same topographical location could be even higher. Generally, species with highly variable phenological behaviour provide evidence for growth plasticity (Devineau, 1999). Studies in the literature report the existence of several competing climate drivers of phenology that are not directly linked to water conditions. This makes their initial interpretation difficult. However, all the drivers cited influence tree water status and support the hypothesis of control by endogenous plant water (Borchert et al., 2002). Leaf flush may be under genetic control, which makes it to depend on day length, hours of sunshine or temperature, while its success may depend on water conditions. Indeed, leaf phenology and shoot growth are increasingly considered to be a way for a plant to fine-tune its water status to atmospheric conditions (Devineau, 1989; Fournier, 1995; Do et al., 2005). The concomitance found by Devineau (1997) between periods of the greatest trunk shrinkage and of leaf flush is in accordance with observations made in other tropical vegetation under contrasting climates (Jeffers & Boaler, 1966; Daubenmire, 1972; Lieberman, 1982; Asthon et al., 1983).). This suggests that defoliation could prevent tissue dehydration.

Indeed, it is increasingly acknowledged that differences in leaf phenology between species are linked to differences in their water regime and in their vulnerability to xylem cavitation, leading to differences in their responses to soil and atmospheric drought (Delzon et al., 2010). These new insights underline the role of leafing rhythms in the internal water balance of woody plants, especially in semi-arid and arid environments. Leaf shedding limits water loss under prevailing evaporative demand and water availability (rainfall, soil water regime and access to the water table). When environmental changes exceed the plasticity of a given species, the species is eliminated. It may or not have been replaced by one or more other species with different phenology and plasticity, and resulted in the distribution of phenological strategies observed along the climate gradient (Fournier, 1991; Fournier et al., 2001). Locally, whatever the location along the climate gradient, similar processes influenced by factors other than climate are thought to be at work. Topography and soil surface crusts, depth, texture, and fertility are probably the main factors that determine spontaneous species distribution through water and nutrient availability, as shown by Fournier & Planchon (1998) in herbaceous communities. The phenological plasticity of woody species and the wide range of strategies at the community scale contribute to the capacity of the woody plants to survive in the heterogeneous and unstable environments of WTA. Species plasticity, like their physiological independence from the rainy season, would be expected to increase with an increase in the length of the leafing period in the driest parts of the gradient.

4.3 Implications in terms of land use and climate change effects: lack of support studies

Phenological diversity between species allows vegetation in WTA to survive heterogeneous and changing conditions. Similarly, within-species phenological variability is a measure of species plasticity, which largely contributes to the survival and expansion of a species in such environments. Consequently, it is logical to claim that the most plastic species are predisposed to be the best candidates to survive global changes.

The increasing dominance of the semi-evergreen species *Guiera senegalensis* in southern Niger (Seghieri & Simier, 2002; Seghieri & Laloë, 2004; Seghieri et al., 2005) can thus be explained as follows. The species was able to fill the niche gaps previously occupied by several deciduous species disappeared under the pressure of crop and wood harvest. With their larger environmental tolerance, semi-evergreens obtained a decisive competitive advantage from the disappearance of less resistant species. The fact they are more flexible than obligate deciduous plants contributes to their survival and expansion in the most exploited agro-systems (Poupon, 1979; Devineau, 1999; Williams et al., 1997; Seghieri & Simier, 2002; Seghieri & Laloë, 2004; Seghieri et al., 2005). However, this makes natural regeneration of the other original native species difficult, as water balance generally deteriorates with vegetation degradation (Casenave & Valentin, 1992; Seghieri et al., 1994, 1995). That is why the WTA vegetation may follow the putative law of the "expansion drive of desert plants" reported in the Middle East and in Mediterranean areas (Blondel & Aronson, 1999): "when following human transformation of ecosystems, plants from more xeric habitats, since they are pre-adapted to a large range of conditions, tend to invade more mesic ones rather than the reverse" (Zohary, 1973). In addition, the plasticity of leafing and flowering appears to be an adaptive trait to resist crop pressure as well as savanna fires (Devineau, 1999). Among species whose leaves persist up to the disturbance, those that should be preferentially selected are those whose leaves rapidly reappear afterwards and are still able to flower.

In WTA, the main climate changes expected are an increase in temperature and in rainfall variability, with an increasing proportion of extreme events (Hély et al., 2006; Frappart et al., 2009; Lebel & Ali, 2009). Lack of water disturbs carbon assimilation most when temperatures are too high. This could explain why in the Sahel, where temperatures can be very high and lack of water is a major limitation, most of the phenological processes were found by Seghieri et al. (submitted) to depend on variations in temperature. But the multiple environmental factors that appear to influence phenology may also contribute to species survival thanks to compensatory effects. For example, the effect of an increase in temperature may be compensated by the beneficial effects of an increase in CO_2 concentrations on carbon gain, especially in C3 plants. High plasticity and the many strategies which enable them to remain photosynthetically active independently of the presence of leaves, also probably buffer the negative effects of severe but temporary droughts on Sahelian species. Indeed, changes in the patterns of density, size and species composition of woody populations were documented in 24 rangeland sites along the north-south bioclimatic gradient in the Sahel monitored from 1984 to 2006 by Hiernaux et al. (2009). In the short term, woody populations were seriously affected by the droughts that occurred in 1983–1984. Drought-induced mortality was not more severe under drier climates but occurred sooner after drought in shallow soils. Active recruitment started immediately, i.e. in 1985, *Acacias* species being among the first pioneers. Thanks to the drought, which reduced competition from the other components of the vegetation cover, recruitment was successful. The resulting species composition was first interpreted as a shift towards more arid tolerant flora (*Balanites aegyptiaca, Maerua crassifolia* and *Commiphora africana)*, but diversification started in the mid 1990s (*Combretum aculeatum, Combretum glutinosum* and *Guiera senegalensis*) supporting the hypothesis of the resilience of Sahelian vegetation. Unfortunately, no phenological observations were reported in this interesting study, which concerned only the Sahel.

At present, woody cover is still an integral part of the whole agro-sylvo-pastoral system in WTA. Parkland systems are increasing in Sahelo-Sudanian areas. Some species of great economic interest are thus protected by farmers even if none are planted. In this way, some local agro-pastoral landscapes in the Sahel have considerably more tree cover than they did 30 years ago (Reij et al., 2009). Vast expanses of Sahelian farmlands which were devoid of uncontrolled vegetation in the early 1980s are now studded with trees, shrubs, and crops (WRI, 2008). Elsewhere, deep disturbances of agro-ecosystems have occurred due to crop pressure, which have 1) reduced the woody cover with major hydrological effects in Sahelo-Sudanian areas (Leblanc et al., 2008), 2) modified fire regimes in Guinean areas (Devineau, 1999), and 3) expanded tree plantations in the wettest areas. Impacts on remote sensing (Brown et al., 2010), and consequences in terms of the carbon and water balances (Boulain et al., 2009) remain to be analysed.

In the past decade, the study of phenology has acquired a new legitimacy in climate change research. A conference entitled "The times they are a-changin'; Climate change, phenological responses and their consequences for biodiversity, agriculture, forestry, and human health" was held in Wageningen in December 2001 (Van Vliet & Schwartz, 2002). It revealed that a change in natural event timing is underway in a wide range of locations and is affecting a wide range of species. The recommendations made on this occasion are still applicable to future studies in WTA where the woody stratum is a key component of the surface-vegetation-atmosphere interface.

5. Challenges for the future

In the context of climate change, the increased interest in phenology sends a clear signal that long term monitoring networks on woody phenology are required in WTA to provide better regional understanding, and hence to improve global forecasting. The main defy at global scale is to foresee the possible consequences of environmental and socioeconomic changes on phenological responses. In WTA, the main challenges are 1) to measure the impacts of land use (wood harvest, pasture, fire, and agriculture) on the phenology of both woody species and canopy cover, 2) to increase our knowledge of the processes involved in their relationships with environmental factors. To take up the challenges, it is extremely important for phenologists to exchange information and ideas with scientists from other disciplines. There are many opportunities for collaborative studies, including methods of data analysis, scenario development, and ecological complexity modelling, as well as monitoring phenology at different scales. Remote sensing approaches, for instance, can provide appropriate tools for upscaling the results of functional (canopy phenology, leaf life span, plant physiological and morphological traits) and demographic (population dynamics) approaches combined. Finally, riparian species have rarely been studied up to now, despite the fact they play a major role in surface-vegetation-atmosphere transfers, given their high density along streams and their probable permanent access to free water.

In this chapter, we have described and explained the distribution of phenological strategies along the WTA gradient in terms of the number of species, and of the dominance of deciduous species. However, replications of environmental situations in which a given species can be found would improve the identification of specific survival strategies in changing environments. We have shed light on the role of leaf phenology in the plant water balance and its adaptation to periodically dry environments. This aspect now needs to be investigated at greater depth. In the future studies on phenology of the woody cover in WTA, we recommend to systematically include knowledge of the relative importance of a given species or of a given strategy in the stand cover. Studies that attempt to quantify relationships between environmental factors and woody phenology are rare and should be encouraged. Implementing these recommendations would provide the essential link with remote sensing data and surface-atmosphere transfer models.

6. Appendix

Phenology of woody species reported in the literature in West Tropical Africa, mostly in savannas and steppes, after Apko, 1992; De Bie et al., 1998; Depommier, 1998; Devineau, 1997; Devineau, 1999; Diouf, 1996; Diouf, 2003; Do et al., 2009; Fournier A., 1991; Grouzis & Sicot, 1980; Hiernaux et al., 1994; Ickowicz et al., 2005; Mahamane et al., 2007; Poupon, 1979; Roupsard et al., 1999; Sanon et al., 2007; Seghieri et al., 1995; Seghieri & Galle, 1999; Seghieri & Simier 2002 (full references in the reference list below).

Deciduous species		Phenology
Acacia adansonii	←	Leafing
	←	Flowering
	←	Fruiting
Acacia ataxacantha	←	Leafing
		Flowering
		Fruiting
Acacia dudgeoni	← ← ← — — — →	Leafing
	← ← ← — — — →	Flowering
	← — — — →	Fruiting
Acacia erythrocalyx	← ← ← — — — — — →	Leafing
	← — →	Flowering
	— — → ← — — —	Fruiting
Acacia gourmaensis	← — — — — — — — →	Leafing
	← — — — — →	Flowering
	← — →	Fruiting
Acacia macrostachya	← — — — — — — — →	Leafing
	← — — — →	Flowering
	← — — — →	Fruiting
Acacia nilotica	←	Leafing
		Flowering
		Fruiting
Acacia polyacantha	←	Leafing
		Flowering
		Fruiting
Acacia senegal	← ← ← — → — → — →	Leafing
	← — →	Flowering
	← — — — — — →	Fruiting
Acacia sieberiana	←	Leafing
		Flowering
		Fruiting
Adansonia digitata	← ← — → → →	Leafing
	↔ ← — → ↔	Flowering
	↔ ↔ ← — — →	Fruiting
Adenium obesum	← — — — →	Leafing
	← — — →	Flowering
	← — — →	Fruiting
Albizia chevalieri	←	Leafing
		Flowering
		Fruiting
Bombax costatum	← — — — — — — — →	Leafing
	→ ←	Flowering
	— — → ←	Fruiting
Boswellia dalziellii	←	Leafing
		Flowering
		Fruiting
Bridelia ferruginea	← — — — — →	Leafing
	← ←	Flowering
	← ← — — →	Fruiting
Burkea africana	←	Leafing
		Flowering
		Fruiting
Cassia sieberana	← — ← — — — — — — — →	Leafing
	↔ ↔ ↔ ← — ← — — → ↔ ↔ ↔	Flowering
	↔ ↔ ↔ ↔ ← — → ↔ ↔ ↔ ↔ ↔	Fruiting
Cassia singueana	←	Leafing
		Flowering
		Fruiting

Deciduous species		Phenology
Cochlospermum planchoni	← — — — →	Leafing
	← — →	Flowering
	← — →	Fruiting
Combretum lecardii		Leafing
	no detail available	Flowering
		Fruiting
Commiphora africana	← — — — →	Leafing
	→ ← —	Flowering
	← — — →	Fruiting
Commiphora pedunculatum	← —	Leafing
		Flowering
		Fruiting
Cordyla pinnata	← —	Leafing
		Flowering
		Fruiting
Crateva adansonii	← —	Leafing
		Flowering
		Fruiting
Dalbergia melanoxylon	← —	Leafing
		Flowering
		Fruiting
Detarium microcarpum	← — — — — →	Leafing
	← →	Flowering
	→ → → ← — — →	Fruiting
Dichrostachys cinerea	← — — ← — → — →	Leafing
	← — — → →	Flowering
	← — →	Fruiting
Dichrostachys glomerata	← — — — →	Leafing
	← →	Flowering
	← →	Fruiting
Dombeya quiqueseta	← — — — →	Leafing
	— → ←	Flowering
	← — →	Fruiting
Entada abyssinica	← —	Leafing
		Flowering
		Fruiting
Entada africana	← ← — — → →	Leafing
	← — ← — →	Flowering
	← — →	Fruiting
Erythrina senegalensis	← —	Leafing
		Flowering
		Fruiting
Euphorbia balsamifera	← — ← — → →	Leafing
	← →	Flowering
	← →	Fruiting
Feretia apodanthera	← — — — ← — → — →	Leafing
	← — ← — →	Flowering
	— — → ← — — → —	Fruiting
Gardenia ternifolia	← — ← — — — — — — →	Leafing
	← → ← →	Flowering
	← ← — ← — — —	Fruiting
Grewia bicolor	← — ← — — → — →	Leafing
	↔ ← → ↔	Flowering
	← →	Fruiting
Grewia flavescens	←	Leafing
		Flowering
		Fruiting

Deciduous species		Phenology
Grewia lasidiscus	← — — — — → — →	Leafing
	↔ ↔	Flowering
	↔ ↔	Fruiting
Grewia tenax	← — — — — — →	Leafing
	← →	Flowering
	↔	Fruiting
Grewia venusta	← — ← — — — → — →	Leafing
	↔ ← — ↔	Flowering
	← → ↔	Fruiting
Hexalobus monopetalus	← —	Leafing
		Flowering
		Fruiting
Lannea humilis	← — — — →	Leafing
	← →	Flowering
	← →	Fruiting
Pericopsis laxiflora	← — — — — — — →	Leafing
	← — →	Flowering
	← →	Fruiting
Pseudocedrela kotschyi	←	Leafing
		Flowering
		Fruiting
Pteleopsis suberosa	← ← — →	Leafing
	— — — — — → ← — —	Flowering
	← →	Fruiting
Pterocarpus erinaceus	← — — — — — — → — →	Leafing
	← →	Flowering
	← →	Fruiting
Pterocarpus lucens	← ← — — — — → — →	Leafing
	← ↔ → →	Flowering
	← — — → →	Fruiting
Sclerocarya birrea	← ← — — — — — → — →	Leafing
	← — → ← →	Flowering
	← ← — — → →	Fruiting
Securidaca longipedunculata	→ ← ← — — — — — → — —	Leafing
	— — — — ← ← → ← —	Flowering
	↔ ↔ ↔	Fruiting
Securigena virosa	← ← — — — — — → — →	Leafing
	← ← — — → →	Flowering
	← ← — — → — →	Fruiting
Sterculia setigera	← ← — — — — — → — →	Leafing
	← — → ← ← →	Flowering
	← → ← — →	Fruiting
Stereospermum kunthianum	← —	Leafing
		Flowering
		Fruiting
Strychnos spinosa	← ← — — → — → — →	Leafing
	← — ← →	Flowering
	← — ← — — — →	Fruiting
Terminalia laxiflora	← — ← — — — — — →	Leafing
	← — →	Flowering
	← — — →	Fruiting
Terminalia macroptera	← —	Leafing
		Flowering
		Fruiting
Xeroderris stühlmannii	← ← — — — → →	Leafing
	← — →	Flowering
		Fruiting
Ziziphus mucronata	← →	Leafing
		Flowering
		Fruiting

Long leafing species		**Phenology**
Acacia albida	— — — — — — — → ← ← — —	Leafing
	→ — → ← — —	Flowering
	← — — →	Fruiting
Acacia gerardii	→ ← — — — — — — — — — —	Leafing
	← — →	Flowering
	← — — →	Fruiting
Annona senegalensis	← — — — ← — — — — — →	Leafing
	← — — — — — — ↔ → ← —	Flowering
	— — — — — — — — ← → ← ←	Fruiting
Anogeissus leiocarpus	→ ← ← — — — — — → — —	Leafing
	— — → ← — — — ← — → — —	Flowering
	↔ — → ← ← —	Fruiting
Combretum aculeatum	← ← — — — — — — →	Leafing
	← ← — ← →	Flowering
	— → ← ← — → —	Fruiting
Combretum micranthum	→ ← — ← ← — — — — — — —	Leafing
	← — ← ← → ← — →	Flowering
	→ — → ← ← — — —	Fruiting
Combretum molle	— — — → ← — — — — — — —	Leafing
	← — — →	Flowering
	← —	Fruiting
Crossopteryx febrifuga	↔ — — ↔ ← — — — — — → — —	Leafing
	— — — ← ← — → ← —	Flowering
	← — — — — →	Fruiting
Gardenia erubescens	← — — → ← — — — — — → —	Leafing
	← — — — — — → ← —	Flowering
	↔ ↔ ↔ ↔ ↔ ↔ ← → ↔ ↔ ↔ ↔	Fruiting
Lannea acida	← — — — — — — — — →	Leafing
	↔ ↔ ← — →	Flowering
	↔ ↔	Fruiting
Lannea microcarpa	← — — — — — — — — →	Leafing
	← ← →	Flowering
	← → →	Fruiting
Lannea vetulina	← — — — — — — — — →	Leafing
	← →	Flowering
	← →	Fruiting
Lonchocarpus laxiflorus	← — — — — — — — — →	Leafing
	← →	Flowering
	← →	Fruiting

Semi-evergreen species		Phenology
Acacia hockii	→ ← — — — — → — →	Leafing
	← — →	Flowering
	← — →	Fruiting
Acacia raddiana	→ — → ← ← — — — — — → —	Leafing
	← — — → ← ← — →	Flowering
	— → ← — — ← →	Fruiting
Acacia seyal	→ ← ← — — — — — → — —	Leafing
	— — → ← ← — — — — — — — —	Flowering
	← — →	Fruiting
Combretum collinum	← — — ← — — — — — → →	Leafing
	— — — ← ← — → ← — —	Flowering
	— — → ← — — ← — — —	Fruiting
Combretum nigricans	→ ← ← — → — — — — — —	Leafing
	← — ← → — →	Flowering
	→ — → ← — — → — —	Fruiting
Cordya mixa		Leafing
		Flowering
		Fruiting
Daniellia oliveri	← — → ← — — — — — — → →	Leafing
	← ← — →	Flowering
	← — — — — →	Fruiting
Diospyros mespiliformis	← — — — — — — — — — →	Leafing
	← → ← →	Flowering
	→ ← — — → ←	Fruiting
Ficus thonningii		Leafing
		Flowering
		Fruiting
Gardenia sokotensis		Leafing
		Flowering
		Fruiting
Guiera senegalensis	→ — ↔ — ← — — ← — — — →	Leafing
	→ — → ← — ← — →	Flowering
	→ — — → → ← — —	Fruiting
Isoberlinia doka	← — — — ← — — — — — — ↔	Leafing
	← — ↔ — — — →	Flowering
	← ← — — →	Fruiting
Ozoroa insignis	→ ← — — — — — — — — — —	Leafing
	← — — →	Flowering
	← — — →	Fruiting
Parkia biglobosa	← — — — — — — — → →	Leafing
	← — → ← —	Flowering
	← — — →	Fruiting
Piliostigma thonningii	→ ← — — — — — — — → → —	Leafing
	↔ ← — → ↔ →	Flowering
	↔ — ← — — — → ← ← — ↔ →	Fruiting
Prosopis africana	↔ — ↔ — — — — — — — — — —	Leafing
	← ← — →	Flowering
	→ ← — — — — → —	Fruiting
Pterocarpus santalinoides		Leafing
		Flowering
		Fruiting
Tamarindus indica	→ ← — — — — — — — — —	Leafing
	← — ← — ← → → ← — —	Flowering
	← → ← — — →	Fruiting
Terminalia avicennioides	← — — ← ← — — — — — → →	Leafing
	← ← — →	Flowering
	← — ← → →	Fruiting
Vittelaria paradoxa	← — — → — ← — — — — →	Leafing
	← — → ← ← → → ← — —	Flowering
	← ← — →	Fruiting
Ximenia americana	↔ — — — — — — — — — — — —	Leafing
	← — — → — — →	Flowering
	← — — ↔ →	Fruiting
Ziziphus mauritiana	→ ← ← — ← → — → — —	Leafing
	→ ← ↔ ↔ → ← — —	Flowering
	← → ← → ← →	Fruiting

Evergreen species		Phenology
Alfezia africana	— — ↔ — — — — — — — — — —	Leafing
	← →	Flowering
	← — →	Fruiting
Balanites aegyptiaca (1)	↔ — — ↔ ↔ — ↔ — — — — —	Leafing
	← — → ← — → ← — →	Flowering
	← — — — ← — → — — →	Fruiting
Berlinia grandiflora		Leafing
	no detail available	Flowering
		Fruiting
Borassus aethiopium		Leafing
	no detail available	Flowering
		Fruiting
Boscia angustifolia		Leafing
	no detail available	Flowering
		Fruiting
Boscia salicifolia		Leafing
	no detail available	Flowering
		Fruiting
Boscia senegalensis	— — ↔ — — — — — — — — ↔ —	Leafing
	→ ← — — — ← ← —	Flowering
	← — — — — — — — — — — — →	Fruiting
Cadaba farinosa (1)	— — ↔ — — — — — — — — — —	Leafing
	← — — — — — →	Flowering
	← — →	Fruiting
Calotropis procera		Leafing
	no detail available	Flowering
		Fruiting
Capparis fascicularis		Leafing
	no detail available	Flowering
		Fruiting
Capparis sepiaria (2)	— — — — — — — — — — — — —	Leafing
	← — — →	Flowering
	← — — →	Fruiting
Celtis integrifolia		Leafing
	no detail available	Flowering
		Fruiting
Cola cordifolia		Leafing
	no detail available	Flowering
		Fruiting
Combretum fragrans	— — — — — ↔ — — — — — — —	Leafing
		Flowering
		Fruiting
Combretum glutinosum	— — ↔ — — ↔ — — — — — — —	Leafing
	← → ↔ — — → ← → ← —	Flowering
	— ← ↔ — — → ← — ← —	Fruiting
Ficus glumosa		Leafing
	no detail available	Flowering
		Fruiting
Ficus ingens		Leafing
	no detail available	Flowering
		Fruiting
Ficus platyphylla		Leafing
	no detail available	Flowering
		Fruiting
Ficus sycomor		Leafing
	no detail available	Flowering
		Fruiting
Flemingia faginea		Leafing
	no detail available	Flowering
		Fruiting
Hippocrates africana		Leafing
	no detail available	Flowering
		Fruiting
Holarrhena floribunda		Leafing
	no detail available	Flowering
		Fruiting

Evergreen species		Phenology
Leptadenia hastata		Leafing
	no detail available	Flowering
		Fruiting
Maytenus senegalensis (1)	↔ — — ↔ — — — — — — — — —	Leafing
	↔ — — ↔ — →	Flowering
	← — →	Fruiting
Nauclea latifolia	↔ ↔ ↔ — — — — — — — — — — —	Leafing
	← — →	Flowering
	← — — — — →	Fruiting
Opilia celtidifolia		Leafing
	no detail available	Flowering
		Fruiting
Parinari curatellifolia	— — — — — ↔ — — — — — — —	Leafing
	← — — — →	Flowering
	← — — — — — — — — — — — →	Fruiting
Piliostigma reticulatum	↔ — ↔ — — — — — — — — — — —	Leafing
	← — — →	Flowering
	→ — → ← — —	Fruiting
Rotula aquatica riparian		Leafing
	no detail available	Flowering
		Fruiting
Saba senegalensis		Leafing
	no detail available	Flowering
		Fruiting
Salix subserrata riparian		Leafing
	no detail available	Flowering
		Fruiting
Sarcocephalus latifolius		Leafing
	no detail available	Flowering
		Fruiting
Syzygium guineense		Leafing
	no detail available	Flowering
		Fruiting
Vitex simplicifolia		Leafing
	no detail available	Flowering
		Fruiting

(1) several periods of leaf turn over and at any times

(2) period of leaf turn-over non mentionned

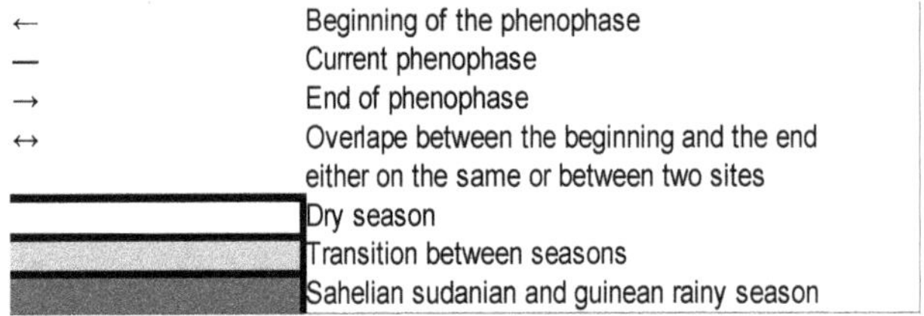

7. Acknowledgements

We are grateful to M-G. Tournoud and A. Gioda for their helpful suggestions.

8. References

Adjanohoun, E.J., Adjakidjè, V., Ahyi, M.R.A., Aké Assi, L., Akoègninou, A., d'Almeida, J., Apovo, F., Boukef, K., Chadare, M., Cusset, G., Dramane, K., Eyme, J., Gassita, J.N.,

Gbaguidi, N., Goudote, E., Guinko, S., Houngnon, P., Lo, I., Keita, A., Kiniffo, H.V., Kone-Bamba, D., Musampa Nseyya, A., Saadou, M., Sodogandji, T., De Souza, S., Tchabi, A., Zinsou Dossa, C., & Zohoun, T. (1989). *Contribution aux études ethnobotaniques et floristiques en République populaire du Bénin. Agence de coopération culturelle et technique (ACCT), coll.* Médecine traditonnelle et pharmacopée, *ISBN 9290281529, Paris.*

AETFAT (1959). Index of the papers on systematic phanerogamy and of the new taxa concerning Africa south of the Sahara and Madagascar. Association for the Taxonomic Study of the Flora of Tropical Africa, Bruxelles.

Agresti, A. (1996). An introduction to categorical data analysis, John Wiley & Sons, ISBN 0-471 11338-7, New York.

Ashton, P.S., Guillaumet, J.L., & Lawton, R.M. (1983). La forêt naturelle : biologie, régénération et croissance des arbres. In: Ecosystèmes forestiers tropicaux d'Afrique, Fournier, F., Sasson, A. (eds.), pp. 162-197, ORSTOM, UNESCO, Recherches sur les Ressources naturelles XIX, ISBN 2-7099-0681-3, Bondy.

Aubréville, A. (1949). Climats, Forêts et Désertification de l'Afrique Tropicale, Société d'Editions Géographiques, Maritimes et Coloniales. Paris.

Batanouny, K.H., & Abdel Wahab, A.M. (1973). Eco-physiological studies of desert plants. VIII. Root penetration of Leptadenia pyrotechnica (Forsk.) Decne. in relation to its water balance. Oecologia, Vol. 11, No 2, (June 1973), pp. 151-161, ISSN 0029-8549.

Bernhard, F., (1970). Etude de la litière et de sa contribution au cycle des éléments minéraux en forêt ombrophile de Côte d'Ivoire. Oecologia Plantarum, T. 5, No 3, pp. 247-266, ISSN 0029-8557.

Bernhard-Reversat, F., Huttel, C., & Lemée, G. (1972). Quelques aspects de la périodicité écologique et de l'activité végétale saisonnière en forêt ombrophile sempervirente de Côte d'Ivoire. In: Proceedings of the symposium on tropical ecology with an emphasis on organic productivity, New-Delhi, 1970.

Blondel, J., & Aronson, J. (1999). Biology and Wildlife of the Mediterranean Region. Oxford University Press, ISBN 0198500351, Oxford.

Borchert, R. (1999). Climatic periodicity, phenology, and cambium activity in tropical dry forest trees. IAWA Journal, Vol. 20, No 3, pp. 239-247, ISSN 0928-1541.

Borchert, R., Rivera, G., & Hagnauer, W. (2002). Modification of vegetative phenology in a tropical semi-deciduous forest by abnormal drought and rain. Biotropica, Vol. 34, No 1, (Mar. 2002) , pp. 27-39, ISSN 1744-7429.

Boulain, N., Cappelaere, B., Ramier, D., Issoufou, H.B.A., Halilou, O., Seghieri, J. ,Guillemin, F., Oï, M., Gignoux, J., & Timouk, F. (2009). Towards an understanding of coupled physical and biological processes in the cultivated Sahel - 2. Vegetation and carbon dynamics. Journal of Hydrology, Vol. 375, No 1-2, (Aug. 2009), pp. 190-203, ISSN 0022-1694.

Bourlière, F., & Hadley, M. (1983). Present-day savannas : an overview. In: Tropical savannas. Ecosystems of the Word 13, Bourlière, F. (ed.) pp. 1-17, Elsevier, ISBN 0444420355, Amsterdam.

Bowman, D.M.J.S. & Prior, L.D. 2005. Why do evergreen trees dominate the Australian seasonal tropics? Australian Journal of Botany, Vol. 53, No. 5, pp. 379-399, (Aug. 2005), ISBN 0067-1924.

Breman, H., & Kessler, J.J. (1995). Woody plants in agro-ecosystems of semi-arid regions. Advanced series in agricultural sciences, vol. 23. Springer-Verlag, ISBN 3-540-58354-8, Berlin.

Brown, M. E., de Beurs, K., Vrieling, A. 2010. The response of African land surface phenology to large scale climate oscillations. Remote Sensing of Environment, Vol. 114, No 10, (Oct. 2010), pp. 2286–2296, ISSN 0031-9228.

Casenave, A., & Valentin, C. (1992). A runoff capability classification system based on surface features criteria in semi-arid areas of West Africa. Journal of Hydrology, Vol. 130, No 1-4, (Jan. 1992), pp. 231–249, ISSN 0022-1694.

Caillaut, S. (2011). Le feu, la brousse et la savane. Modélisation spatiale de la dynamique des paysages soudaniens (Burkina Faso). PhD, Caen University, France.

Chuine, I, & Beaubien, E. G. (2001). Phenology is a major determinant of tree species range. Ecology Letters, Vol. 4, No 5, (Aug. 2001), pp. 500-510, ISSN 1461-023X.

Clements, F.E. (1936). Nature and Structure of the Climax. Journal of Ecology, Vol. 24, No 1, (Feb. 1936), pp. 252-284, ISSN 1365-2745.

Daubenmire, R. (1972). Phenology and other characteristics of tropical semi-deciduous forest in north-western Costa Rica. Journal of Ecology, Vol. 60, No 1, (Mar 1972), pp. 147-170, ISSN 1365-2745.

De Bie, S., Ketner, P., Paase, M., & Geerling, C. (1998). Woody plant phenology in the West Africa savanna. Journal of Biogeography, Vol. 25, No 5, (Sep. 1998), pp. 883-900, ISSN 0305-0270.

Delabre, E. (1998). Caractérisation et évolution d'écosystèmes anthropisés sahéliens: les milieux post-culturaux du sud-ouest nigérien. Doctoral thesis, Paris VI University, Paris.

Delzon, S., Douthe, C., Sala, A., & Cochard, H. (2010). Mechanism of water-stress induced cavitation in conifers: bordered pit structure and function support the hypothesis of seal capillary-seeding. Plant, Cell and Environment, Vol. 33, No 12, (Dec. 2010), pp. 2101-2111, ISSN 0140-7791.

Depommier, D. (1998). Etude phénologique de *Faidherbia albida* : effet de l'émondage, du site et de la dimension de l'arbre sur les phénophases de l'espèce au Burkina Faso. In: L'Acacia au Sénégal, Campa, C., Grignon, C., Gueye, M., Hamon, S. (eds.), pp. 159-179, ORSTOM éditions, coll. Colloques et Séminaires, ISBN 2-7099-1423-9, Paris.

Devineau, J.L. (1976). Données préliminaires sur la litière et la chute des feuilles dans quelques formations forestières semi-décidues de moyenne Côte-d'Ivoire. Oecologia Plantarum, T. 11, No 4, pp. 375-395, ISSN 0029-8557.

Devineau J.L. (1982). Étude pondérale des litières d'arbres dans deux types de forêts tropophiles en Côte d'Ivoire. Annales de l'Université d'Abidjan, série E (écologie), Vol. 15, pp. 27-62.

Devineau, J.L. (1984). Structure et dynamique de quelques forêts tropophiles de l'ouest africain (Côte d'Ivoire). Thèse de Doctorat d'Etat, Paris VI University, Paris.

Devineau, J.L. (1989). Accroissements en surface terrière dans des forêts semi-caducifoliées de Côte-d'Ivoire. Comptes Rendus de l'Académie des Sciences de Paris, t. 309, Série III, (Sep 1989), pp. 423-428, ISSN 0764-4469.

Devineau, J.L. (1991). Variabilité de la croissance en circonférence des arbres dans les forêts semi-décidues de Lamto (Côte d'ivoire). Revue d'Ecologie : La Terre et la Vie, Vol. 46, No 2, (Apr-Jun 1991), pp. 95-124, ISSN 0249-7395.

Devineau, J.L. (1997). Évolution saisonnière et taux d'accroissement des surfaces terrières des ligneux dans quelques peuplements savanicoles soudaniens de l'ouest burkinabé. Ecologie, t. 28, No 3, pp. 217-232, ISSN 1259-5314.

Devineau, J.L. (1999). Seasonal rhythms and phenological plasticity of savanna woody species in a fallow farming system (south-west Burkina Faso). Journal of Tropical Ecology, Vol. 15, No 4, (Jul. 1999), pp. 497-513, ISSN 0266-4674.

Devineau, J.L., Fournier, A., & Nignan, S. (2010). Savanna fire regimes assessment with MODIS fire data: Their relationship to land cover and plant species distribution in western Burkina Faso (West Africa). Journal of Arid Environments, Vol. 74, No 9, (Sep. 2010), pp. 1092-1101, ISSN: 0140-1963.

Do F., Rocheteau A., Diagne A.L. & Grouzis, M. (1998). Flux de sève et consommation en eau d'Acacia tortilis dans le Nord Ferlo. In: L'Acacia au Sénégal, Campa, C., Grignon, C., Gueye, M., Hamon, S. (eds.), pp. 63-80, ORSTOM éditions, coll. Colloques et Séminaires, ISBN 2-7099-1423-9, Paris.

Do, F., Goudiaby, V.A., Gimenez, O., Diagne, A.L., Diouf, M., Rocheteau, A. & Akpo, L.E. (2005). Environmental influence on canopy phenology in the dry tropics. Forest Ecology and Management, Vol. 215, No 1-3, (Aug. 2005), pp. 319-328, ISSN: 0378-1127.

Do, F., Rocheteau, A., Diagne, A.L, Goudiaby, V.A., Granier, A. & Lhomme, J.P. (2008). Stable annual pattern of water use by Acacia tortilis in Sahelian Africa. Tree Physiology, Vol. 28, No 1, (Jan. 2008), pp. 95-104, ISSN 0829-318X.

Diouf, M. (2003). Caractéristiques fondamentales de la feuillaison d'une espèce ligneuse sahélienne: Acacia tortilis (Forsk) Hayne. Variations selon les microsites topographiques au Ferlo (Nord-Sénégal). PhD, Cheikh Anta Diop University, Dakar.

Eamus, D. & Prior L. (2001). Ecophysiology of trees of seasonally dry tropics: comparisons among phenologies. In: Advances in Ecological Research, Vol. 32, Caswell, H. (ed.), pp. 113-197, Academic Press-Elsevier, ISBN 978-0120139323, London.

Fournier, A. (1991). Phénologie, croissance et production végétales dans quelques savanes d'Afrique de l'Ouest. ORSTOM éditions, coll. Etudes et Thèses, ISBN : 2-7099-I 052-7, Bondy.

Fournier, A., & Planchon, O. (1998). Link of vegetation with soil at a few meter-scale: herbaceous floristic composition and infiltrability in a Sudanian fallow-land. Acta Oecologica, Vol. 19, No 3, (May-Jun 1998), pp. 215-226, ISSN 1146-609X.

Fournier, A., Floret, Ch., & Gnahoua, G.M. (2001). Végétation des jachères et succession post-culturale en Afrique tropicale. In: La jachère en Afrique tropicale. Vol. 2 - De la jachère naturelle à la jachère améliorée, Floret, Ch., & Pontanier, R. (eds), pp. 123-168, J. Libbey Eurotext éditions, ISBN 2-7099-1442-5, Montrouge.

Fournier, C. (1995). Fonctionnement hydrique de six espèces ligneuses coexistant dans une savane sahélienne (région du Ferlo, nord Sénégal). ORSTOM éditions, coll. Travaux et Documents Microédités, ISBN 2-7099-1 239-2, Bondy.

Frappart, F., Hiernaux, P., Guichard, F., Mougin, E., Kergoat, L., Arjounin, M., Lavenu, F., Koité, M., Paturel, JE. & Lebel , T. (2009). Rainfall regime across the Sahel band in the Gourma region, Mali. Journal of Hydrology, Vol. 375, No 1-2, (Aug. 2009), pp. 128–142, ISSN 0022-1694.

Funk, C.C., & Brown, M.E. (2009). Declining global per capita agricultural production and warming oceans threaten food security. Food Security, Vol. 1, No 3, (Sep. 2009), pp. 271–289, ISSN 1876-4517.

Garrity, D.P., Akinnifesi, F.K., Ajayi, O.C., Weldesemayat, S.G., Mowo, J.G., Kalinganire, A., Larwanou, M., & Bayala, J. (2010). Evergreen Agriculture: a robust approach to sustainable food security in Africa. Food Security, Vol. 2, No 3, (Sep. 2010), pp. 197–214, ISSN 1876-4517.

Giorgi, F. (2002). Variability and trends of sub-continental scale surface climate in the twentieth century. Part I: observations. Climate Dynamics, Vol. 18, No 8, (Apr. 2002), pp. 675-691, ISSN 0930-7575.

Givnish, T.J. *(2002).* On the adaptive significance of evergreen vs. deciduous leaves: solving the triple paradox. Silva Fennica, Vol. 36, No 3, (Aug. 2002), pp. 703-743, ISSN 0037-5330.

Goudiaby, V.C.A. (2003). Impact d'un déficit pluviométrique sur la feuillaison d'Acacia tortilis (Forsk.) Hayne subsp. raddiana (Savi) Brenan var. raddiana dans le Nord-Ferlo au Sénégal. PhD, Cheikh Anta Diop University, Dakar.

Goutorbe, J.-P., Lebel, T., Tinga, A., Bessemoulin, P., Brouwer, J., Dolman, A.J., Engman, E.T., Gash, J.H.C., Hoepffner, M., Kabat, P., Kerr, Y.H., Monteny, B., Prince, S., Said, F., Sellers, P. & Wallace, J.S. (1994). HAPEX-Sahel: a large scale study of land-atmosphere interactions in the semi-arid tropics. Annales Geophysicae, Atmospheres, Hydrospheres and Space Sciences, Vol. 12, No 1, (Feb. 1994), pp. 53-64, ISSN 0992-7689.

Grouzis, M. & Sicot, M. (1980). A method for the phenological study of browse populations in the Sahel : the influence of some ecological factors, In: Browse in Africa, the Current state of Knowledge. International Symposium. Addis Abeba, 8-12 avril 1980, Le Houerou ed., pp. 233-240, ILCA, Addis Ababa.

Haarsma, R. J., Selten, F. M.,Weber, S. L. & Kliphuis, M. (2005). Sahel rainfall variability and response to greenhouse warming, Geophysical Research Letters, Vol. 32, L17702, ISSN 0094-8276.

Hall, J.B., & Swaine, M.D. (1981). Distribution and ecology of vascular plants in a tropical rain forest. : Forest Vegetation in Ghana, W. Junk publishers, ISBN 9061936810, The Hague.

Hallé, F., Oldeman, R.A. A., & Tomlinson, P.B. (1978). Tropical trees and Forest: an architectural analysis, Springer Verlag, ISBN 0387084940, Berlin.

Harrison, S. P., Prentice, I. C., Barboni, D., Kohfeld, K.E., Ni, J., & Sutra, J-P. (2010). Ecophysiological and bioclimatic foundations for a global plant functional classification, Journal of Vegetation Sciences, Vol. 21, No 2, (Apr. 2010), pp. 300-317 , ISSN 1100-9233.

Hély, C., Bremond, L., Alleaume, S., Smith, B., Sykes, M. T. & Guiot , J. (2006). Sensitivity of African biomes to changes in the precipitation regime. Global Ecology and Biogeography, Vol. 15, No 3, (May 2006), pp. 258–270, ISSN 1466-822X.

Heubes, J. , Kühn, I. , König, K., Wittig, R., Zizka, G., Hahn, K. (in press). Modelling biome shifts and tree cover change for 2050 in West Africa. Journal of Biogeography, (Jun 2011), DOI: 10.1111/j.1365-2699.2011.02560.x, ISSN 1365-2699.

Hiernaux, P.H.Y, Cissé, M.I., Diarra, L., & de Leeuw, P.N. (1994). Fluctuations saisonnières de la feuillaison des arbres et des buissons sahéliens. Conséquences pour la quantification des ressources fourragères. Revue d'Elevage et de Médecine Vétérinaire dans les Pays tropicaux, Vol. 47, No 1, pp. 117.125, ISSN 1951-6711.

Hiernaux, P.H.Y, Diarra, L., Trichon, V., Mougin, E., Soumaguel, N., Baup, F. (2009). Woody plant population dynamics in response to climate changes from 1984 to 2006 in Sahel (Gourma, Mali). Journal of Hydrology, Vol. 375, No 1-2, (Aug. 2009), pp 103-113, ISSN 0022-1694.

Ickowicz, A., Friot, D. & Guérin, H. (2007). Acacia senegal, arbre fourrager sahélien? Bois et Forêt des Tropiques, Vol. 284, No 2, pp. 59-69, ISSN 0006-579X.

IPCC (2007). Climate change 2007: the physical science basis. Contribution of the WG 1 to the fourth assessment report of the Intergovernmental Panel on Climate Change, Solomon, S., Qin, D., Manning, M., Chen, Z., Marquis,M., Averyt, K.B., Tignor, M., & Miller, H.L. (eds.), Cambridge University Press, ISBN 978 0521 88009-1, Cambridge.

Jeffers, J. N. R., & Boaler, S. B. (1966). Ecology of a miombo site, Lupa North Forest Reserve, Tanzania. Weather and plant growth 1962-64. Journal of Ecology, Vol. 54, No 2, (Jul. 1966), pp. 447-463, ISSN 1365-2745.

Koch, G.W., Vitousek, P.M., Steffen, W.L., & Walker, B.H. (1995). Terrestrial transects for global change research. Vegetatio, Vol. 121, No 1–2, (Dec. 1995), pp. 53–65, ISSN 00423106.

Kochmer, J.P. & Handel, S.N. (1986). Constraints and competition in the evolution of flowering phenology. Ecological Monographs, Vol. 56, No 4, (Dec. 1986), pp. 303-325, ISSN 0012-9615.

Lebel, T., & Ali, A. (2009). Recent trends in the Central and Western Sahel rainfall regime 4 (1990–2007). Journal of Hydrology, Vol. 375, No 1-2, (Aug. 2009), pp. 52-64, ISSN 0022-1694.

Leblanc, M.J., Favreau, G., Massuel, S.,Tweed, S.O., Loireau, M., Cappelaere, B. (2008). Land clearance and hydrological change in the Sahel: SW Niger. Global and Planetary Change, Vol. 61., No 3-4 , (Apr. 2008), pp. 135-150, ISSN 0921-8181.

Le Houerou, H.N. (1980). Browse in Africa. The Current State of Knowledge. In: Browse in Africa, the Current state of Knowledge. International Symposium. Addis Abeba, 8-12 avril 1980, Le Houerou ed., pp. 83–102, ILCA, Addis Ababa.

Le Houerou, H.N. (1989). The Grazing Land Ecosystems of the African Sahel. Springer-Verlag, Ecological Studies vol 75, ISBN 3540507914, Berlin.

Lieberman, D. (1982). Seasonality and phenology in a dry tropical forest in Ghana. Journal of Ecology, Vol. 70, No 3, (Nov. 1982), pp. 791-806, ISSN 1365-2745.

Liming, F.G. (1957). Home made dendrometers. Journal of Forestry, Vol. 55, No 8, (Aug. 1957), pp. 575-577, ISSN: 0022-1201.

Mahamane, A., Saadou, M., & Lejoly, J. (2007). Phénologie de quelques espèces ligneuses du parc national du "W" (Niger). Sécheresse, Vol. 18, No 4, (Oct. Nov. Dec. 2007), pp. 354-358, ISSN 1147-7806.

Mearns, L.O., Hulme, M., Carter, T.R., Leemans, R., Lal, M. & Whetton, P. (2001). Climate scenario development. In: Climate change 2001: the scientific basis, contribution of the WG1 to the IPCC third assessment report, Houghton, J.T., Ding, Y., Griggs, D.J., Noguer, M., van der Linden, P.J., Dai, X., Maskell, K. & Johnson, C.A. (eds.), pp. 739-768, Cambridge University Press, ISBN 0521 80767 0, Cambridge.

Menaut, J.C. (1971). Études de quelques peuplement ligneux d'une savane guinéenne de Côte d'Ivoire. PhD., Paris Science University, Paris.

Menaut, J.C. (1983). The vegetation of african savannas. In: Tropical savannas. Ecosystems of the Word 13, Boulière F. (ed.), pp 109-149, Elsevier Scientific Publishing Company, ISBN 978-0444420350, Amsterdam.

Menaut, J.C. & César, J. (1979). Structure and primary productivity of Lamto savannas, Ivory Coast. Ecology, Vol. 60, No 6, (Dec. 1979), pp. 1197-1210, ISSN: 0012-9658.

Migahid, A.M., Abdel Wahad, A.M. & Batanouny, K.H. (1972). Eco-physiological studies of desert plants. VII. Water relations of Leptadenia pyrotechnica (Forsk.) Decne. Growing in the egyptian desert. Oecologia, Vol. 10, No 1, (Mar. 1972), pp. 79-91,

Neelin, J.D., Munnich, M., Su, H., Meyerson, J.E., & Holloway, C.E. (2006). Tropical drying trends in global warming models and observations. Proceedings of the National Academy of Sciences of the United States of America, Vol. 103, No 16, (Apr. 2006), pp. 6110-6115, ISSN-0027-8424.

Pau, S., Wolkovich, E.M., Cook, B.I., Davies, T.J., Kraft, N.J.B., Bolmgren, K. , Betancourt, J. L., & Cleland, E.E. (2011). Predicting phenology by integrating ecology, evolution and climate science. Global Change Biology, (Aug. 2011), DOI 10.1111/j.1365-2486.2011.02515.x, ISSN 1365-2486.

Peyre de Fabregues, B. (1980). Végétation. In: Atlas du Niger, Bernus, E., Sidikou, A.H., Laclavère, G. (eds.), pp 20-21, Jeune Afrique éditions, coll. Les Atlas Jeune Afrique, ISBN 2852581515, Paris.

Poupon, H. (1979). Etude de la phénologie de la strate ligneuse à Fété Olé de 1971 à 1977. Bulletin de l'IFAN, Tome. 41, ser. A, No 1, pp. 44-91, ISSN 0018-9634.

Reij, C., Tappan, G., & Smale, M. (2009). Agroenvironmental Transformation in the Sahel: Another Kind of "Green Revolution". IFPRI Discussion Paper, ser. 00914, (Nov. 2009), International Food Policy Research Institute, Washington.

Redelsperger, JL., Thorncroft, C., Diedhiou, A., Lebel, T., Parker, D.J., & Polcher, J. (2006). African Monsoon Multidisciplinary Analysis (AMMA): An International Research Project and Field Campaign. Bulletin of the American Meteorological Society, Vol. 87, No 12, (Dec. 2006), pp. 1739–1746, ISSN 0003-0007.

Roupsard, O., Ferhi, A., Granier, A., Pallo, F., & Depommier, D. (1999). Reverse phenology and dry-season water uptake by Faidherbia albida (Del.) A. Chev. in an agroforestry parkland of Sudanese west Africa. Functional Ecology, Vol. 13, No 4, (Aug. 1999), pp. 460–472, ISSN 0269-8463.

Saadou M. (1990). La végétation des milieux drainés nigériens à l'est du fleuve Niger. PhD, Abdou Moumouni University, Niamey.

Sanon, H.O., Kaboré-Zoungrana, C., & Ledin, I. (2007). Edible biomass production from some important browse species in the Sahelian zone of West Africa. Journal of Arid Environments, Vol. 71, No 4, (Dec. 2007), pp. 376–392, ISSN 0140-1963.

Scheiter, S., & Higgins, S.I. (2009) Impacts of climate change on the vegetation of Africa: an adaptive dynamic vegetation modelling approach. Global Change Biology, Vol. 15, No 9, (Sep. 2009), pp. 2224-2246, ISSN 1365-2486.

Seghieri, J., Carreau, J., Boulain, N., (de) Rosnay, P., Arjounin, M., & Timouk, F. (submitted). Predictive power of physical factors for phenological responses of woody plants in the central Sahel (Mali). Plant Ecology.

Seghieri, J., Floret, C. & Pontanier, R. (1994). Development of an herbaceous cover in a Sudano-Sahelian savanna in North Cameroon in relation to available soil water. Vegetatio, Vol. 114, No 2, (Oct. 1994), pp. 175-184, ISSN 1385-0237.

Seghieri, J., Floret, C., & Pontanier, R. (1995). Plant phenology in relation to water availability : herbaceous and woody species in the savannas of northern Cameroon. Journal of Tropical Ecology, vol. 11, No 2 , (May 1995), pp. 237-254, ISSN: 0266-4674.

Seghieri, J., & Galle, S. (1999). Runoff contribution to a sahelian two-phase mosaic system: soil water regime and vegetation life cycles. Acta Oecologica, Vol. 20, No 3, (May 1999), pp. 209–218, ISSN: 1146-609X.

Seghieri, J., & Laloë, F. (2004). Characterization of the variability of the daily course of leaf water potential in the dominant shrub species within Sahelian fallows in south-west Niger. Ecological Modelling, Vol. 173, No 2-3, (Apr. 2004), pp. 271–281 , ISSN 0304-3800.

Seghieri, J., & Simier, M. (2002). Variations in phenology of a residual invasive shrub species in Sahelian fallow savannas, south-west Niger. Journal of Tropical Ecology, Vol. 18, No 6, (Nov. 2002), pp. 897- 912, ISSN: 0266-4674.

Seghieri, J., Simier, M., Mahamane, A., Hiernaux, P., Rambal, S. (2005). Adaptative above-ground biomass, stand density and leaf water potential to droughts and clearing in Guiera senegalensis, a dominant shrub in Sahelian fallows (Niger). Journal of Tropical Ecology, Vol. 21, No 2, (Mar 2005), pp. 203-213, ISSN 0266-4674.

Seghieri, J., Vescovo, A., Padel, K., Soubie, R., Arjounin, M., Boulain N., (de) Rosnay, P., Galle, S., Gosset, M., Mouctar, A. H., Peugeot, C., & Timouk, F. (2009). Relationships between climate, soil moisture and phenology of the woody cover in two sites located along the West African latitudinal gradient. Journal of Hydrology, Vol. 375, No 1-2, (Aug. 2009), pp. 78- 89, ISSN 0022-1694.

Sokal, R.R., & Michener, C. D. (1958). A statistical method for evaluating systematic relationships. University of Kansas Science Bulletin, Vol. 38, Pt. 2, (Mar. 1958), pp. 1409-1438, ISSN 0022-8850.

Thornton, P.K., Jones, P. G., Owiyo, T., Kruska, R. L., Herrero, M., Orindi, V., Bhadwal, S., Kristjanson, P., Notenbaert, A., Bekele, N., Omolo, A. 2008. Climate change and poverty in Africa: Mapping hotspots of vulnerability. African Journal of Agricultural and Resource Economics, Vol. 2, No 1, (Mar 2008), pp. 24-44, ISSN 1993-3738.

van Vliet, A.J.H., & Schwartz M.D. (2002). Phenology and climate : the timing of life cycle events as indicators of climatic variability and change. International Journal of Climatology, Vol. 22, No 14, (Nov. 2002), pp. 1713–1714, ISSN 0899-8418.

Williams, R.J., Duff, G.A., Bowman, D.M.J.S., & Cook., A. (1996). Variation in the composition and structure of tropical savannas as a function of rainfall and soil texture along a large-scale climatic gradient in the Northern Territory. Journal of Biogeography, vol. 23, No 6, (Nov. 1996), pp. 747-756, *ISSN 0305-0270.*

Williams, R. J., Myers, B. A., Muller, W. J., Duff, G. A. & Eamus, D. (1997). Leaf phenology of woody species in a North Australian Tropical Savanna. Ecology, Vol. 78, No 8, (Dec. 1997) pp. 2542–2558, ISSN: 0012-9658.

WRI (World Resources Institute) (2008). Chapter 3, Roots to Resilience: Case Studies, Turning back the desert, How farmers have transformed Niger's landscapes and livelihoods. In: World Resources 2008, Roots of resilience: Growing the wealth of the poor, WRI publisher, pp. 142-157, ISBN 1569736006, Washington.

White, F. (1983). The vegetation of Africa, a descriptive memoir to accompany the UNESCO/AETFAT/UNSO vegetation map of Africa. UNESCO, ser. Natural Resources Research 20, ISBN: 9231019554, Paris.

Wright, H.E. (1974). Landscape development, forest fires, and wilderness management. Science, Vol. 186, No 4163, (Nov. 1974), pp. 487-495, ISSN 0036-8075.

Zohary, M.J.O. (1973). Geobotanical foundations of the Middle East. Gustav Fischer Verlag. ISBN 3347301454 , Stuttgart.

8

Phenology of Species in the Cerrado of Mato Grosso State, Brazil – *Brosimum gaudichaudii* Trécul (Moraceae)

Rozilaine Aparecida Pelegrini Gomes de Faria,
Maria de Fatima Barbosa Coelho,
Maria Cristina de Figueiredo e Albuquerque
and Rodrigo Aleixo Brito de Azevedo
Universidade Federal de Mato Grosso,
Brasil

1. Introduction

The roots of the *Brosimum gaudichaudii* Trécul., a plant of the Moreaceae family, frequent in the cerrado region and Northeast of Brazil, which is popularly known as "mamica de cadela", "mamacadela" and "algodão" (cotton), are employed by popular medicine against vitiligo and skin deseases (Agra et al., 2008). Its fruit, of a yellow-orange color, is chewed as chewing gum, due to the latex it contains and due to its flavor, while the powder obtained of roots is used as a fragrance (Hoehne, 1946).

Usually plants used for fighting vitiligo (leucomelanodermia), contain furocoumarins. Furocoumarins are of a broad spectrum of application in the treatment of leucomelanodermias, with photosensitizing activity upon the skin (Alchorne, 2001). Many studies have been carried out about this type of activity of the furocoumarins (Jimbow, 1998).

The growth of most species of the Cerrado is related to seasonality, with frequent renewal of the leaves and flowering in the dry season. The study of phenology not only contributes to our understanding of the dynamics of the community, but also about the interaction with abiotic factors such as precipitation, photoperiod, radiation, soil quality or biotic factors such as presence / absence of predators, which are inextricably linked with the phenological pattern of species (Pedroni et al., 2002).

Seasonality of fruiting is due to the seasonality of flowering. The great concentration of fruit in synchronized phenophase attracting disperser favoring seed dispersal in space, while asynchrony favors the dispersion over time (Pina-Rodrigues & Piratelli, 1993).

The next occurrence of fruiting individuals may be related to their habit of scattering, however, each plant stage seems to be linked to the strategy of the plant to establish and perpetuate in the middle (Pina-Rodrigues & Piratelli, 1993). Early or late flowering can be a strategy linked to the present pollinators and seed dispersal by animals or mode anemocoric fruiting is linked to either the wet season, both in the dry season (Oliveira 1998).

In general in the Cerrado, the pattern of seed dispersal of species is constant, occurring in the dry season, ensuring the germination of seeds at the beginning of the rainy season (Bulhão & Figueiredo, 2002). Flowering in this dry season occurs after the fall of leaf pattern generally observed in the deciduous species typical of the Cerrado region (Sarmiento & Monasterio, 1983).

Thus, to know the phenology of a species is of fundamental importance for the programming of gathering fruits and seeds in order for an integrated management programs for reforestation and restoration of degraded areas. Whereas there may be variations between phenophases as the region where the species is included, the objective of this study was to evaluate the phenology of *B. gaudichaudii* and relate the phenology with environmental factors such as rainfall and temperature.

2. Material and methods

2.1 Study area

The study was conducted from 2006 to 2008, in an area of 2264m^2 with Cerrado in the community of Mata Cavalos, coordinates 15°50'34,9"S and 56°24'03,0"W and 10 km away from the town of Nossa Senhora do Livramento, Mato Grosso, Brazil.

The climate is tropical and warm sub-humid and average annual temperature of 24°C, the peak of 42°C and less than 0°C (Arruda et al., 2008). The study area lies in different levels of conservation due to human influence. The monthly data of precipitation and average temperature were obtained from records of agrometeorological Padre Ricardo Remetter in the Experimental Farm of Universidade Federal de Mato Grosso, located in Santo Antonio do Leverger, distant 33 km from Cuiaba city.

2.2 Phenological data

The phenological data were collected in a sample of 50 plants that were randomly selected, identified and numbered. As Fournier and Charpantier (1975), the number of 10 individuals per species is sufficient to monitor the phenological stages. However, some authors used the number of individuals ranging from 2 to 41 (Mantovani et al., 2003). The observations occurred every 20 days by registering the presence of phenophase flowering, fruiting, and proportion of budding leaves in the canopy.

The intensity of the events was estimated for each individual according to the criteria of Fournier (1974), (0) absence of the phenological event; (1) presence of an event in a range of 25%, (2) presence of event in a range of 26 to 50% (3) presence of event ranging from 51 to 75% and (4) presence of event ranging from 76 to 100%.

The botanical material was identified by taxonomist of Central Herbarium of the Universidade Federal de Mato Grosso, Dr Rosilene Rodrigues, and is deposited under the number 37693.

2.3 Data analysis

We used two methods of data analysis: a) percentage of intensity of phenophase where individuals were analyzed in a semi-quantitative scale ranging from 0 to 4, by calculating the percentage of each plant stage (Fournier, 1974), b) activity index indicating the proportion of individuals who express specific phenology. Through these indices can estimate the synchrony of phenophase (Bencke & Morellato, 2002). To check synchrony (Z) between individuals of the sample was applied to the formula:

$$Z = \frac{\sum X_i}{N} \qquad (1)$$

and

$$X_i = \sum_{i=1}^{n} e_j (N-1) f_i \qquad (2)$$

where,

Xi = individual i synchrony with their conspecifics;
N = total number of individuals in the sample
e_j = number of records in which individuals are in the same phenophase j, i ≠ j;
f_i = the number of reports in which individual i is on phenophase considered.

This index ranges from zero (no synchrony), when there is no synchrony between individuals in the population, a (perfect synchrony), the event of all individuals of the species would have occurred in the same period during the year.

3. Results and discussion

During the 18 months of observations, the period of increased rainfall that characterizes the rainy season, occurred between October to May. It was found that in the month of September was one of the lowest rainfall in the region in October and the highest average monthly temperature, 28.1 ° C (Figure 1).

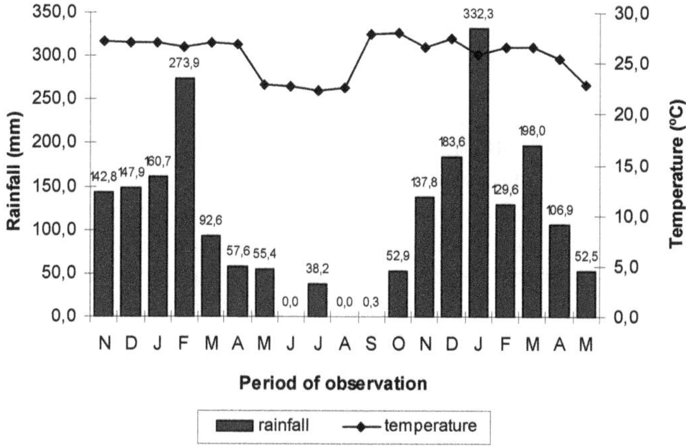

Fig. 1. Meteorological characterization of the study area: precipitation and monthly mean temperatures for the years 2006-2008.

The presence of phenophases is very characteristic among individuals. The sprouting may be present in both apical buds as the whole length of stem and branches. The reproductive phenology, apparently did not affect the leaf fall, since individuals with the presence of leaves in the canopy showed shoots and inflorescences.

In the dry season, abiotic factors such as photoperiod, thermoperiod and irradiance vary with seasonality in spite of their lower amplitudes. The length of day and the maximum and minimum temperatures also tend to increase this time of year (Bulhão & Figueiredo, 2002).

In the period that coincided with the rainy season, from November 2006 to May 2007, a greater number of individuals at the stage of leaf (100% of subjects), but with intensity around 50%. The temperature rise in the period between September and October brought the issue of leaves and shoots, and number of individuals in the phenology of leaf increased by intensifying the phenological pattern, with synchrony index equal to 0.96, indicating high synchrony within the population (Figure 2).

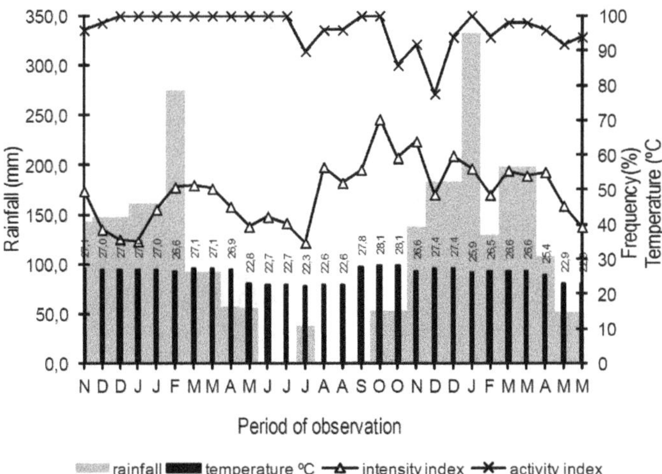

Fig. 2. Activity index and intensity of leaf phenophase of *B. gaudichaudii* in the region of Nossa Senhora do Livramento, Mato Grosso, Brasil in 2006-2008.

Nunes et al. (2005) found that the falling leaves of *G. ulmifolia* happened in the months of dry and cold season, and observed that this occurs for species that are embedded in environments marked seasonality, which concentrate the rain in a restricted period of the year. However, for species in the study of leaf fall occurred in July, the month when we registered the lowest average temperature, 22.3°C (Figure 1), indicating that the low temperature effect was more striking in this event. In *Heteropterys aphrodisiaca* (Coelho & Spiller, 2008), and *Lafoensia pacari* (Santos et al., 2009) the drought period was more relevant to leaf fall.

For the reproductive phase, the index of synchrony was equal to 0.69 and can be classified as average. Flowering occurs between June and December 2007, however, the peak intensity occurred in August with 52% of phenology among individuals while a larger number of plants (92%) developed stage in September 2007 (Figure 3).

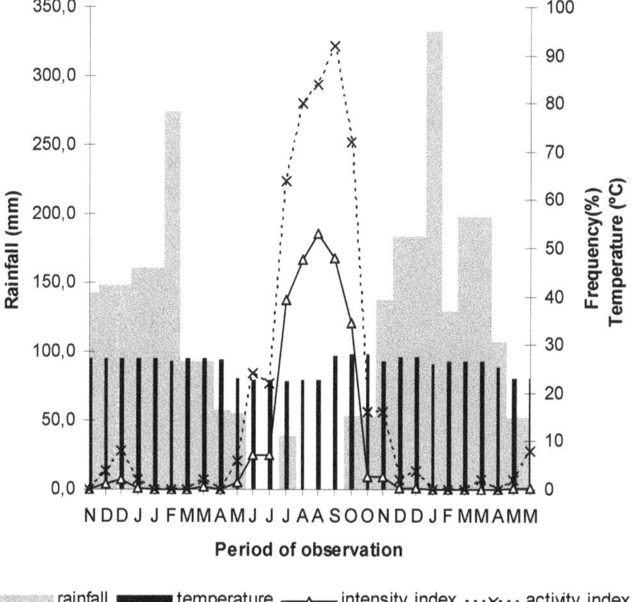

Fig. 3. Activity index and intensity of flowering phenophase of *B. gaudichaudii* in the region of Nossa Senhora do Livramento, Mato Grosso, Brasil in 2006-2008.

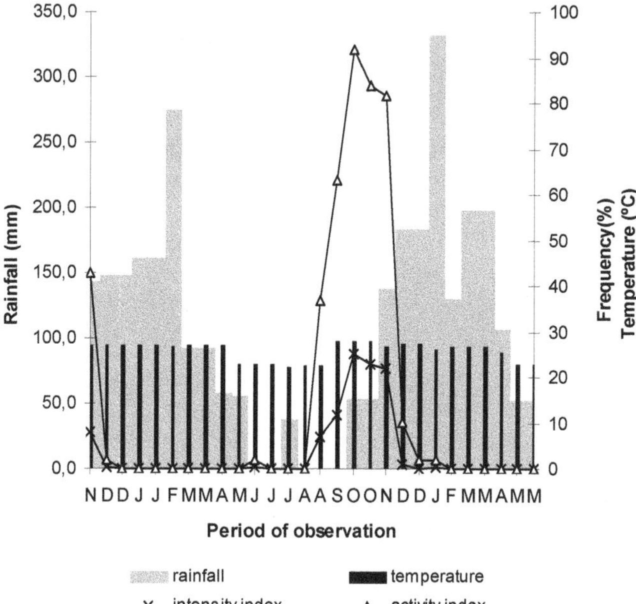

Fig. 4. Activity index and intensity of fruiting phenophase of *B. gaudichaudii* in Nossa Senhora do Livramento, Mato Grosso, Brasil in 2006-2008.

For Bulhão and Figueiredo (2002), flowering in the dry season and / or early rainy season is an indication that the bloom, as well as issuance of sprouts probably depend on rainfall, a factor relative to other environmental factors that do not rains. Emission was observed in flowers and fruits to *B. gaudichaudii* during periods of absence of rain, suggesting a relative independence of the rainy season, as observed by Almeida et al. (1998). But Nunes et al. (2005) suggested that for *Guazuma ulmifolia*, the flowering phenology tended to occur early in the warm season, indicating that these events happen at specific times of year.

The highest average temperature for the year 2007, the area surveyed, was recorded between the months of September (27.8°C) and October (28.1°C), the month when the highest indexes of coincident activity for both clones sprouting (95%) and flowering (92%) of subjects analyzed. One may suggest that these events would be more interconnected with thermoperiod than with the onset of the rainy season.

However, fruiting phenophase in synchrony was greater among individuals in the event population index calculated as ($Z = 0.72$). This phenophase occurred between August and December 2007, coinciding for the month of October 2007 a number of higher frequency in individuals with fruits (91%), however with low intensity resulting in a lower production of fruits (25%) (Figure 4).

Factors such as photoperiod and intensity of solar radiation also could influence the duration and intensity of phenological patterns evaluated. Mantovani and Martins (1988) observed that species that thrive in the rainy season are related to higher temperatures and longer photoperiods. The existence of xylopodium for support the period where there is lack of rain, the renewal of foliage and flowering in the dry season are indications that phenophases of woody species of Cerrado are not limited by the reduction of soil water as suggested Lenza and Klink (2006). As the photoperiod and leaf abscission could induce flowering in the dry season.

Seasonality of fruiting species of the Cerrado would be related to the dispersal agent. Thus wind dispersed species produce fruits during the dry season (June to August) while the zoochorous tend to produce throughout the year would be conditioned or to biotic factors as the most appropriate time for the establishment of the plant (Pina-Rodrigues & Piratelli, 1993). Ressel et al. (2004) observed that the dispersal of fruits of *B. gaudichaudii* is zoochorous and peak fruiting occurs at the beginning of the rainy season, as observed in this study area (Figure 4).

The seed germination during the rainy season favors the establishment of seedlings. Species of the Cerrado as *Copaifera langsdorfii*, even dispersal of fruit occurs in the dry season, seedling growth begins in the next rainy season (Felfilli et al., 1999). In *B. gaudichaudii*, the fruiting period is classified as short-lived and occurs during the rainy season. The classification of short period for fruiting period less than five months in a study of phenology of species in gallery forest was used by Antunes and Ribeiro (1999).

4. Conclusions

Phenology of the studied population of *Brosimum gaudichaudii* are asynchronous. The temperature factor is more relevant to the beginning of the reproductive phenology of the rainfall factor. The most appropriate period for collecting the fruits of *Brosimum gaudichaudii* in Mato Grosso State is from October to December.

5. Acknowledgment

The community of Mata Cavalos for allowing the conduct of work in his field and his help in data collection. At National Development Council (CNPq) for the Productive Grant's first author.

6. References

Agra, M.F.; Silva, K.N.; Basílio, I.J.L.D; França, P.F.; Barbosa-Filho, J.M. (2008). Survey of medicinal plants used in the region Northeast of Brazil. *Revista Brasileira de Farmacognosia,* Vol.18, No. 3, (Setembro), pp. 472-508, ISSN 0102-695X.

Alchorne, M.M.A. (2001). Discromias 176-177. In: Prado F.C.; Ramos J. & Valle, J.R. *Atualização Terapêutica 2001 - Manual Prático de Diagnóstico e Tratamento,* Ed. Artes Médicas, ISBN 8573791489, São Paulo, Brasil, pp. 1630

Almeida, S.P.; Proença, C.E.B.; Sano, S.M. & Ribeiro, J.F. (1998). *Cerrado: espécies vegetais úteis.* Embrapa, ISBN 8586764027, Planaltina, Brasil, 464p.

Antunes, N.B. & Ribeiro, J.F. (1999). Aspectos fenológicos de seis espécies vegetais em matas de galeria do Distrito Federal. *Pesquisa Agropecuária Brasileira,* Vol.34, No.9, (Setembro) pp. 1517-1527, ISSN 1678-3921

Arruda, C.A.S.; Vilanova, S.R.F.; Chichorro, J.F. (2008).Turismo rural e agricultura familiar: o caso de Nossa Senhora do Livramento-MT. *Interações,* Vol.9, No.2, (Dezembro) pp. 149-157, ISSN 1518-7012.

Bencke, C.S.C. & Morellato, L.P.C. (2002). Comparação de dois métodos de avaliação da fenologia de plantas, sua interpretação e representação. *Revista Brasileira de Botânica,* Vol.25, No.3, (Setembro), pp. 269-275, ISSN 0100-8404.

Bulhão, C.F. & Figueiredo, P.S. (2002). Fenologia de leguminosas arbóreas em uma área de cerrado marginal no nordeste do Maranhão. *Revista Brasileira de Botânica,* Vol.25, No.3, (Setembro), pp. 361-369, ISSN 0100-8404.

Coelho, M.F.B. & Spiller, C. (2008). Fenologia de *Heteropterys aphrodisiaca* O. Mach.- Malpighiaceae, em Mato Grosso. *Revista Brasileira de Plantas Medicinais,* Vol.10, No.1, (Outubro), pp. 1-7, ISSN 1516-0572

Felfili, J.M.; Silva Junior, M.C.S.; Dias, B.J.; Rezende, A.V. (1999). Estudo fenológico de *Stryphnodendron adstringens* (Mart.) Coville no cerrado *senso strictu* da Fazenda Água Limpa no Distrito Federal, Brasil. *Revista Brasileira de Botânica,* Vol. 22, No.1 (Abril), pp. 83-90, ISSN 0100-8404.

Fournier, L.A. & Charpantier, C. (1975). El tamaño de la muestra y la frecuencia de las observaciones en el estudio de las caracteristicas fenológicos de los árboles tropicales. *Turrialba,* Vol.25, No.3, (Setembro), pp. 45-48, ISSN 0300-4317.

Fournier, L.A. (1974). Um método cuantitativo para la medición de características fenológicas em árboles. *Turrialba,* Vol.24, No.4, (Dezembro), pp. 422-423, ISSN 0300-4317.

Hoehne, F.C. (1946). *Frutos Indígenas do Brasil,* Secretaria Agricultura, Indústria e Comércio - Instituto de Botânica, São Paulo, Brasil, pp. 88.

Jimbow, K. (1998). Vitiligo. Therapeutic advances. *Dermatologic Clinics*, Vol.16, No.2, (April), pp. 399-407, ISSN:0733-863.

Lenza, E. & Klink, C.A. (2006). Comportamento fenológico de espécies lenhosas em um cerrado sentido restrito de Brasília-DF. *Revista Brasileira de Botânica*, Vol.29, No.4, , (Dezembro), pp. 627-638, C

Mantovani, M.; Ruschel, A.R.; Reis, M.S.; Puchalski, A. & Nodari, R.O. (2003). Fenologia reprodutiva de espécies arbóreas em uma formação secundária da Floresta Atlântica. *Revista Árvore*, Vol.27, No. 4, (Dezembro), pp. 451-458, ISSN 0100-6762.

Mantovani, W. & Martins, F.R. (1988). Variações fenológicas das espécies do Cerrado da Reserva Biológica de Moji Guaçu, Estado de São Paulo. *Revista Brasileira de Botânica*, Vol.1, No.1, (Fevereiro), pp. 101-112, ISSN 0100-8404.

Nunes, Y.R.F.; Fagundes, M.; Santos, R.M.; Domingues, E.B.S.; Almeida, H.S. & Gonzaga, A.P.D. (2005). Atividades fenológicas de *Guazuma ulmifolia* Lam. (Malvaceae) em uma floresta estacional decidual no norte de Minas Gerais. *Lundiana*, Vol.6, No.2, (Fevereiro), pp. 99-105, ISSN 1676-6180.

Oliveira, P.E. (1998). Fenologia e biologia reprodutiva das espécies de Cerrado. In: Sano, S.M. & Almeida, S.P. (Eds.) *Cerrado: ambiente e flora*. EMBRAPA-CPAC, ISBN 8570750080, Planaltina, Brasil, pp. 167-191.

Pedroni, F.; Sanchez, M. & Santos, F.A.M. (2002). Fenologia de copaíba (*Copaifera langsdorfii* Desf. Leguminosae, Caesalpinoideae) em uma floresta semidecídua no sudeste do Brasil. *Revista Brasileira de Botânica*, Vol.25, No.2, (Junho), pp. 183-194, ISSN 0100-8404.

Piña-Rodrigues, F.C.M. & Piratelli, A.J. (1993). Aspectos ecológicos da produção de sementes. In: Aguiar, I.B.; Piña-Rodrigues, F.C.M. & Figliolia, M.B. (Eds.). *Sementes florestais tropicais*, ABRATES, ISBN 3-900051-07-0, Brasília, Brasil, pp.47-81.

Ressel, K.; Guilherme, F.A.G.; Schiavini, I. & Oliveira, P. (2004). Ecologia morfofuncional de plântulas de espécies arbóreas da Estação Ecológica do Panga. *Revista Brasileira de Botânica*, Vol.27, No.2 (Junho), pp. 311-323, ISSN 0100-8404.

Santos, L.W.; Coelho, M.F.B. & Pirani. (2009). Fenologia de *Lafoensia pacari* St. Hil. (Lythraceae) em Barra do Garças, Mato Grosso, Brasil. *Revista Brasileira de Plantas Medicinais*, Vol.11, No.1, (Março), pp. 12-17, ISSN 1516-0572.

Sarmiento, G. & Monasterio, M. (1983). Life forms and phenology. In: Bouliere, F. (Ed). *Ecosystems of the world: tropical savannas*. Elsevier Science, Amsterdan, Netherlands, pp. 79-108. ISBN: 0-444-42035-5.

9

Phenology of Tree Species Populations in Tropical Dry Forests of Southeastern Brazil

Yule Roberta Ferreira Nunes,
Giovana Rodrigues da Luz and Lílian de Lima Braga
*Universidade Estadual de Montes Claros,
Brazil*

1. Introduction

Tropical dry forests (TDFs) are forest formations that occur in regions with two marked seasons (dry and wet) and are characterized by some degree of foliar deciduousness (Murphy & Lugo, 1986). Recent estimates show that 1,048,700 km^2 of TDFs exist worldwide (Miles et al., 2006). The two largest continuous areas of this physiognomy are located in South America and include northeastern Brazil and the region incorporating southeastern Bolivia, Paraguay and northern Argentina (Miles et al., 2006).

In Brazil, these forests, usually called dry forests (Seasonally Deciduous Forests *sensu* Veloso et al., 1991), lose over 50% of their leaf area (Nascimento et al., 2004) as a physiological response to water shortages (Zalamea & González, 2008). Thus, the structural and ecophysiological properties of TDFs are significantly determined by the duration and timing of the dry season, which selects for adaptations associated with avoidance, resistance or tolerance to water stress (Singh & Kushwaha, 2005). Therefore, the floristic composition of TDFs is diverse, and they contain numerous liana, shrub and large tree species (Madeira et al., 2009). In particular, these forests show a great diversity of phenological patterns and large inter-annual variations, particularly at the level of individuals (Singh & Kushwaha, 2005).

In general, Brazilian TDFs are naturally fragmented and occur in different biomes: Cerrado and Caatinga in the Southeast, Central West and Northeast, and they are interspersed with evergreen formations, such as the Amazon and Atlantic Forest, in the North and South, respectively (Scariot & Sevilha, 2005). TDFs also occur in the transition zones of different biomes. However, according to the Brazilian classification of vegetation, Seasonally Deciduous Forests are included as part of the Atlantic Forest biome, specifically that found in Minas Gerais State (Veloso et al., 1991), and this biome is one of the most endangered in Brazil, with only 8% its original forest cover remaining. Therefore, human use is restricted in TDFs and is legally enforced by Brazilian environmental laws (Lima, 2001).

Although protected by law, TDFs are continuously converted to agricultural uses (Maass, 1995), as they generally occur on high fertility soils (Ribeiro & Walter, 1998). Moreover, different political, social and economic pressures have increased the human impact on these forests, leading to severe disturbance and intense deforestation (Murphy & Lugo, 1986; Maass, 1995). In fact, tropical dry forests are the most threatened tropical ecosystems (Miles et al., 2006; Pezzini, 2008). Additionally, most studies in tropical forests were conducted in tropical humid or wet forests, and only 14% of the published articles on tropical forests focused on dry forests (Sánchez-Azofeifa et al., 2005). Therefore, given their relevance and degree of threat, studies on TDFs at the population and community level are essential to understanding their dynamics, maintaining their biodiversity and developing conservation policies to protect these ecosystems (Pezzini, 2008).

In this context, phenological studies of plant species in TDFs may help to understand the dynamics of plant communities (Spina et al., 2001) and act as an indicator of plant responses to climate and soil conditions in a region (Fournier, 1974). Information gathered from phenology studies can be used as a tool to support forest management when developing conservation strategies for the regional fauna and flora (Fournier, 1974).

The phenology of plants in TDFs has been largely associated with exogenous factors, such as precipitation, temperature and photoperiod (Bullock & Solís-Magallanes, 1990), which mainly influence vegetative phenophases (Lobo et al., 2003), and endogenous factors, such as competition for pollinators and dispersers and pressure from flower, fruit and seed predators (Lobo et al., 2003). Thus, because seasonal cues are more pronounced in TDFs, phenological patterns are particularly important to understanding how dry forests function (Justiniano & Fredericksen, 2000). The mix of evergreen and deciduous species gives the forest a phenological complexity that is not found in other tropical forest formations (Burnham, 1997). Nevertheless, information on the phenology of TDF species is scarce (however, see Borchert, 1994; Borchert et al., 2002; Borchert et al., 2004; Nunes et al., 2005; Nunes et al., 2008). This study addresses two basic questions aimed at increasing knowledge regarding the phenology of TDF tree species populations: (i) whether there any reproductive and vegetative phenological patterns in TDF tree species populations and (ii) whether the phenological responses of these populations are determined primarily by precipitation and temperature.

2. Methods

2.1 Study sites

This study was conducted in three sites located in northern Minas Gerais State (Fig. 1): Parque Estadual da Mata Seca (Mata Seca State Park, hereafter PEMS) (Site 1), CROS Reserve (Site 2) and Juramento Reservoir Reserve (Site 3). According to the Thornthwaite climate classification (Thornthwaite, 1948), the climate in northern Minas Gerais is dry sub-humid (C1), with two marked seasons (dry and wet) during the year (Antunes, 1994). The average annual precipitation ranges from 733-1,305 mm, with average annual temperatures between 16.8°C and 26.2°C. Rain usually falls from October to March, with November, December and January as the wettest months, while the driest period occurs from June to August (Nimer & Brandão, 1989). The study sites are located in the transition between the Cerrado and Caatinga biomes (Rizzini, 1997).

The PEMS was created in 2000 and currently has an area of 15,360 ha. The park is located in the city of Manga (14°97′02″S and 43°97′02″W) and represents the only conservation area on the left margin of the São Francisco River with dry forests growing in flat soils (Madeira et al., 2009). Based on its forest structure, the PEMS is a mosaic of different successional stages (Madeira et al., 2009). The average annual temperature in the PEMS area is 24.4°C, and the average annual precipitation is 871 mm (Antunes, 1994), with less than 60 mm of this rain falling in a seven month period (Espírito-Santo et al., 2008).

The CROS Reserve, owned by the Rocha & Souza Mining Construction Company, is located in the city of Montes Claros (16°38′53″S and 43°53′30″W) approximately 10 km from the urban area. The total area of the property is 49 ha, with 20% (9.8 ha) assigned as a permanent protection area (PPA). Dry forests and cerrado (savanna) are the main vegetation types in this area. Exposed rocks can be found on the soil surface and are typical of soils under limestone outcrops of basaltic origin (Silva & Scariot, 2004). The average annual precipitation is 1,000 mm, and the average temperature is 23°C.

The Juramento Reservoir Reserve is located in the city of Juramento (16°46′20″S and 43°39′56″W) and has an area of 3,100 ha. The reserve, owned by the Minas Gerais State Water Company (COPASA), has a 7.63 km² reservoir. This reservoir was formed by the damming of the Juramento, Saracura and Canoas Rivers, which are part of the Verde Grande River sub-basin and the São Francisco River basin. The area surrounding the reservoir is 23.37 km² and has been under protection since the reservoir's construction in 1986 (Nunes et al., 2005). The average annual temperature is 23°C, and precipitation is approximately 1,000 mm/year (Nunes et al., 2005).

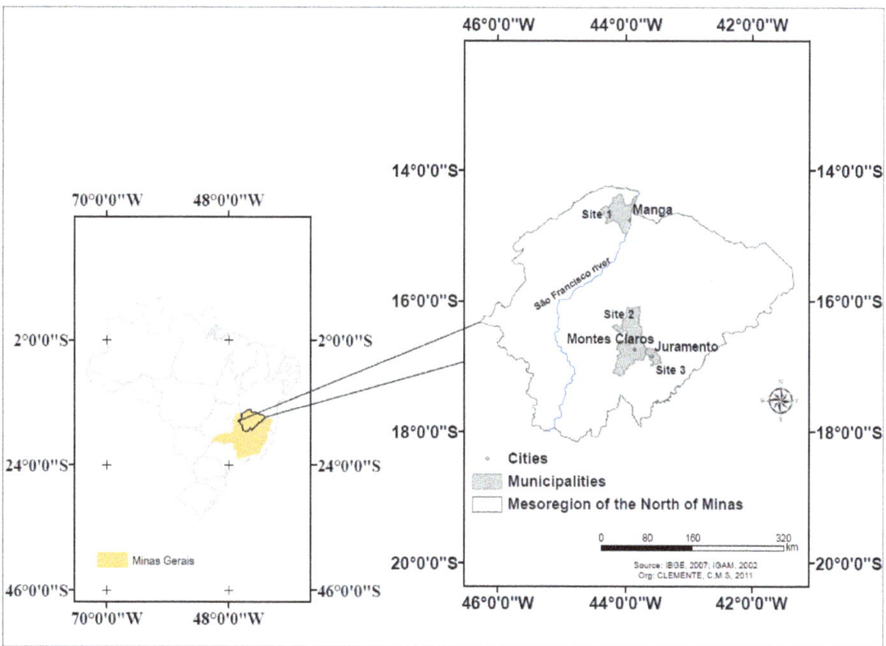

Fig. 1. Geographical location of the study sites in southeastern Brazil (Minas Gerais State). Site 1: Parque Estadual da Mata Seca, PEMS (Manga); Site 2: CROS Reserve (Montes Claros); Site 3: Juramento Reservoir Reserve (Juramento).

2.2 Sampling of phenological data

Populations from 20 TDF tree species were selected for monitoring of phenological patterns (Table 1). The circumference of marked trees at breast height (cbh, 1.3 m above ground) was greater than 15 cm. Each species had 20 individuals marked (Nunes et al., 2005; 2008), except for *Cedrela odorata* L. and *Sapium glandulosum* (L.) Morong, which had 14 and 15 individuals

Family	Species	Dispersion syndrome	Study Site	Source
Anacardiaceae	*Astronium fraxinifolium* Schott ex Spreng.	Ane	3	Maragon et al. 2010
	Myracrodruon urundeuva Allemão	Ane	1	Barbosa et al. 2005
	Spondias tuberosa Arruda	Zoo	1	Maia 2004
Burseraceae	*Commiphora leptophleos* (Mart.) J. B. Gillet	Zoo	1	Barbosa et al. 2005
Cactaceae	*Pereskia bahiensis* Gurke	Zoo	1	Britton & Rose 1963
	Pilosocereus pachycladus Ritter	Zoo	1	Obs. pes. [1]
Euphorbiaceae	*Jatropha mollissima* (Pohl) Baill.	Auto	1	Santos et al. 2005
	Sapium glandulosum (L.) Morong	Zoo	1	Maia 2004
Fabaceae	*Anadenanthera colubrina (Vell.) Brenan*	Auto	3	Barbosa 2005
	Chloroleucon tortum Mart. Pittier	Zoo	3	Obs. pes. [1]
	Copaifera langsdorffii Desf.	Zoo	3	Pereira et al. 2008
	Senna spectabilis DC. H. S. Irwin & Barneby	Auto	1	Barbosa et al. 2005
	Senegalia polyphylla DC.	Auto	2	Obs. pes. [1]
	Mimosa hostilis Benth.	Auto	1	Barbosa et al. 2005
	Pseudopiptadenia contorta (Benth.)	Auto	1	Obs. pes. [1]
Malvaceae	*Guazuma ulmifolia* Lam.	Zoo	3	Moreira et al. 2009
	Heteropterys byrsonimifolia Lam.	Ane	3	Stefanello et al. 2009
Meliaceae	*Cedrela odorata* L.	Ane	2	Silva e Rodal 2009
Polygonaceae	*Ruprechtia fagifolia* Meisn.	Ane	2	Obs. pes.[1]
Sapindaceae	*Dilodendron bipinnatum* Radlk.	Zoo	2	Obs. pes.[1]

Table 1. Species selected for phenological survey at three tropical dry forest sites in Minas Gerais, southeastern Brazil. (Ane = anemochorous, Auto = autochorous, Zoo = zoochorous). 1- Personal observation.

marked, respectively. In each field trip, phenophases were recorded through careful observation of the canopy with binoculars. The phenology of individual trees was recorded for 12 months, beginning in July and ending in June of the following year. The phenological survey at Site 1 was conducted in 2008 and 2009, at Site 2 in 2007 and 2008 and at Site 3 in 2002 and 2003.

Fig. 2. Phenophases observed in the species studied: flower buds of *Spondias tuberosa* Arruda (a) and *Pereskia bahiensis* Gurke (b); flower opening in *Senna spectabilis* DC. H. S. Irwin & Barneby (c); ripe fruits of *S. tuberosa* (d); fruiting in *Mimosa hostilis* Benth. (e), *P. bahiensis* (f), *Pilosocereus pachycladus* Ritter (g), *Jatropha molissima* (Pohl) Baill. (h), *Myracrodruon urundeuva* Allemão (i) and *Pseudopiptadenia contorta* (Benth.) (j); total deciduousness (k) and young leaves (l) in *M. urundeuva*.

The following reproductive phenophases were recorded: (1) flower buds, from the formation of the floral structure to the opening of flower buds; (2) flower opening (anthesis), from the opening of flower buds to flower fall; (3) fruiting, from fruit formation until ripening, represented by the presence of green fruits; and (4) ripe fruits, represented by the presence of dark-brown, yellow or yellowish-green fruits, depending on species. The following vegetative phenophases were also recorded: (5) leafing, represented by emergence of small leaves of different coloration (bright green or red); (6) leaf fall, characterized by leaves falling under a soft breeze and fallen leaves under the tree canopy; and (7) full deciduousness, characterized by total absence of leaves (Nunes et al., 2005; 2008) (Fig. 2).

2.3 Data analyses

The activity index observation method was used to quantify phenological events of the species (Bencke & Morellato, 2002). The activity index refers to the percentage of individuals displaying an event at each sampling date, with the presence or absence of the particular phenophase being recorded. In addition to indicating the occurrence of an event, this method also estimates the synchrony among individuals in a population (Pedroni et al., 2002).

To test whether species populations are differently affected by climate variables, the phenological data of each phenophase were correlated with total precipitation and average temperature, in the same period (month), using Spearman correlation tests (Zar, 1996) and were qualitatively compared with the climatological norms for each site (Fig. 3). The average temperature and total precipitation data for each month were collected at three weather stations located near the study sites in the cities of Itacarambi (Site 1), Montes Claros (Site 2) and Juramento (Site 3). Climatological norms for each city from 1991-2007 (Site 1), 2001-2006 (Site 2) and 1987-2002 (Site 3) were also used to discriminate between dry and wet seasons. So, for the climatological norms we used total precipitation and mean temperature of each year and extract the mean of these variables for all the period, 17 years to Site 1, five years to Site 2 and 16 years to Site 3. These differences in data period are associated to the time of station installation and the availability of the data.

To verify whether populations of the different species exhibit similar reproductive and vegetative phenophases at the different sites, a cluster analysis of the activity index of each species was performed, with Euclidean distance as the coefficient of association (Ludwig & Reynolds, 1988). Moreover, the species studied were also classified according to seed dispersal syndrome (Table 1), following van der Pilj (1982) classification, in zoochorous, when the dispersal is done by animals; anemochorous, when the dispersion is carried by the wind; and autochorous, when the dispersion occurs by explosive dehiscence or gravity. The species classification was made according to information on the biology of the species found in literature (Briton & Rose 1963; Maia 2004; Barbosa 2005; Barbosa et al. 2005; Santos et al. 2005; Pereira et al. 2008; Moreira et al. 2009; Silva & Rodal 2009; Stefanello et al. 2009; Maragon et al. 2010), expert consultation and increased knowledge of the authors. This classification was intended to detect patterns related to reproductive strategies of species.

3. Results and discussion

3.1 Reproductive phenology

Synchronous flowering within populations and among species is more frequent in dry areas (Bullock, 1995). In general, flower bud (late dry season) and anthesis (late wet season) phenophases were concentrated in the wet season. Flower buds occurred in September and October for 50% of the species at Site 1, from August to November for 50% of the species at Site 2 and in October for 67% of the species at Site 3 (Fig. 4a). For most species, anthesis was recorded from September to November; 60% of the species displayed this event in September at Site 1, 50% from September to November at Site 2 and 50% in September and October at Site 3 (Fig. 4b). This pattern seems to be common in seasonal tropical forests with well-defined dry and wet seasons, as several authors (Morellato et al., 1989; Morellato & Leitão-Filho, 1990; 1992; 1996) have also observed flowering in the wet season or during the

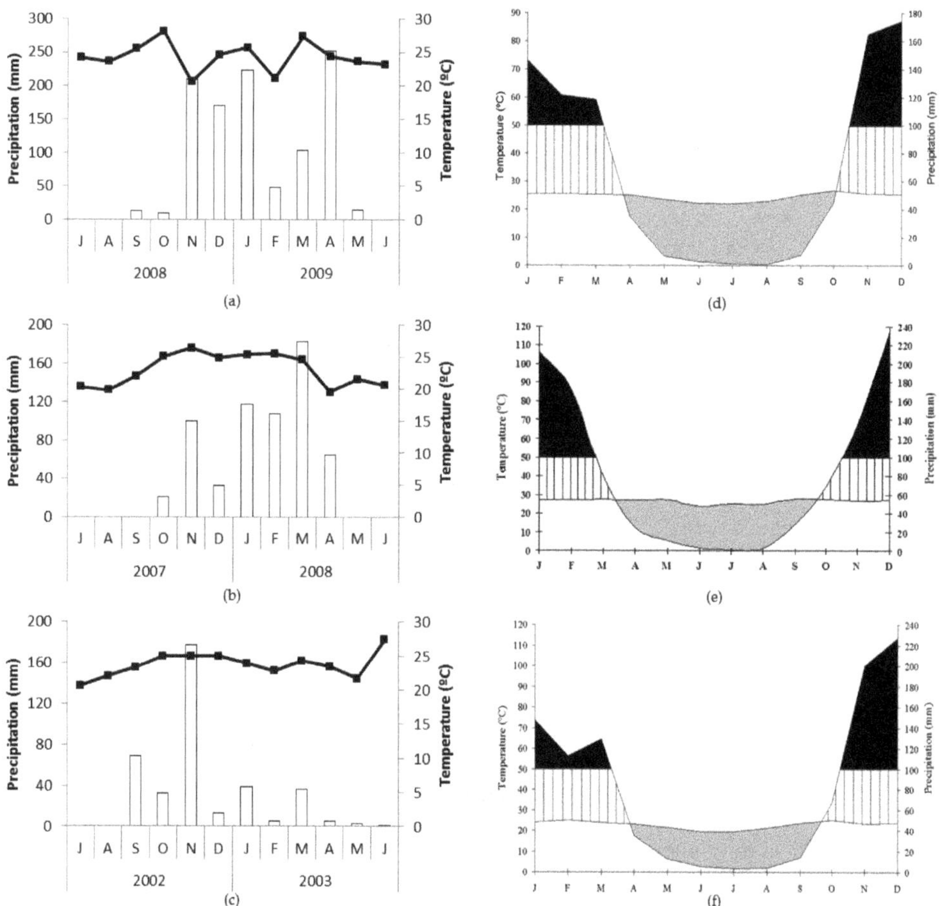

Fig. 3. Climate description of the study sites in southeastern Brazil: total precipitation (bars) and average monthly temperatures (line) for (a) Site 1 (Parque Estadual da Mata Seca, Manga: 2008-2009), (b) Site 2 (CROS Reserve, Montes Claros: 2007-2008) and (c) Site 3 (Juramento Reservoir Reserve, Juramento: 2002-2003); climatological norms for (d) Site 1 (Itacarambi, 1991-2007), (e) Site 2 (Montes Claros, 2001-2006) and (f) Site 3 (Juramento, 1987-2002). For d, e and f the black color refer to the monthly rainfall greater than 100 mm (hydric surplus), the gray color to the drought period and the ratched the humid period.

transition from the dry to wet season. According to Morellato et al. (1989), most species produce flowers during this period because the first rains and the increasing photoperiod stimulate the onset of flowering. This notion has also been advanced by Pereira et al. (2008) who suggested that, prior to anthesis, flower primordia are in a dormant state imposed by drought, which can be interrupted when moisture levels in the environment increase. Moreover, higher precipitation levels and temperatures in the wet season lead to increased litter decomposition and nutrient availability to plants (Morellato, 1992), which in turn are able to invest in reproductive phenophases. During this period, population density and pollinator activity are also higher (Wikander, 1984).

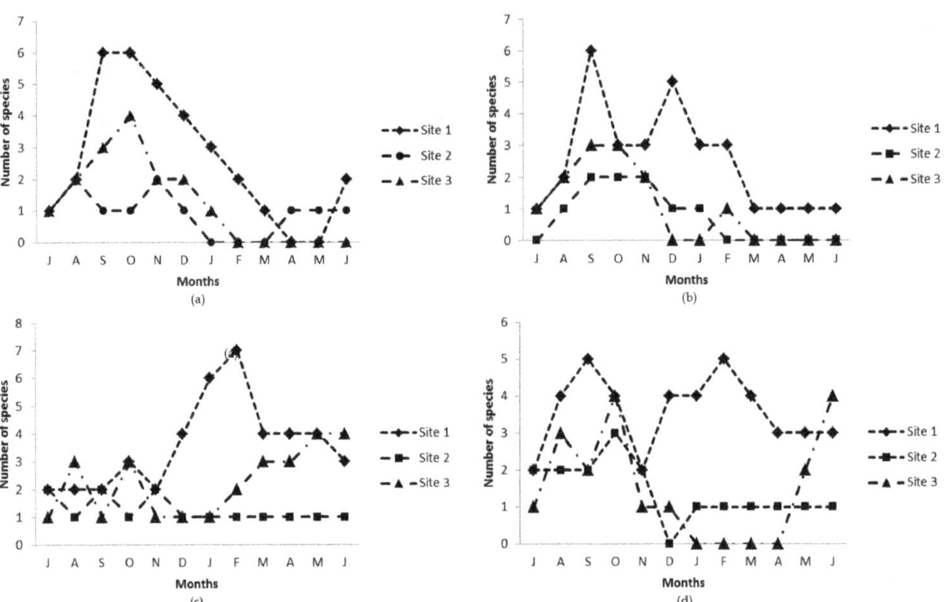

Fig. 4. Reproductive phenology of 20 tropical dry forest tree species in southeastern Brazil from July to June (12 months). Number of species displaying phenophase (a) flower buds, (b) anthesis, (c) green fruits and (d) ripe fruits at the three study sites in Minas Gerais (Site 1 = Parque Estadual da Mata Seca - Manga; Site 2 = CROS Reserve - Montes Claros; and Site 3 = Juramento Reservoir Reserve - Juramento).

Fruiting varied among sites, with 50% of the species fruiting in February at Site 1, 50% between July and November at Site 2 and 67% in May and June at Site 3 (Fig. 4c). The emergence of ripe fruits occurred from September to February at Site 1 for 50% of the species, in October at Site 2 for 75% of the species and at from October to June Site 3 for 67% of the tree species analyzed (Fig. 4d). Thus, unlike flowering, green fruits and ripe fruits appear at different times during the year, both in the dry and wet seasons. The timing and duration of reproduction may differ greatly among plant groups with different pollination or dispersal vectors (Bullock, 1995). The fruiting patterns observed in this study are likely associated with the seed dispersal mode of each species. In TDFs, fleshy fruits are abundant in the wet season and rare in the dry season, while anemochorous and autochorous fruits are almost inconspicuous in the wet season but abundant in the dry season (Frankie et al., 1974). In fact, five species fruiting in the dry season and five in the wet season. The majority of species that fruiting in the dry season (60%) was wind-dispersed (anemochorous), while the fruiting for most species in the wet season (80%) was dispersed by animals (zoochorous). Another interesting pattern, several species (45%) had fruits in both season, most of these (44.4%) belonging to autochorous syndrome. However, one species (*Dilodendron bipinnatum* Radlk.) not fruiting during the study period.

In tropical forests, the most frequent seed dispersal syndrome is zoochory, followed by anemochory and autochory (Howe & Smallwood, 1982). Nevertheless, wind dispersal is relatively common in deciduous forests (Van Schaik et al., 1993) and is largely present in TDFs (Bullock, 1995). In these forests, plants lose leaves seasonally, allowing for greater wind movement not only in the canopy but also in lower vegetation strata (Griz & Machado, 2001). Thus, reduction in physical obstacles resulting from greater deciduousness of tree species promotes the longer distance transport of anemochorous seeds (Janzen, 1967). Conversely, as activity of dispersers is higher in wet seasons (Smythe, 1970), the fruiting of most zoochorous species was concentrated in this season. It should also be noted that autochorous species (e.g., *Mimosa hostilis* Benth. and *Senna spectabilis* DC. H. S. Irwin & Barneby) with fruiting distributed throughout the year do not rely on any single dispersal agent and maximize seed dispersal over time (Piña-Rodrigues & Aguiar, 1993). Although dispersal is not primarily zoochorous, fruiting spread throughout the year provides an important food source to fauna. As food availability is limited during dry seasons, the dispersal of several species relies on secondary zoochorous dispersal.

3.2 Vegetative phenology

All of the species studied produced new leaves in the wet season. Leafing pick was recorded from October to February at Site 1, in October and November at Site 2 and from October to December at Site 3 (Fig. 5a). Conversely, leaf fall occurred from March to May for 90% of the species at Site 1, from July to October and in May and June for 100% of the species at Site 2 and from July to September and in May and June for 100% of the species at Site 3 (Fig. 5b). Full deciduousness was recorded for 90% of the species in August and September at Site 1, for all of the species from July to October at Site 2 and for 66% of the species in October at Site 3 (Fig. 5c). Thus, the onset of leafing at all sites occurred during the dry/wet season transition, whereas leaf fall was strongly associated with dry seasons, with the full absence of leaves observed for some species during these periods.

A widespread feature of tropical dry forests is the seasonal production of new leaves (Longman & Jenik, 1987), which is stimulated by increased water and light availability (Lieberman & Lieberman, 1984; Borchert, 1994). Besides, the existence of synchrony in leafing, in tropical dry forests, has been suggested as a strategy to avoid the emergence of young leaves at unfavorable times (van Schaik et al., 1993). This synchrony, observed in the present study, affects herbivore communities leading to fluctuations in population size, as most food resources are present for only a few months (Peters et al., 2001). Therefore, leaves produced during flushing peaks sustain less damage than those produced out of synchrony because herbivores are more easily satiated when food availability increases (Janzen, 1975).

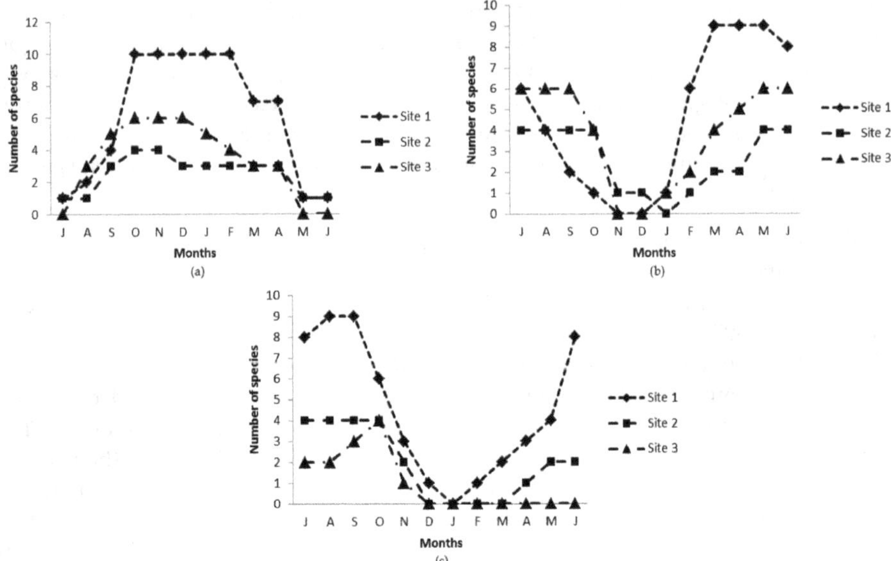

Fig. 5. Vegetative phenology of 20 tropical dry forest tree species in southeastern Brazil. Number of species displaying phenophase (a) leafing, (b) leaf fall and (c) full deciduousness at the three study sites in Minas Gerais (Site 1 = Parque Estadual da Mata Seca - Manga; Site 2 = CROS Reserve - Montes Claros; and Site 3 = Juramento Reservoir Reserve – Juramento).

Leaf fall, on the contrary, is likely a consequence of increased water stress and leaf ageing (Elliott et al., 2006). Leaf fall is intense during the early dry season because evapotranspiration increases, leading to greater nutrient removal from soil by roots (Morellato, 1992). At the end of the dry season, the air is drier and water availability decreases, nutrient uptake is inhibited and most species shed their leaves, reducing evapotranspiration (Morellato, 1992). Synchronous leaf fall, showed in this study, may be the result of various processes (Janzen, 1980), for example, for many deciduous species in tropical forests, the retention of leaves may be costly when water availability is reduced; besisde, as trees shed leaves with the arrival of the dry season, the need for any individual tree to retain its leaves for canopy space is reduced; and also, when green foliage disappears from the community, plant species without major anti-herbivore defenses will likely sustain greater damage to the rest of their leaves.

3.3 Phenophases versus climate variables

Despite the variations in phenological behavior among species, the climate variables influenced more the vegetative phenology than the reproductive phenology. There was a correlation between the production of flower buds and precipitation in 20% of the species and with temperature only in *C. odorata* and *D. bipinnatum*. Precipitation and temperature also affected the opening of flowers in three (15%) species. These data show that both climate variables affect the flowering of tropical dry forest species. According to Mantovani & Martins (1988), these environmental variables affect pollinator availability, which are directly related to the flowering patterns of plant species. These individuals likely share the same pollinator agents (mostly insects), which have high population densities at that time of year (Souza-Silva et al., 2001). Conversely, fruiting was only correlated with precipitation, and 20% of the species were influenced by this variable. Ripe fruits were negatively correlated with precipitation in *C. odorata* and *M. hostilis* and positively with temperature in *Ruprechtia fagifolia* Meisn. Water availability is an important factor in the production of fleshy, succulent fruits that are attractive to dispersers (Piña Rodrigues & Piratelli, 1993). Among the dispersal guilds of tree species, zoochory is positively correlated with precipitation (Bullock, 1995). Therefore, there is a connection between the production of energy-rich fruits and the attraction of effective dispersers. However, in the present study there is no association among dispersion guilds, climate variables and fruit production, because the association of precipitation and fruiting and ripe fruits phenophases occurred in different dispersion guilds.

Precipitation and temperature are important environmental variables that determine tree stem water stress in the dry tropics (Singh & Kushwaha, 2005). Precipitation positively influenced leafing in 55% of the species, while temperature positively influenced occurrence of this phenophase only in *Anadenanthera colubrina* (Vell.) Brenan. Leaf fall was also negatively correlated with precipitation in eight species (40%) and with temperature in *R. fagifolia*. There was a negative correlation between full deciduousness and precipitation in 50% of the species. Negative correlations between leaf fall and environmental variables indicate that this phenophase is associated with a reduction in temperature and precipitation (Nunes et al., 2005). Leaf fall during drier periods represents an economization of water by plants (Borchert et al., 2002), as it results in large nutrient losses. Full deciduousness is particularly evident in nutrient-rich soils, in which plants can retrieve nutrients that are lost in this process from the soil (Jaramillo & Sanford, 1995). Therefore, the soil conditions of TDFs enable plants to shed leaves because they grow in high fertility soils (Oliveira-Filho et al., 1998). A marked seasonality in vegetative phenological patterns has been observed in several studies focusing on dry forests (Lieberman, 1982; Bullock & Solís-Magallanes, 1990; Machado et al., 1997). Tree species in this study behaved consistently with this phenological pattern, which evolved through leaf fall to avoid water stress during dry periods, thus reducing transpiration and pronounced dehydration (Bullock & Solís-Magallanes, 1990).

Site	Species	Climate variable	Phenophase	Activity	
				r	p
1	*Commiphora leptophleos*	P	Full deciduousness	-0.63	0.03
1	*Jatropha mollissima*	P	Fruiting	0.58	0.05
		P	Leafing	0.67	0.02
		P	Full deciduousness	-0.72	0.01
1	*Mimosa hostilis*	P	Ripe fruits	-0.77	0.00
		P	Leafing	0.74	0.01
1	*Myracrodruon urundeuva*	P	Flower buds	-0.88	0.00
		P	Anthesis	-0.72	0.01
		P	Full deciduousness	-0.72	0.01
1	*Pereskia bahiensis*	P	Fruiting	0.61	0.04
		P	Full deciduousness	-0.75	0.01

1	*Pseudopiptadenia contorta*	P	Leafing	0.64	0.02
		P	Full deciduousness	-0.67	0.02
1	*Pilosocereus pachyladus*	P	Flower buds	-0.65	0.02
		P	Leafing	-0.63	0.03
1	*Sapium glandulosum*	P	Full deciduousness	-0.75	0.00
1	*Senna spectabilis*	P	Full deciduousness	-0.80	0.00
1	*Spondias tuberosa*	P	Full deciduousness	-0.74	0.01
2	*Cedrela odorata*	T	Flower buds	0.60	0.04
		P	Ripe fruits	-0.71	0.01
		P	Leafing	0.87	0.00
		P	Leaf fall	-0.86	0.00
		P	Full deciduousness	-0.86	0.00
2	*Dilodendron bipinnatum*	P	Flower buds	-0.67	0.02
		T	Flower buds	-0.82	0.00
		P	Leafing	0.79	0.00
		P	Leaf fall	-0.79	0.00
		P	Full deciduousness	-0.76	0.00
2	*Senegalia polyphylla*	P	Leafing	0.93	0.00
		P	Leaf fall	-0.82	0.00
2	*Rupretchia fagifolia*	T	Ripe fruits	0.67	0.02
		T	Leaf fall	-0.78	0.00
		P	Leaf fall	-0.74	0.01
3	*Anadenanthera colubrina*	T	Leafing	0.58	0.05
		P	Fruiting	-0.60	0.04
		P	Leaf fall	-0.65	0.02
3	*Astronium fraxinifolium*	T	Anthesis	-0.60	0.04
		P	Leafing	0.74	0.02
		P	Leaf fall	-0.68	0.02
3	*Chloroleucon tortum*	P	Fruiting	-0.59	0.04
		P	Leafing	0.59	0.05
		P	Leaf fall	-0.63	0.03
3	*Copaifera langsdorffii*	P	Leaf fall	-0.63	0.03
3	*Guazuma ulmifolia*	P	Flower buds	0.81	0.00
		P	Anthesis	0.63	0.03
		P	Leafing	0.70	0.01
		P	Leaf fall	-0.62	0.03
3	*Heteropterys byrsonimifolia*	P	Leafing	0.67	0.02

Table 2. Correlation between the climate variables precipitation (P) and temperature (T) and the reproductive and vegetative phenophase activity of 20 tropical dry forest tree species in southeastern Brazil (Site 1 = Parque Estadual da Mata Seca - Manga; Site 2 = CROS Reserve - Montes Claros; and Site 3 = Juramento Reservoir Reserve - Juramento).

3.4 Phenophases versus sites

Few phenological studies have compared populations located in sites under different environmental conditions (Frankie et al., 1974; Borchet, 1980). This experimental approach, however, is valid for understanding the extent of phenological variability as a survival strategy and to determine how environmental factors influence these patterns (Goulart et al., 2005). Although this study did not address variation between populations of the same species in different regions, phenological patterns generated by different populations may indicate local environmental variation, both in climate and soil conditions. In general, there

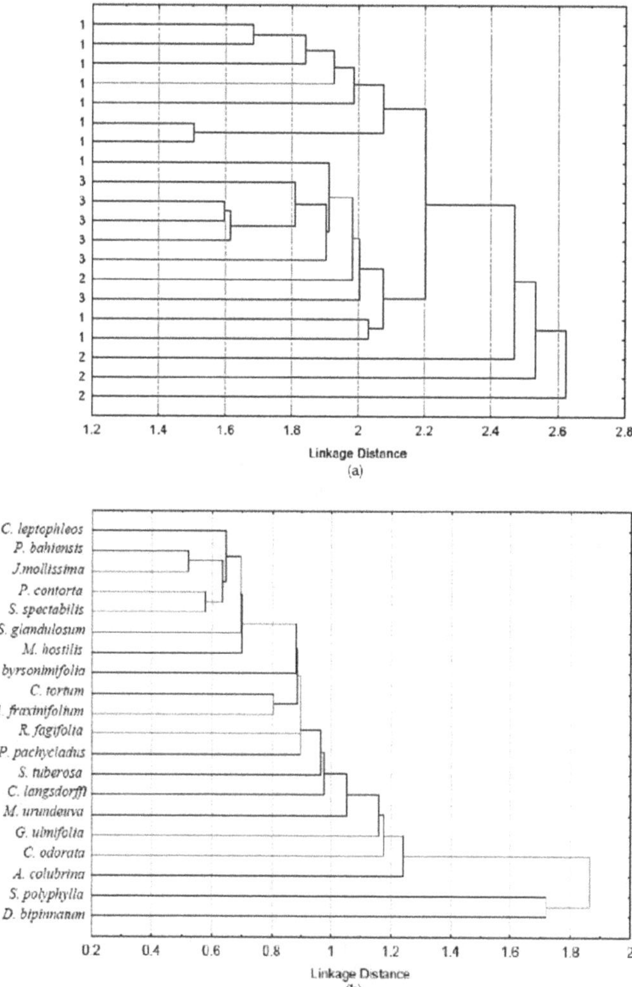

Fig. 6. Similarity dendrogram of the reproductive and vegetative phenophase activity of 20 tropical dry forest tree species in southeastern Brazil. (a) Species and (b) species location in the different study sites (Site 1 = Parque Estadual da Mata Seca - Manga; Site 2 = CROS Reserve - Montes Claros; and Site 3 = Juramento Reservoir Reserve – Juramento).

was a tendency of species to group by site (Fig. 6). This result suggests that there is a difference between phenological patterns of species in different sites and that these patterns are grouped by site. In fact, two large groups composed primarily of species from Sites 1 (PEMS, Manga) and 3 (Juramento Reservoir Reserve, Juramento) were formed in the cluster analysis, while species from Site 2 (CROS Reserve, Montes Claros) were not included in any of these groups.

The extent and intensity of seasonal drought may differ due to the geographical location of the study sites (Singh & Kushwaha, 2005). For instance, Site 1 is at a northern location, and this location determines its climatic differences to the other sites. According to the climatological norms of the study sites (Fig. 2), there are differences in the average temperature (Site 1 = 24.2°C, Site 2 = 23°C and Site 3 = 23°C) and precipitation (Site 1 = 871 mm, Site 2 = 1,000 mm and Site 3 = 1,000 mm) among sites, especially between Site 1 and the others. In addition to climate variables, soil conditions may also determine the phenological behavior of plants (Morellato, 1992). The differences between Sites 2 and 3 are likely caused by such factors. In fact, the TDF at Site 3 is growing in limestone outcrops (slope seasonal dry forest or dry calcareous forest), while at Site 2, the TDF is growing in flat soils. Therefore, although the proximity of the sites results in similar climates, differences in soil and topography may determine the occurrence of similar phenological patterns in different species. The water availability in the soil is considered to be the determining factor of leaf dynamics in the dry tropics (Borchert, 1994). The amount and timing of water that is available to plants on rocky outcrops is likely different from that of plants growing in deep soils.

4. Conclusion

This study showed different reproductive and vegetative phenological patterns among TDF tree species. Flowering and fruiting occurred in the wet season, whereas leaf fall was concentrated in the dry season. Nevertheless, fruiting seemed to be associated with the dispersal guilds of the different species. Temperature and precipitation were correlated with all phenophases analyzed, and the correlations were more robust with vegetative phenology. Species occurring in the same site showed phenological patterns that were more similar than species from different sites.

Climate and soil conditions may influence the occurrence of adaptations that result in similar phenological patterns among different species. Despite strong biotic interactions with processes such as pollination, propagule dispersal, predation and herbivory, the environmental factors that act directly upon the phenological behavior of species also affect these biotic agents and determine the occurrence of both vegetative and reproductive phenological patterns in TDF species.

5. Acknowledgments

The authors thank the Fundação de Amparo à Pesquisa de Minas Gerais (FAPEMIG, DEG 2393/03 and DEG 2583/05), the Collaborative Network Research Tropi-dry and the Interamerican Institute for Global Change Research (IAI, CRN II-21) for funding this research. Giovana R. Luz is grateful for a scholarship from FAPEMIG (10.186/08). The Instituto Nacional de Meteorologia (INMET), Empresa de Pesquisa Agropecuária de Minas Gerais (EPAMIG), Companhia de Saneamento de Minas Gerais (COPASA) and Universidade Estadual de Montes Claros (UNIMONTES) provided climatic data. We thank

UNIMONTES for logistical support and especially the students of the Laboratório de Ecologia e Propagação Vegetal for their field assistance. The Instituto Estadual de Florestas de Minas Gerais (IEF/MG), Construtora Rocha & Souza (CROS) and COPASA gave permission to work at the study sites and provided logistical support. We thank Carlos Magno Santos Clemente for preparing the map of the study area and Saimo Rebleth de Souza for helping with the graphs.

6. References

Antunes, F.Z. (1994). Caracterização climática, *Informe Agropecuário* 17: 15-19.

Barbosa, D.C.A., Barbosa, M.C.A. & Lima, L.C.M. (2005). Fenologia de espécies lenhosas da caatinga, *in* Leal, I.R., Tabarelli, M. & Silva, J.M.C. (ed.), *Ecologia e conservação da caatinga*, Editora Universitária da UFPE, Recife, pp. 657-694.

Barbosa, D.C.A. (2005). Estratégia de germinação e crescimento de espécies lenhosas da caatinga com germinação rápida, *in* Leal, I.R., Tabarelli, M. & Silva, J.M.C. (ed.), *Ecologia e conservação da caatinga*, Editora Universitária da UFPE, Recife, pp.625-656.

Bencke, C.S.C. & Morellato, L.P.C. (2002). Estudo comparativo da fenologia de nove espécies arbóreas em três tipos de floresta atlântica no sudeste do Brasil, *Revista Brasileira de Botânica* 25(2): 237-248.

Borchet, R. (1980). Phenology and ecophysiology of tropical trees: *Erythrina poeppigiana* O. F. Cook, *Ecology* 61: 1065-1074.

Borchert, R. (1994). Soil and stem water storage determine phenology and distribution of tropical dry forest trees, *Ecology* 75(5): 1437-1449.

Borchert, R., Rivera, G. & Hagnauer, W. (2002). Modification of Vegetative Phenology in a Tropical Semi-deciduous Forest by Abnormal Drought and Rain, *Biotropica* 34(1): 27-39.

Borchert, R., Meyer, S.A., Felger, R.S. & Porter-Bolland, L. (2004). Environmental control of flowering periodicity in Costa Rican and Mexican tropical dry forests. *Global Ecology and Biogeography* 13: 409–425.

Britton, N.L. & Rose, J.N. (1963). *The Cactaceae*, General Publishing Company ,New York.

Bullock, S.H. (1995). Plant reproduction in neotropical dry forests, *in* Bullock, S.H., Mooney, H.A. & Medina, E. (ed.), *Seasonally dry tropical forests*, Cambridge University Press, Cambridge, pp. 277-303.

Bullock, S.H. & Solis-Magallanes, J.A. (1990). Phenology of canopy trees of a tropical deciduous forest in México, *Biotropica* 22: 22-35.

Burnham, R.J. (1997). Stand characteristics and leaf litter composition of a dry forest hectare in Santa Rosa National Park, Costa Rica, *Biotropica* 29: 384-395.

Elliott, S., Baker, J.P. & Borchert, R. (2006). Leaf flushing during the dry season: the paradox of Asian monsoon forests, *Global Ecology and Biogeography* 15: 248-257.

Espírito-Santo, M.M., Fagundes, M., Sevilha, A.C., Scariot, A.O., Azofeifa, G.A.S, Noronha, S.E. & Fernandes, G.W. (2008). Florestas estacionais deciduais brasileiras: distribuição e estado de conservação, *MG Biota* 1(2): 5-13.

Fournier, L.A.O. (1974). Un método cuantitativo para la medición de características fenológicas en árboles, *Turrialba* 24(4): 422-423.

Frankie. G.W., Baker, H.G. & Opler, P.A. (1974). Comparative phonological studies of trees in tropical wet and dry forests in the lowlands Costa Rica, *Journal of Ecology* 62: 881-919.

Goulard, M.F., Lemos-Filho, J.P. & Lovato, M.B. (2005). Phenological variation within and among populations of *Plathymenia reticulata* in brazilian Cerrado, the Atlantic Forest and transitional sites, *Annals of Botany* 96: 445-455.

Griz, L.M.S. & Machado, I.C.S. (2001). Fruiting phenology and seed dispersal syndromes in Caatinga, a tropical dry forest in the northeast of Brazil, *Journal of Tropical Ecology* 17: 303 - 321.

Howe, H.F. & Smallwood, J. (1982). Ecology of seed dispersal, *Annual Review of Ecology and Systematics* 13: 201-228.

Janzen, D.H. (1980). *Ecologia vegetal nos trópicos*, EPU - Editora da Universidade de São Paulo, São Paulo.

Janzen, D.H. (1975). Behaviour of *Hymenaea courbaril* when its predispersal seed predator is absent, Science 189:145-147.

Janzen, D.H. (1967). Synchronization of sexual reproduction of trees within the Dry Season in Central America, *Evolution* 21: 620-637.

Jaramillo, V.J. & Sanford Jr., R. (1995). Nutrient cycling in tropical deciduous forests, *in* Bullock, S.H., Mooney, H.A. & Medina E. (ed.), *Seasonally dry tropical forests*, Cambridge University Press, Cambridge, pp. 346–361.

Justiniano, M.J. & Fredericksen, T.S. (2000). Phenology of Tree Species in Bolivian Dry Forests, *Biotropica* 32: 276-281.

Lieberman, D. (1982). Seasonality and Phenology in a Dry Tropical Forest in Ghana, *The Journal of Ecology* 70: 791-806.

Lieberman, D. & Lieberman, M. (1984). The causes and consequences of synchronous flushing in a dry tropical forest, *Biotropica* 16: 193-201.

Lima, A. (2001). *Aspectos jurídicos da proteção da Mata Atlântica*, Instituto Socioambiental, São Paulo.

Lobo, J.A., Quesada, M., Stoner, K.E., Fuchs, E.J., Herrerias-Diego, Y., Rojas, J. & Saborio, G. (2003). Fators affecting phenological patterns of Bombacaceae trees in seasonal forests in Costa Rica and México, *American Journal of Botany* 90(7): 1054–1063.

Longman, K.A. & Jenik, J. (1987). *Tropical forest and its enviroments*. Longman Scientific & Technical, New York.

Ludwig, J.A. & Reynolds, J.F. (1988). *Statistical ecology: a primer on methods and computing*, John Wiley & Sons, New York.

Maass, J.M. (1995). Conversion of tropical dry forest to pasture and agriculture, *in* Bullock, S.H., Mooney, H.A. & Medina, E. (ed.), *Seasonally Dry Tropical Forests*, Cambridge University Press, New York, pp. 399–422.

Machado, I.C.S., Barros, L.M. & Sampaio, E.V.S.B. (1997). Phenology of Caatinga species at Serra Talhada, PE, northeastern Brasil, *Biotropica* 29: 57-68.

Madeira, B.G., Espírito-Santo, M.M., D'Ângelo-Neto, S., Nunes, Y.R.F., Sánchez-Azofeifa, G.A., Fernandes, G.W. & Quesada, M. (2009). Changes in tree and liana communities along a successional gradient in a tropical dry forest in south-eastern Brazil, *Plant Ecology* 201: 291-304.

Maia, G.N. (2004). Catingueira, *in* Maia, G.N. (ed.), *Caatinga: árvores e arbustos e suas utilidades*. Leitura e Arte, São Paulo, pp. 159-169.

Mantovani, W. & Martins, F.R. (1988). Variações fenológicas das espécies do cerrado da Reserva Biológica de Mogi Guaçu, Estado de São Paulo, *Revista Brasileira de Botânica* 11: 101-112.

Maragon, G.P., Cruz, A.F., Barbosa, W.B., Loureiro, G.H. & Holanda, A.C. (2010). Dispersão de sementes de uma comunidade arbórea de um remanescente de Mata Atlântica, município de Bonito, PE. *Revista Verde* 5(5): 80-87.

Miles, L., Newton, A.C., Defries, R.S., Ravilious, C., May, I., Blyth, S., Kapos, V. & Gordon, J.E. (2006). A global overview of the conservation status of tropical dry forests, *Journal of Biogeography* 33(3): 491–505.

Moreira, A.L., Queiroz, E.P. & Pigozzo, C.M. (2009). Síndromes de dispersão de frutos e sementes do fragmento urbano (19° BC) de Mata Atlântica, Cabula, Salvador, Bahia, *Candombá - Revista Virtual* 5(1): 13-25.

Morellato, L.P.C., Rodrigues, R.R., Leitão Filho, H.F. & Joly, C.A. (1989). Estudo comparativo da fenologia de espécies arbóreas de floresta de altitude e floresta mesófila semidecídua na serra do Japi, Jundiaí, São Paulo, *Revista Brasileira de Botânica* 12(1/2): 85-98.

Morellato, L.P.C. (1992). Sazonalidade e dinâmica de ecossistemas florestais na Serra do Japi, *in* Morellato, L.P.C. (ed.), *História natural da Serra do Japi: ecologia e preservação de uma área florestal no Sudeste do Brasil,* Universidade Estadual de Campinas/Fundação de Amparo à Pesquisa do Estado de São Paulo, Campinas, pp. 98-110.

Morellato, L.P.C. & Leitão-Filho, H.F. (1992). Padrões de frutificação e dispersão na Serra do Japi, *in* Morellato, L.P.C. (ed.), *História natural da Serra do Japi: ecologia e preservação de uma área florestal no sudeste do Brasil,* Unicamp, Campinas, pp. 112-138.

Murphy, P.G. & Lugo, A.E. (1986). Ecology of tropical dry forest. *Annual Review of Ecology and Systematics* 17(1): 67-88.

Nascimento, A.R.T., Felfili, J.M. & Meirelles, E.M. (2004). Florística e estrutura da comunidade arbórea de um remanescente de Floresta Estacional Decidual de encosta, Monte Alegre, GO, Brasil, *Acta Botanica Brasilica* 18: 659-669.

Nimer, E. & Brandão, A.M.P.M. (1989). *Balanço hídrico e clima da região dos cerrados,* IBGE, Rio de Janeiro.

Nunes, Y.R.F., Fagundes, M., Almeida, H.S., Veloso, M.D.M. (2008). Aspectos ecológicos de Aroeira (*Myracrodruon urundeuva* Allemão - Anacardiaceae): fenologia e germinação de sementes, *Revista Árvore* 32(2): 233-243.

Nunes, Y.R.F., Fagundes, M., Santos, R.M., Domingues, E.B.S., Almeida, H.S. & Gonzaga, A.P. (2005). Atividades fenológicas de *Guazuma ulmifolia* Lam. (Malvaceae) em uma floresta estacional decidual no norte de Minas Gerais, *Lundiana* 6(2): 99-105.

Oliveira-Filho, A.T, Curi, N., Vilela, E.A., & Carvalho, D.A. (1998). Effects of canopy gaps, topography and soils on the distribution of woody species in a central Brazilian deciduous dry forest, *Biotropica* 30: 362–375.

Pedroni, F., Sanchez, M. & Santos, F.A.M. (2002). Fenologia da copaíba (*Copaifera langsdorffii* Desf. - Leguminosae, Caesalpinioideae) em uma floresta semidecídua no sudeste do Brasil, *Revista Brasileira de Botânica* 25(2): 183-194.

Pereira, T.S., Costa, M.L.M.N., Moraes, L.F.D. & Luchiari, C. (2008). Fenologia de espécies arbóreas em Floresta Atlântica da Reserva Biológica de Poço das Antas, Rio de Janeiro, Brasil, *Iheringia* Sér. Bot. 63(2): 329-339.

Peters, P.J., Read, J. & Sanson, G.D. (2001). Variation in the guild composition of herbivorous insect assemblages among co-occuring plant species, *Austral Ecology* 26: 385-399.

Pezzini, F.F. (2008). Fenologia e características reprodutivas em comunidades arbóreas de três estágios sucessionais em Floresta Estacional Decidual do norte de Minas Gerais. *Dissertação de Mestrado,* Universidade Federal de Minas Gerais, Belo Horizonte, 130p.

Piña-Rodrigues, F.C.M. & Aguiar, I.B. (1993). Maturação e Dispersão de Sementes, *in* Aguiar, I.B., Piña-Rodrigues, F.C.M., Figliolia, M.B. (ed.), *Sementes Florestais Tropicais,* Abrates, Brasília, pp. 215-274.

Piña-Rodrigues, F.C.M. & Piratelli, A.J. (1993). Aspectos ecológicos da produção de sementes, *in* Aguiar, I.B., Piña-Rodrigues, F.C.M. & Figliolia, M.B. *Sementes florestais tropicais,* Abrates, Brasília, pp. 47-81.

Ribeiro, J.F. & Walter, B.M.T. (1998). Fitofisionomias do bioma cerrado, *in* Sano, S.M. & Almdeida, S.P. (ed.), *Cerrado: ambiente e flora*, EMBRAPA-CPAC, Planaltina, pp.89-166.

Rizzini, C.T. (1997). *Tratado de fitogeografia do Brasil*. Âmbito Cultural Edições Ltda, Rio de Janeiro.

Sanchez-Azofeifa, G.A., Quesada, M., Rodriguez, J.P., Nassar, J.M., Stoner, K.E. Castillo, A., Garvin, T., Zent, E.L., Calvo-Alvarado, J.C. & Kalacska, M.E.R. (2005). Research priorities for Neotropical Dry Forests, *Biotropica* 37(4): 477-485.

Santos, M.J., Machado, I.C. & Lopes, A.V. (2005). Biologia reprodutiva de duas espécies de *Jatropha* L. (Euphorbiaceae) em Caatinga, Nordeste do Brasil, *Revista Brasileira de Botânica* 28(2): 361-373.

Scariot, A. & Sevilha, A.C. (2005). Biodiversidade, Estrutura e Conservacão de Florestas Estacionais Deciduais no Cerrado, *in* Scariot, A., Felfili, J.M. & Sousa-Silva, J.C. (ed.), *Cerrado: Ecologia, Biodiversidade e Conservação*, Ministério do Meio Ambiente, Brasília, pp. 123–139.

Silva, M.C.N.A. & Rodal, M.J.N. (2009). Padrões das síndromes de dispersão de plantas em áreas com diferentes graus de pluviosidade, PE, Brasil, *Acta Botanica Brasilica* 23(4): 1040-1047.

Silva, L.A. & Scariot, A. (2004). Composição e estrutura da comunidade arbórea de uma floresta estacional decidual sobre o afloramento calcário no Brasil Central, *Revista Árvore* 28: 69-75.

Singh, K.P. & Kushwaha, C.P. (2005). Emerging paradigms of tree phenology in dry tropics. *Current Science* 89(6): 964-975.

Smythe, N. (1970). Relationships between fruiting seasons and seed dispersal methods in a Neotropical forest, *American Naturalist* 104: 25-35.

Souza-Silva, M., Fontenelle, J.C.R., Martins, R.P. (2001). Seasonal abundance anda species composition of flower-visiting flies, *Neotropical Entomology* 30(3): 351-359.

Spina, P.A., Ferreira, W.M. & Leitão-Filho, H.M. (2001). Floração, frutificação e síndromes de dispersão de uma comunidade de floresta de brejo na região de Campinas (SP), *Acta Botanica Brasilica* 15: 349-368.

Stefanello, D., Fernandes-Bulhão, C. & Martins, S.V. (2009). Síndromes de dispersão de sementes em três trechos de vegetação ciliar (nascente, meio e foz) ao longo do rio Pindaíba, MT, *Revista Árvore* 33(6): 1051-1061.

Thornthwaite, C.W. (1948). An approach toward a rational classification of climate, *Geographical Review* 38(1): 55-94.

van der Pijl, L. (1982). *Principles of dispersal in higher plants*, Springer-Verlag, Berlim.

van Schaik, C.P., Terborgh, J.W. & Wright, S.J. (1993). The phenology of tropical forest: adaptative significance and consequences of consumers, *Annual Review of Ecology and Systematics* 24: 353-377.

Veloso, H.P., Rangel Filho, A.L.R. & Lima, J.C.A. (1991). *Classificação da vegetação brasileira adaptada a um sistema universal*, Instituto Brasileiro de Geografia e Estatística, Rio de Janeiro.

Wikander, T. (1984). Mecanismos de dispersión de diasporas de una selva decidua en Venezuela, *Biotropica* 16: 276-283.

Zalamea, M. & González, G. (2008). Leaffall phenology in a subtropical wet forest in Puerto Rico: from species to community patterns, *Biotropica* 40(3): 295-304.

Zar, J.H. (1996). *Biostatistical analysis*, Prentice-Hall, New Jersey.

10

Flowering Phenology in Teak Seed Orchards – Genetic Variation, Measurement and Implications to Seed Orchard Fertility

Vasudeva Ramesh[1] and Rajesh Gunaga[2]
[1]*University of Agricultural Sciences, Dharwad, Karnataka,*
[2]*Dr. B.S.K. Krishi Vidyapeeth, Dapoli, Maharashtra,*
India

1. Introduction

Teak (*Tectona grandis.* Linn. f.) is an important timber species preferred all over the world because of its versatile range of uses (Katwal, 2005). Its matchless timber properties include strength, workability, attractiveness, lightness, resistance to termite, fungus and weather, as well as seasoning capacity without splitting or cracking. Teak occurs in natural forests between 9° to 26° N latitude and 73° to 104° E longitude, which includes southern and central India, Myanmar, Laos People's Democratic Republic and northern Thailand (White, 1991). The species is a dominant component of diverse forest types. For instance in India, teak forests are spread over a large geographic area of 8.9 million ha which range from very-dry to very-moist types. As a consequence, natural populations exhibit great genetic variability for economically important traits (Katwal, 2005). In fact the Indian subcontinent is considered the center of diversity for teak because of the huge genetic variation for economically important traits such as bole form, timber quality, biochemical traits and others (Anmol Kumar et al., 1997). Further teak is emerging as one of the predominant plantation species in the Indian subcontinent as well as in 30 other countries of tropical Asia, Africa and Latin America. Though teak plantations account for 5-8 per cent of the total forest area in the tropics (Ball et al. 1999), about 90 per cent of the quality hardwood plantations for timber production belongs to only teak (Granger, 1998). As per 1990 data, globally, area under teak plantations was 2.2 million ha, with 94 per cent in Tropical Asia, especially in India and Indonesia, followed by Tropical Africa (about 4.5 %) and remaining area in Tropical America (FAO, 1995). Presently about 1.5 million ha of teak plantations exist in India and around 50,000 ha are raised annually (Subramanian et al., 2000). This has resulted in a huge demand for quality planting stocks in astronomical quantities.

Genetic improvement of teak in India, started in the year 1954, has focused mainly on identifying phenotypically superior trees from diverse growing regions and deploying them as vegetatively propagated clones in seed orchards (Gunaga & Vasudeva, 2005). Seed orchard is essentially a collection of phenotypically superior and diverse individuals of a species, which is silviculturally managed to produce genetically superior seed crop through the process of open pollination (Askew, 1986). Establishment of seed orchards using superior clones from diverse regions has been an important strategy of genetic improvement programme of teak. It is assumed that the offspring developed through a random mating among the superior types would also be genetically superior. The main purpose of establishing seed orchard is to mass-produce such genetically superior seeds, which are easily accessible and collectable. Thus, seed orchards form an important link between on-going tree improvement programme and commercial planting activity. The establishment of seed orchards would be a genetic dead-end unless full potential of an orchard is realized through harvesting of genetically improved seed crops. Unfortunately, low fruit production in clonal seed orchards has been an important limitation in teak improvement programmes of India and worldwide (Gunaga & Vasudeva, 2005; Indira, 2005).

1.1 Importance of flowering in seed orchards and the basic assumptions

The assumption made while establishing a seed orchard is that diverse genotypes are highly compatible and completely overlap in their flowering phenology and the process would lead to big genetic gain. Synchronous flowering among different clones helps to achieve random mating (panmixis) and hence good seed output. Further, it also reduces the foreign pollen contribution in the process of pollination, if any. In general the following requirements are assumed to be fulfilled in an idealized seed orchard (Askew, 1986, Vasudeva et al. 1999)

- The orchard is completely isolated from the influence of undesirable pollen from outside
- Natural self-pollination occurs only in insignificant amounts
- Clones are equally productive in male and female flowers (in case of unisexual species)
- Pollen flight and female flower receptivity coincides especially in monoecious types
- The gamete contributions of the parents are in the same proportions as they were in the orchard
- The pollen contributions to the seed crop are either uniform for all parents or that they are in similar proportions to the seed crop.

Unfortunately, the above assumptions have not been tested among established clonal seed orchards. It is well known that in a seed orchard, the levels of variation for reproductive traits among the constituent genotypes influence fruit production. Further, the variation in flowering behavior among the clones influence the extent of gene exchange between clones and consequently the genetic composition of the seeds produced (Gunaga & Vasudeva, 2002; Vasudeva et al., 2005). Hence, understanding variation for reproductive phenology is fundamental to the successful operation of any seed orchard. Since superior genotypes identified from diverse regions are used in a clonal seed orchard (CSO), understanding the flowering phenology of the constituent clones becomes imperative to achieve maximum synchrony. A large of number of reports is available for temperate species, which document asynchronous flowering among the clones in a seed orchard, especially among monoecious species (Gunaga, 2000; Vasudeva et al., 2001). Any factor that affects the seed production in seed orchards would be a hindrance for logical end of tree improvement programmes (Griffin, 1984). Therefore any variation in flowering phenology within a seed orchard can potentially alter the quantity as well as genetic quality of fruits (Sedgley & Griffin, 1989). Although, there seems to be a direct association between the extent of flowering synchrony among the constituent clones and the fruit production in a seed orchard, surprisingly, there are very few reports in India that focus on these issues. A few authors, however, have also identified this gap of information earlier (Subramanian et al., 1994; Nagarajan et al., 1996; Radhamani et al., 1998, Gunaga & Vasudeva, 2002). Anmol Kumar (1992) has reported that flowering among local clones belonging to the Maharashtra province of India was early as compared to the clones from other parts (provenance) of India. Rawat et al. (1992) have reported that flowering in teak seed orchard established at New forest, Dehra Dun, Northern India was asynchronous and recognized that these patterns may cause lower fruit production in the seed orchard. A few reports on these lines are also available in south east Asia. For instance, Palupi & Owens (1998) have shown asynchronous flowering among clones of teak at clonal seed orchard established at East Java of Malaysia. A lack of a suitable tool for an objective measurement of flowering synchrony among different clones has also contributed to the negligence!! Hence there is a need for development of objective criteria to evaluate flowering phenological events. In this chapter we firstly bring out patterns genetic variation for reproductive phenology in teak seed orchards; review a novel method of computing the overlap of flowering among individuals in an orchard and discuss its implications to seed orchard fertility from a management perspective.

2. Flowering phenology in teak and its genetic control

Flowering phenological patterns are most diverse and least understood in the tropics (Corlett, 1998). Pattern of flowering is well studied in most of the temperate and sub-temperate species like radiata pine (Griffin,1984), loblolly pine (Askew, 1986), douglas fir (EL-Kassaby & Askew, 1991), black pine (Matziris, 1994) and *Eucalyptus citriodora* (Verma et al., 1989). Studies of the reproductive biology and pollination in teak (Hedegrat, 1973; Kaosa-ard, 1991) have shown that teak is predominantly an out crossed species and pollinated by insects. Isozyme analysis of seeds from different clones by Kjaer & Suangtho (1995) has confirmed these results. Self-incompatibility and short stigmatic receptive period (which ranges for just about few hours in one morning) are the main reasons for higher out crossing rates (Hedegrat, 1973).

2.1 Variation for flowering phenology among teak clones

Despite extensive planting programmes and the importance of increasing fruit production in seed orchards, little is known about the clonal variation for reproductive biology of teak. To use floral traits as a criterion while selecting the clones, it is essential to understand their genetic control. Hence, in order to achieve better genetic gain, it is imperative to assess variation and genetic control of the floral and fecundity traits in an orchard. However, in general, reproductive traits have been consistently ignored while selecting plus trees of teak as well as while upgrading the existing seed orchards. There is only one report on the extent of genetic control of floral features in teak (Hanumatha et al., 2001). Perhaps the only study that focuses on estimating the genetic parameters of flowering phenophases in teak was published by Gunaga & Vasudeva (2005). In their study the authors have assessed the clonal variation for flowering phenology and estimated the extent of genetic control over flowering phenophases. Phenophases of 407 rametes of 24 teak clones arising from diverse regions of southern India were monitored every week in a 19-year-old-teak CSO of southern India. The passport data on the clones used in the study are provided in the Table 1. The time taken for commencement of flower buds, flowering and peak flowering were calculated as the number of days from 1st January to the date of their first appearance on every tree. Peak flowering was defined as the time when approximately 75 per cent of the flowers on a tree are in bloom. Duration of all phenophases was considered as total number of days of respective events.

Sl. No	Teak Clone I.D.	Provenance of the clone	Latitude (North)	Longitude (east)	Altitude (meters a msl)
1	MYHD1	North	15° 17′	74° 38′	573
2	MYHD2	North	15° 17′	74° 38′	573
3	MYHD3	North	15° 17′	74° 38′	573
4	MYHD4	North	15° 17′	74° 38′	573
5	MYHV1	North	15° 06′	74° 36	570
6	MYHV3	North	15° 06′	74° 36	570
7	MYHV4	North	15° 06′	74° 36	570
8	MYHV5	North	15° 06′	74° 36	570
9	MYHV6	North	15° 06′	74° 36	570
10	MYHV7	North	15° 06′	74° 36	570
11	MyHaK1	North	15° 11′	74° 41′	500
12	MyHaK2	North	15° 11′	74° 41′	500
13	MyHaK3	North	15° 11′	74° 41′	500
14	MYSA1	Central	13° 53′	74° 28	571
15	MYSA2	Central	13° 53′	74° 28	571
16	MYSS2	Central	13° 53′	74° 28	571
17	MyBL1	Central	13° 40′	75° 39′	571
18	MYHuT1	South	12° 13′	76° 00′	850
19	MYHuT2	South	12° 13′	76° 00′	850
20	MYHuT3	South	12° 13′	76° 00′	850
21	MYHuT6	South	12° 13′	76° 00′	850
22	MYHuT7	South	12° 13′	76° 00′	850
23	MYHuT8	South	12° 13′	76° 00′	850
24	MyMK3	South	11° 55′	76° 11′	690

Table 1. Passport data of teak clones of a Clonal Seed Orchard (CSO) in Karnataka, southern India.

2.2 Influence of provenance on flowering phenology among teak clones

Generally in South Indian conditions, the period of 'flowering' to 'fruit maturity' in teak takes about nine months. Flower buds appear during April, coinciding with the on-set of monsoon showers, and continue through August. However, Palupi & Owens (1998) have shown that among the teak clones of East Java, flower initiation started from January through middle of February, while the initiation of peak flowering varied from mid January to March first week. Blooming in a clonal seed orchard of south India started in the second week of May and continued until fourth week of September, with peak flowering from fourth week of May to second week of September (Gunaga & Vasudeva, 2005). Thus huge variations for flowering phenology naturally occur in the populations of Teak.

Reports from a wide range of commercial tree species of temperate regions has shown that the variation for time of flowering is genetically based (Griffin, 1984 in Radiata pine; Askew, 1986 in loblolly pine; EL-Kassaby & Askew, 1991 in Douglas fir; Matziris, 1994 in Black pine and Verma et al., 1999 in *Eucalyptus citriodora*). Similarly, as shown by Gunaga & Vasudeva (2005), large genetic variation occurs among teak clones for different flowering phenophases (Figure 1). A few other authors have also reported clonal variation in flowering pattern of teak. Anmol Kumar (1992) reported that flowering among local clones belonging to Maharashtra, Central India, was early when compared to the clones from other states. Drastic variation for flowering among clones in a 10-year old model teak seed orchard in Dehra Dun, Northern India was reported by Rawat et al. (1992).

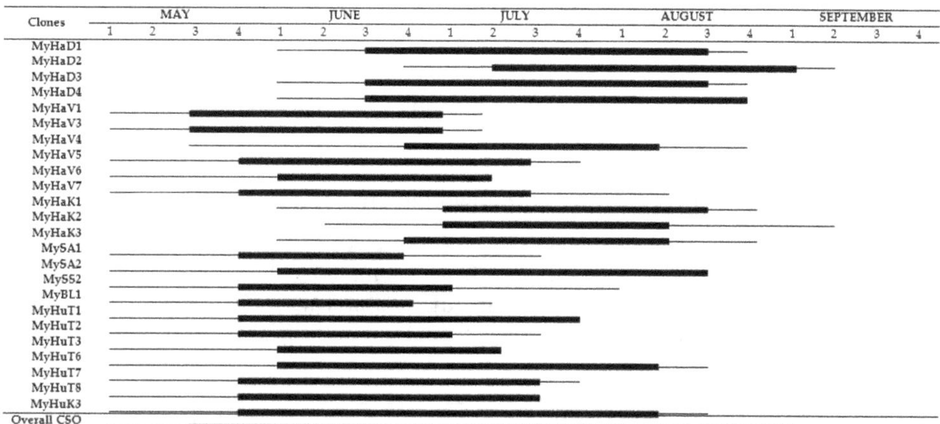

Fig. 1. Phenogram showing clonal variation for duration of flowering (thin line) and peak flowering (bold line) in a clonal seed orchard (CSO) of teak in southern India.

Provenance	Flowering initiation		Peak flowering initiation		Duration of flowering		Duration of peak flowering	
	Mean	**±SD**	**Mean**	**±SD**	**Mean**	**±SD**	**Mean**	**±SD**
Northern	184.78	24.04	207.11	21.78	51.79	15.89	15.57	7.56
Central	147.50	10.15	174.78	13.55	57.09	12.11	16.24	8.9
Southern	152.18	15.05	178.27	18.29	54.75	11.58	17.23	9.26
F Ratio	114.85		97.30		3.05		1.24	
P. level	<0.001		<0.01		NS		NS	
CV (%)	12.25		10.26		27.86		50.56	

Table 2. Provenance variation for time of budding, flowering, peak flowering (expressed as number of days from 1st January 1999) and duration of flowering and peak flowering (number of days) in the Clonal Seed Orchard of teak. NS= non-significant. CV = Coefficient of variation.

Interestingly, Gunaga & Vasudeva (2005) have also shown strong provenance influence on flowering phenology among the clones of a CSO. Teak clones from central and southern Indian provenances were early in commencement of flower buds, flowers, and peak flowering (by about 30-40 days) when compared to clones from northern provenance (Table 2; Figure 2). In their study, clearly two peaks for flowering among teak clones were observed -the first, during May to June corresponding mainly to the Central and Southern clones; the second, during July corresponding to those from northern clones (Figure 2). However, duration of flowering and peak flowering among clones of different provenances did not differ. Clearly, a significant proportion of rametes of clones from southern and central regions escaped peak rainfall period in the orchard when compared to those from northern region (Figure 2). Perhaps this is the first empirical evidence among the CSOs of teak in India confirms that clones of diverse origin do not contribute equally to the resultant seed crop. Such genetically based geographical variation in phenology is shown to occur in several plant species. For instance, Yanninen et al. (1988) have shown that northern and southern clones of *Viccinium myrtillus* differ in their spring and autumn phenology. Female flower production in loblolly pine varied between Texas and Arkansas sources (Richmond & Mc Kinley, 1986). Weinstein (1989) reported that the development of flower and the vegetative bud in *Pinus halepensis* and *P. brutia* showed differences among altitudinal races.

2.3 Estimation of genetic parameters for flowering phenophases

Though flowering phenology has been the most conserved life history trait, estimates of genetic parameters such as heritability has been attempted only for few timber species. Gunaga & Vasudeva (2005) have adopted an Analysis of Variance (ANOVA) approach to decompose total variability in flowering phenology into genetic and environmental components using the following model:

Sl.no	Source	d.f.	Expected mean square
1	Between clones	(c-1)	$\sigma^2 + n_0 \sigma^2_C$
2	Error	Σ (n_i-1)	σ^2

Where,

c = number of clones
σ^2 = sum of squares due to error
σ^2_c = sum of squares due to clones (genotypic variance)
n_0 = $\Sigma n_i - \{\Sigma (n_i-1)/ n_i\}$ as suggested by Kedharnath (1982)
n_i = number of rametes of ith clone

From this analysis, following genetic parameters were estimated:

Heritability

Broad sense heritability $(H_1^2) = \frac{\sigma^2_c}{(\sigma^2_c + \sigma^2)}$ (Individual ramete basis)

Broad sense heritability $(H_2^2) = \frac{\sigma^2_c}{(\sigma^2_c + \sigma^2 / n_0)}$ (Clonal mean basis)

following Matziris (1994).

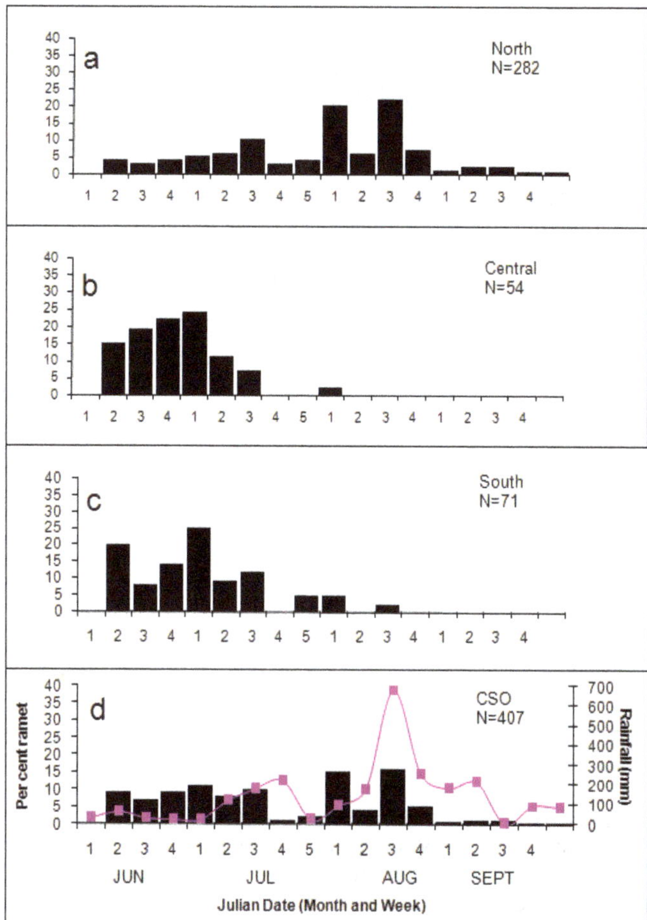

Fig. 2. Distribution of teak ramets of 24 clones from different provenances (a-c) and from overall CSO (d) with respect to initiation of flowering across time. Y- axis represents the per cent ramets in each class. The red line in panel (d) suggests the rainfall pattern in the CSO located in southern India. N= total number of ramets.

Standard error for heritability on individual ramete basis was calculated as follows (Kedharnath, 1982):

$$SEH_1^2 = \frac{2(n-1)\left(1-H_1^2\right)^2\left\{1+(n_0-1)H_1^2\right\}}{n_0^2(n-c)(c-1)}$$

Where,

c = number of clones

n_0 = Coefficient of σ^2_c

n = Total number of observation

The authors have shpwn that among the different flowering phenophases commencement of flower buds, flowering and that of peak flowering recorded high heritabilities on individual tree basis (h^2_1= 0.734, 0.637 and 0.709, respectively; Table 4). The values on clonal mean basis were much higher. The low values of standard errors of h^2_1 indicate that these estimates were reliable. However, the duration of flowering and peak flowering were strongly influenced by environmental factors and hence weakly inherited (h^2_1 = 0.03 and 0.004, respectively). These results are in complete agreement with Matziris (1994) who reported that in black pine, the heritabilities on individual tree basis were high for the initiation of male flowering and female receptivity, while that for the duration of flowering stages were lower. Matziris (1997) reported that heritability for female flowering initiation in Aleppo Pine was high.

Variable	Variance		h^2_1	±SE (h^2_1)	h^2_2
	σ^2_P	σ^2_c			
Commencement of flower buds	676.09	496.57	0.734	0.0171	0.979
Commencement of flowering	830.39	529.21	0.637	0.0218	0.967
Commencement of peak flowering	629.97	446.88	0.709	0.0184	0.976
Duration of Flowering	224.22	6.80	0.030	0.0213	0.344
Duration of peak flowering	65.09	0.27	0.004	0.0187	0.064

σ^2_P = Phenotypic variance
σ^2_c = Variation due to clones
h^2_1 = Heritability on individual tree basis
h^2_2 = Heritability on clonal mean basis
SE(h^2_1) = Standard error for h^2_1

Table 4. Estimation of components of variance for flowering phenology in clones of teak in a CSO, southern India.

These results suggests that teak clones are genetically predisposed to initiate flowering during the time to which their respective mother plants have been adapted to, while the duration of these events will be under the control of local environmental factors of the orchard. This also suggests that while the time of initiation of phenophases is highly conserved, the duration may be shaped to be plastic such that maximum reproductive success could be obtained. Hence, the results of the present study could be extrapolated and used as indicator to select clones for future seed orchards such that their flowering time could be properly matched. Further, it also hints that the duration of flowering is more amenable for modification through silvicultural practices to suit to the needs of the orchard manager such that maximum flowering synchrony can be obtained.

3. Measures of synchronous flowering

Till recently, reproductive phenologies were qualitatively described as phenograms developed by Mahall & Bormann (1978) which were nothing but bands on a time line. This technique was readily adaptable to display the synchrony patterns male and female flowering in conifers. However it only provided a qualitative idea about synchrony and was impractical to be adopted whenever large number of clones is to be included. Askew and Blush (1990) have developed an index to measure the overlap between male and female flowering synchrony in monoecious loblolly pine. However, their measure doesn't correspond to species with bisexual perfect flowers such as teak. Evaluating the effectiveness of gamete exchange in a cross pollinated species requires a standard measure of the relative probabilities of crossing between all possible pairs of parents or clones. A quantitative index should have the following criteria:

1. The index should be based on simple observations that could be taken even by a field level worker and it should be simple to compute as well as to interpret
2. The index should be able to characterize a teak clone or a group clones with a single value which truthfully depicts its flowering overlap with either one or a group of clones over a time window.
3. The index should be able to be used as a tool routinely even in orchards with large number of clones to track development of young orchards as well as in monitoring the stability of established orchards.
4. The index must reach its maximum value of 1 if the phenograms of two clones are identical.
5. The index must show its minimum value of 0 if the phenograms of two clones are completely dissimilar.
6. If the phenograms of two clones show some overlap the value of the index must lie between 0 and 1.

Gunaga & Vasudeva (2009) have developed a general overlap index to quantify the exact overlap in flowering phenology among different clones. The authors have adopted the principle of computing similarity in species richness proposed by Morisita (1959) and Horn (1966) and extended to the teak seed orchard scenario. The following is the general formula for computing overlap in any phenophase. The time of peak flowering has been shown an example:

$$\text{Overlap Index } (C) = \frac{2\sum_{t=i}^{n}\left(P_{ij} \times P_{ik}\right)}{\left(\sum P^2{}_{ij} + \sum P^2{}_{ik}\right)}$$

Where,

P_{ij} = is the proportion of ramets of jth clone in peak flowering for a given period i

P_{ik} = is the proportion of ramets of kth clone in peak flowering for a given period i

n = number of weeks in which the flowering was observed.

$P_{ij} \times P_{ik}$ = the joint probability of flowering by two clones

The index is a simple, useful quantitative expression of phenological relationship among clones. The value of overlap index ranges from 0 *i.e.* when there is no overlap between two clones to 1 *i.e.* when there is full overlap between two clones. A higher value of the overlap index is suggestive of greater overlap. This index can be conveniently adopted to compute overlap of flowering between any two clones or between groups of clones belonging to different provenances or to know the flowering synchrony between a clone and rest of the clones in a CSO. The overlap index of a clone with all other clones of the CSO indicates its potentiality to contribute effectively to the ensuing seed crop.

An experimental verification of the new method was also reported by Gunaga & Vasudeva (2009). In their study synchrony among clones for peak flowering was assessed through phenograms (Figure 1) as well as through a novel overlap index (Table- 5). The perusal of trend in the phenogram suggests that generally there are two peak periods in flowering. The first peak during early May to July corresponds mainly to the clones of central and southern

Sl. No	Clone ID	With all other clones	1	2	3	4	5	6	7	8	9	10	11	12	13	14	15	16	17	18	19	20	21	22
1	MyHaD1	**0.798**																						
2	MyHaD2	**0.554**	0.953	-																				
3	MyHaD3	**0.701**	0.989	0.953	-																			
4	MyHaD4	**0.702**	0.989	0.969	0.995	-																		
5	MyHaV1	**0.889**	0.624	0.424	0.603	0.560	-																	
6	MyHaV3	**0.830**	0.719	0.532	0.695	0.661	0.983	-																
7	MyHaV4	**0.546**	0.924	0.794	0.910	0.883	0.851	0.906	-															
8	MyHaV5	**0.751**	0.740	0.550	0.715	0.681	0.973	0.996	0.921	-														
9	MyHaV6	**0.862**	0.669	0.460	0.641	0.603	0.985	0.982	0.878	0.981	-													
10	MyHaV7	**0.714**	0.725	0.531	0.697	0.661	0.976	0.988	0.916	0.991	0.985	-												
11	MyHaK1	**0.734**	0.970	0.991	0.974	0.982	0.504	0.604	0.843	0.617	0.532	0.596	-											
12	MyHaK2	**0.626**	0.983	0.965	0.990	0.993	0.583	0.680	0.890	0.694	0.615	0.672	0.984	-										
13	MyHaK3	**0.720**	0.991	0.952	0.998	0.995	0.622	0.715	0.917	0.733	0.659	0.712	0.975	0.991	-									
14	MySA1	**0.733**	0.719	0.532	0.700	0.660	0.986	0.992	0.907	0.989	0.976	0.983	0.607	0.678	0.717	-								
15	MySA2	**0.893**	0.738	0.564	0.714	0.676	0.975	0.983	0.922	0.977	0.967	0.983	0.636	0.701	0.731	0.983	-							
16	MySS2	**0.483**	0.602	0.400	0.573	0.530	0.991	0.968	0.839	0.962	0.981	0.975	0.477	0.550	0.592	0.971	0.969	-						
17	MyBL1	**0.597**	0.663	0.469	0.631	0.595	0.991	0.983	0.871	0.974	0.986	0.983	0.544	0.616	0.653	0.981	0.986	0.989	-					
18	MyHuT1	**0.775**	0.728	0.551	0.791	0.680	0.974	0.985	0.915	0.983	0.967	0.976	0.629	0.699	0.736	0.988	0.981	0.963	0.968	-				
19	MyHuT2	**0.637**	0.584	0.382	0.553	0.512	0.994	0.975	0.821	0.965	0.977	0.971	0.460	0.536	0.573	0.973	0.967	0.991	0.988	0.956	-			
20	MyHuT3	**0.586**	0.619	0.413	0.587	0.547	0.979	0.970	0.850	0.971	0.985	0.983	0.480	0.560	0.603	0.962	0.959	0.986	0.979	0.950	0.984	-		
21	MyHuT6	**0.851**	0.774	0.608	0.757	0.723	0.962	0.976	0.932	0.970	0.966	0.968	0.679	0.743	0.775	0.980	0.980	0.947	0.970	0.980	0.940	0.937	-	
22	MyHuT7	**0.730**	0.734	0.539	0.704	0.668	0.972	0.986	0.919	0.993	0.983	0.995	0.604	0.678	0.720	0.985	0.978	0.973	0.978	0.977	0.965	0.979	0.967	-
23	MyHuT8	**0.898**	0.793	0.640	0.766	0.740	0.944	0.981	0.933	0.978	0.940	0.963	0.705	0.764	0.786	0.975	0.975	0.926	0.954	0.971	0.934	0.922	0.970	0.966
24	MyHuK3	**0.726**	0.718	0.528	0.691	0.652	0.985	0.986	0.913	0.987	0.983	0.990	0.599	0.667	0.708	0.993	0.987	0.983	0.987	0.985	0.975	0.976	0.978	0.922

Table 5. Overlap index for peak flowering among 25 clones in a Teak clonal seed orchard at Manchikere, Karnataka. The third column represents the overlap index of that clone with all other clones of the CSO (except with itself).

provenances; the second peak during July to August corresponds to those from northern provenance (also see Figure 2). Table 5 shows the overlap index for peak flowering for all clone pairs of 25 clones in a clonal seed orchard of teak in southern India, Karnataka. It also shows the overlap index of a clone with all other clones of the CSO (except with itself). In nearly a fifth of the bi-clonal interaction with respect to peak flowering period, there was less than 0.6 overlap suggesting a lack of synchrony. In fact in 12.55 per cent of interactions, the overlap was less than 0.5. This suggests a gross violation of basic assumption, of random mating among clones, made while establishing a CSO.

Strong provenance influences on flowering phenology was also observed (Table 6). Comparison of overlap index between groups of clones from different provenances yielded several interesting results (Table 6). The overlap index between northern clones and those from southern was least (0.778); while it was 0.798 between northern clones and those of central. Highest overlap of 0.997 was found between clones of southern and central. Clones from central and southern provenances were early in commencement of flower buds, flowers and peak flowering (by about 30-40 days) when compared to clones from northern provenance (Figure 1). It clearly indicates that a significant proportion of ramets of clones from southern and central provenances escape early showers in the orchard when compared to those from northern provenance.

Overlap of clones of	Index Value
Northern provenances and central provenance	0.778
Northern provenances and southern provenance	0.798
Southern provenances and central provenance	0.997

Table 6. Overlap index for peak flowering between provenances.

Genetically based geographical variation was recorded in the present study and it was also shown to occur in several plant species. Hence, simple geographical continuity of the clonal origin could be considered as thumb rule while selecting right clones for seed orchard. Effimov (1987) has also suggested the selection of plus trees based on synchronized flowering while establishing second-generation clonal seed orchards. In fact in Australia constraints for *Pinus caribaea* seed production has prompted a co-operative arrangement wherein special clonal seed orchards would be established using local clones alone. The overlap index between 'early flowering clones' (those clones blooming during May-June month corresponding to Figure 2 b, c) with 'late flowering clones' (those bloomed during July to August months corresponding to Figure 2a) was 0.739. This suggested that about 70 per cent synchrony can be expected between clones which flowered early *i.e.* during June month, with clones that flowered late (during July to August; Figure 1). In order to improve the seed production level in CSO, seed orchards need to be established considering the flowering synchrony of the constituent clones.

4. Management implications

Having a measure of flowering synchrony of each clone with the reminder of the clones in the orchard plays an important role in assessing the genetic value of the orchard crop (Askew & Blush, 1990). Rating an entire orchard for its relative degree of flowering synchronization is effective with this new measure. This helps to track the changes in flowering on a quantitative scale through out the orchard life and hence help in evaluating the panmixia/ genetic gain. The index is a simple, useful quantitative expression of phenological relationship among clones. The overlap index of a clone with all other clones of the CSO indicates its potentiality to contribute effectively to the ensuing seed crop. Assuming there are no variations in the pollen grain fertility, it may also represent a value akin to the general combining ability of a parent. Flowering synchrony and parental balance would influence on panmictic equilibrium (random mating). It may be one of the causes for low fruit production in seed orchards (Gunaga & Vasudeva, 2005). For production of genetically superior seed, abundant and synchronized flowering of all clones in a seed orchard is very necessary, other wise the seed produced cannot guaranteed as genetically superior (Rawat et al., 1992). To overcome these problem, selecting a clone with high overlap index to establish future CSOs could be a good practice to increase the panmixis and hence the genetic quality of the seed crop from a CSO.

5. Conclusion

Genetic improvement of teak has focused mainly on identifying phenotypically superior trees from diverse growing regions and deploying them as vegetatively propagated clones in seed orchards. One of the most important aspects in a seed orchard is the synchrony among the clones for reproductive phenology. This will decide the extent of random mating among the constituent clones and hence the genetic gain in the resultant progeny. Although, there seems to be a direct association between the extent of flowering synchrony among the constituent clones and the fruit production in a seed orchard, surprisingly, there are very few reports in India that focus on these issues. A lack of a suitable tool for an objective measurement of flowering synchrony among different clones has also contributed to the negligence!! Hence there is a need for development of objective criteria to evaluate flowering phenological events. The patterns genetic variation for phenology in teak seed orchards have shown that the reproductive initiation events have strong genetic basis and show moderately high heritable values. A novel index to measure flowering among individuals in an orchard has been developed and the experimental verification of the same has shown that there is a large asynchrony among the clones guided by the provenance influence.

6. Acknowledgement

We thank Karnataka Forest Department for kind permission and co-operation. Authors are also thankful to Professor K.N. Ganeshaiah, School of Ecology and Conservation, College of Agriculture, GKVK Campus, Bangalore for his kind suggestions on the earlier version.

7. References

Anmol Kumar, A. (1992). Teak seed improvement - Achievement and problems, *Indian Forester*, 116(3), pp. 525-533.

Anmol Kumar, A., Gogate, M. G., Sharma, R. and Mandal, A. K. (1997). Genetic evaluation of teak clones of Allapalli region of Maharastra. *Indian Forester*, 123(3), pp. 187-189.

Askew, G.R. (1986). Implication of non-synchronous flowering in clonal seed orchards. In: *IUFRO Conference on a Joint meeting of working parties on breeding theory, progeny testing and seed orchards*, pp. 182-191.

Askew, G.R. and Blush, Th. D. (1990). Short Note: An index of phonological overlap of flowering for clonal coniferous seed orchards. *Silvae Genetica*, 39, 168-172.

Ball, B., Pandey, D. and Hirai, S. (1999). Global overview of teak plantations, In: *Proceedings of 'Site technology and Productivity of Teak plantations'*, 26-29 Jan. 1999,

Chiang Mai, Thailand, pp. 17.

Corlett, R.T. (1998). Potential impacts of climate changes on tropical Asian forests through an influence on phenology. *Climatic Change*, 39(2/3), pp. 439-453.

Effimov, Yu. P. (1987). Seed production and growth of scots pine in orchards of vegetative and seed origin. *Lesnoe Khozyaistvo*, 12, pp. 38-41.

El-Kassaby, Y.A. and Askew, G.R. (1991). The relation between reproductive phenology and reproductive output in determining the gametic pool profile in a Douglas Fir seed orchard. *Forest Science*, 37, pp. 827-835.

FAO, (1995). Forest Resources Assessment 1990: Tropical Forest Plantation Resources (ED. Pandey, D.). FAO Forestry Paper No. 128. Published by FAO, Rome.

Granger, A. (1998). Future supplies of high grade tropical hard woods from intensive plantations. *Jour. of World Forest Resource Management*, 3, pp. 15-29.

Griffin, A.R. (1984). Clonal variation in Radiata pine seed orchards. II. Flowering phenology. *Australian Forest Research*, 14, pp. 271-281.

Gunaga, R. P. and Vasudeva, R. (2005). Causes for low fruit production in clonal seed orchards of teak (*Tectona grandis* Linn. f): A special references to India. *In: Quality Timber Products of Teak from Sustainable Forest Management* (Eds.) Bhat, K. M., Nair, K. K. N. Bhat, K. V., Muralidharan, E. M. and Sharma, J. K. (EDs.). Published by KFRI, Peechi, pp. 352-358.

Gunaga, R. P. (2000). Genetic Variation for Phenology in a Clonal Seed Orchards of Teak (*Tectona grandis* Linn.f). A Thesis submitted to the Department of Forest Biology and Tree Improvement, College of Forestry, Sirsi campus, University of Agricultural Sciences, Dharwad, pp. 128.

Gunaga, R. P. and Vasudeva, R. (2002). Variation in flowering phenology in a clonal seed orchard of teak. *Jour. of Tree Sciences*, 21 (1&2), pp. 1-10.

Gunaga, R.P. and Vasudeva, R. (2009). Overlap index: a measure to assess flowering synchrony among teak (*Tectona grandis* Linn. f) clones in seed orchards, *Current Science*, 97(6), pp. 941-946.

Hanumantha, M., Vasudeva, R., R.P. Gunaga and Swaminath, M.H. (2001). Genetic variation for reproductive success in a clonal seed orchard of teak, *My Forest*, 37 (1), pp. 373-387.

Hedegart, T. (1973). Pollination of teak (*Tectona grandis*). *Silvae Genetica*, 22(4), pp. 124-128.

Horn, H.S. (1966). Measurement of "overlap" in comparative ecological studies. *Am Natur.*, 100, pp. 419-424.

Indira, E.P. (2005). Why teak seed orchards are low productive?, In: *Quality Timber Products of Teak from Sustainable Forest Management*, (EDs.) Bhat, K.M., Nair, KKN, Bhat, K.V., Muralidharan, E.M. and Sharma, J.K., Published by KFRI, Peechi. 347-351.

Kaosa-Ard, A. (1981). Teak (*Tectona grandis* Linn. f.): its natural distribution and related factors. *Natural History Bulletin of Siamese Society*, 29, 55-74.

Katwal, R. P. S. (2005). Teak in India: Status, Prospectus and Perspectives, In: Quality timber products of teak from sustainable forest management, Bhat, K. M., Nair, K. K. N. Bhat, K. V., Muralidharan, E. M. and Sharma, J. K. (EDs.). Published by KFRI, Peechi, pp. 1-18.

Kedharanth, S. (1982) Genetic variation and heritability of juvenile height growth in *Eucalyptus grandis*. *Journal of Tree Sciences*, 6, pp. 46-49.

Kjaer, E. D. and Suangtho, V. (1995). Out crossing rate of teak (*Tectona grandis* L. f.). *Silvae Genetica*, 44, pp. 175-177.

Mahall and Bormann (1978). A quantitative description of the vegetative phenology of herbs in a northern hardwood forest. Bot. Gaz. 139 (4): 467.

Matziris, D. (1994) Genetic variation in the phenology of flowering in Black pine. *Silvae Genetica,* 43, 321-328.

Matziris, D. (1997) Variation in growth, flowering and cone production in a clonal seed orchard of Aleppo pine grown in Greece. *Silvae Genetica,* 46, pp. 224-228.

Morisita, M. (1959). Measuring inter-specific association and similarity between communities. *Mem Fac. Sci. Kayushu Univ. Ser. (Biol),* 3, pp. 65-80.

Nagarajan, B., Mohan Varghese, A. Nicodemus, K.R. Sashidharan, S.S.R. Benet and Kannan, C.S. (1996). Reproductive biology of teak and its implication in tree improvements. In: *Proc. QFRI-IUFRO Conference on Tree Improvement for Sustainable Tropical Forestry,* (Eds). Dieters, M.J., A.C. Matheson, D.G. Nikeles, C.E. Harwood, S.M. Walker), Caloundra, Queensland, Australia.

Palupi, E.R. and Owens, J.N. (1998). Reproductive phenology and reproductive success of teak (*Tectona grandis*). *International Journal of Plant Science,* 159(5), pp. 833-842.

Radhamani, A., Nicodemus, A., Nagarajan, B. and Mandal, A. K. (1998). Reproductive biology of tropical tree species. In: Mandal AK and Gibson GL (Ed) *Forest Genetics and Tree Breeding.* New Delhi. CBS-Publishers and Distributors, 1998, pp. 1-268.

Rawat, M. S., Uniyal, D. P. and Vakshasya, R. K. (1992). Variation in the model teak seed orchard. New Forest, Dehra Dun. *Indian Journal of Forestry,* 118 (1-4), pp. 60-65.

Richmond, G.B. and MC-Kilney, C.R. (1986). An experimental seed orchard in South Texas. *Journal of Forestry,* 84, pp. 19-43.

Sedley, M. and Griffin, A.R. (1989). *Sexual Reproduction of Tree Crops,* Australia Academic Press.

Subramaniam, K.N., Nicodemus, A. and Radhamani, A. (1994). Teak improvement in India, *For. Gen. Resour.,* 22, pp. 33-36.

Subramanian, K., Madal, A.K., Ram Babu, N., Chundamannil, M. and Nagarajan, B. (2000). Site, Technology and Productivity of teak plantations in India, In *Site, Technology and Productivity of teak plantations.* Enters, T. and Nair, C.T.S. (EDs.), FORSPA Publication No. 24, FAO, Bangkok, pp. 51-68.

Vasudeva, R., Gunaga, R.P and Hanumantha, M. (2001). Implication of variation in reproductive traits among teak clones to management and genetics of seed orchards. *Advances in Forestry Research in India,* 24, pp. 120-135.

Vasudeva, R., R. P. Gunaga and Hanumantha, M. (2005). Implications of clonal variation for reproductive traits to improvement of teak. *In: Quality Timber Products of Teak from Sustainable Forest Management,* Bhat, K. M., Nair, K. K. N. Bhat, K. V., Muralidharan, E. M. and Sharma, J. K. (EDs.). Published by KFRI, Peechi, Published by KFRI and ITTO, pp. 417-422.

Vasudeva, R., Rajesh P Gunaga, Hanumantha, M. and Swaminath, M.H. (1999). Non synchronous flowering in teak seed orchards: A cause for low fruit production. In: *Proceedings of National Symposium on Forestry Towards 21st century, Sept., 1999, TNAU, Coimbatore.*

Verma SK, Sharma V. K. and Bagchi SK. (1989). The phenology of flowering of reciprocal F_1 hybrids (*Eucalyptus citriodora* Hook. X *E. torelliana* F.V. Mvell.), F_2 and F_3 segregates and parent species at New Forest, Dehra Dun. *Annals of Forestry,* 7, pp. 120-124.

Weinstain, A. (1989). Geographical variation and phenology of *Pinus halepensis, P. brutia* and *P. eldarica* in Israel. *Forest Ecology and Management,* 27(2), 99-108.

White, K. J. (1991). *Teak: some aspects of research and development.* FAO, Regional Officer for Asia and the Pacific (RAPA), pp. 17.

Yanninen, I., Laakso, S. and Raatikainen, M. (1988) Geographical variation in the phenology and morphology of bilberry in Finland. *Acta Botanica Fennica,* 136, pp. 49-59.

Part 3

Phenology of Fruiting Plants

11

Apple Phenology in Subtropical Climate Conditions

José Luiz Petri, Fernando José Hawerroth, Gabriel Berenhauser Leite, Marcelo Couto and Poliana Francescatto
Epagri/Estação Experimental de Caçador, Caçador, Santa Catarina, Embrapa Agroindústria Tropical, Fortaleza, Ceará, UFSC/Universidade Federal de Santa Catarina, Florianópolis, Santa Catarina, Brazil

1. Introduction

There are two phases that characterize the annual cycle of apple (*Malus domestica* Borkh.): hibernal period (dormancy) and shoot growth. Temperature is the main factor related to dormancy, specifically low temperatures. The evolution of the apple phenological stages during the vegetative phase: induction and flower differentiation, budbreak, full bloom, fruit development and maturation, and yield and production quality, may be visually detected and are affected by climate conditions, mainly by temperature.

Plant-related factors such as cultivar and rootstock also influence this evolution dynamic. The type of the fruit structure and its foliar area affect the fruit formation, and are important parameters to the phenological evolution. Growing apples in the warm winter regions where the chilling requirement is not adequate, can cause the trees to develop a series of anomalies in the phenology referent to budbreak, flowering, growth and development of both fruit and tree (Skinner, 1964; Saure, 1985; Chariani & Stebbins, 1994; Bulon & Faust, 1995; Petri et al., 1996; Iuchi, 2002; Petri & Leite, 2004).

At the end of the dormancy, vegetative and reproductive apple buds evolve and pass through different external phases that are known as phenological stages. According to Saraiva (1973), the study of phenology is based on observations and measurements of a series of plant organs in a determined period. This constitutes an essential element to evaluate the adaptability of fruits species under particular environmental conditions.

The knowledge of the phenological stages is important under a practical viewpoint to manage some cultural practices such as fruit thinning and phytosanitary treatments. These phenological changes have been intensely studied and classified according their development from dormancy to fruitset. Flower differentiation, phenological stages, growth and fruit maturation could be correlated with the climate variables, making possible a forecast of each evolution by mathematical equations. The study of the variability indexes of the evolution of the flower phenological stages are also important, since may indicate the reproductive regularity of cultivars as well as to identify cultivars with coincidence of flowering to be able to recommend pollinators.

The methodology of correlations between chilling units and date of flowering allows to predict in advance the flowering date and consequently to estimate the date of maturation. Estimation of the growth and development of fruits is fundamental in warm winter regions because it is not only related to fruit caliber, but also to alterations that environmental conditions could cause regarding visual aspects mainly fruit form.

2. Flower induction and flower bud development in apple trees

Apple trees have two distinct phases regarding their metabolism: the period of dormancy or vegetative phase. Both are affected by climate conditions, mainly by temperature which leads to an alteration in the phenological phases, including induction and flower differentiation.

Induction and differentiation of the apple flower buds take place just after the last year's flowering. Although it is possible to distinguish the different fruiting organs for the external appearance sometimes buds might not differentiate into flower bud. The flower formation coincides temporally with the shoot and fruit formation. So, that means fruitset affects the metabolism of young shoots and furthermore can inhibit the flowering the following year.

Analysis of bud permits to define the flower percentage and quality. The predicted knowledge of the flower bud formation will give important information about pruning intensity, fruit thinning, fertilization and pollination. In addition, the retrospective of last year's production also aids in the cultural practices to be adopted.

Flowering intensity is one of the parameters to set pruning. Consequently, to minimize the possible faults of this practical, however it does not full assure the fruitset since the latter is year dependent. The quantification of the flower buds will permit to make an equilibrated pruning, avoiding a winter heavy pruning what would promote a large vegetative growth and therefore nutrients competition with the new fruits. Excessive tree vigor reduces fruit set and flower induction.

The process of induction and flower differentiation is affected by several factors such as climatic, nutritional, cultural, physiological and genetic. Flower induction is privileged by the presence of leaf area and negative influenced by excessive fruits on tree. Tree defoliation before induction, excessive fruits and later fruit thinning prejudice the flower induction process.

The period of the apple flower induction occurs in the beginning of shoot growth which means around 45 to 60 days after full bloom. However, there are evidences that induction may occur later, even after fruit harvest in warm regions. As reported by Petri (2002) the period of induction can vary according to cultivar, bud position on tree, climate conditions and nutritional factors. Most flower induction occurs in the beginning of summer but may be extended up to begin of fall under subtropical conditions.

After flower induction flower differentiation takes place. It proceeds throughout vegetative cycle until the next year flowering. The sequence of the differentiation process begins with the appearance of sepals, then stamens, ovary, anthers, pollen and ovule. Since ovary and anthers are formed it is possible to distinguish the flower buds using a stereoscopic microscope.

Though the external appearance of most apple fruit organs is an indicative of flower bud, precise information about the real percentage of flower bud and on which organ or branch it is located is important to the pruning purposes. Old spurs tend to produce small fruits of low quality in relation to new spurs developed from three or younger year old branches

The knowledge of the fruiting habit is important to the definition of cultural practices such as pruning, thinning and the use of growth regulators, aiming higher productions and fruit quality. Several studies have been developed on which type of flower structure is responsible for the formation of the largest fruit with commercial quality (Madail et al., 2010). Apple fruit buds are mixed buds, which means presence of flowers and leaves in the same bud. They are borne terminally or laterally on fruiting structure classified as brindle, and terminally on spurs. Brindles (10-40cm) are formed in the previous year and spurs are two or more year old buds. Axillar buds are located in the insertion of leaves along the brindles (Petri & Leite, 2006).

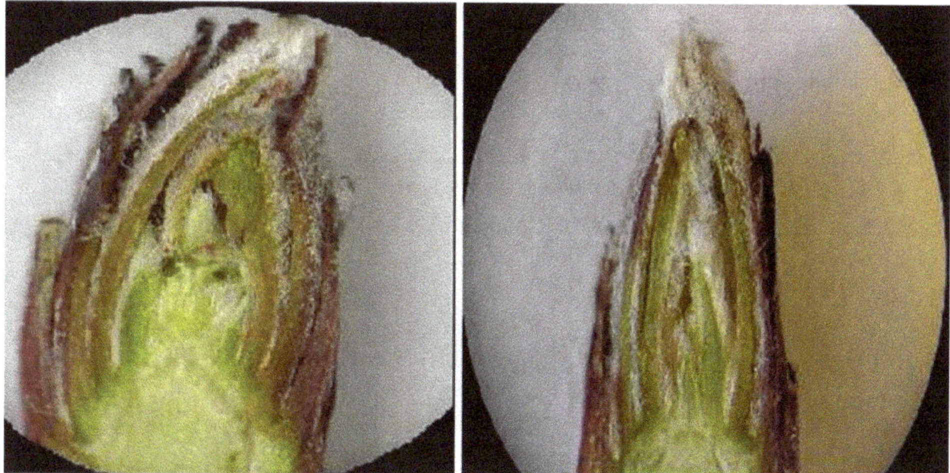

Fig. 1. Flower bud (left) and vegetative bud (right) of apple trees.

Leaf area is a factor of great importance to the fruit development, to the cell division process and to the fruitset. Fruiting structure with larger leaf area shows higher photosynthetic capacity, higher production of assimilates and lesser competition among the different tree organs (Costes, 2003). Nevertheless, other authors affirm that the number of leaves is not sufficient to determine the bud performance (Lauri & Trottier, 2004; Lauri et al., 2006). The bud position in the canopy influences the fruit form. Webster (1976) observed that terminal buds of brindles of 'McIntosh' apples provided more elongated fruits than fruits from axillary buds. Rodrigues & Rodrigues (1977) verified better fruits formed in the basal portion of the brindle, since at this position the size of the fruiting structure and leaf area are larger than in the brindle tip.

From 15th January under the climate conditions of Southern Brazil, it was possible to identify flower and vegetative buds using a stereoscopic microscope (Fig. 1). In this case, the cultivar Gala already had showed well developed flowers, whereas in the cultivar 'Fuji' there was a delay in flower development. There is little difference among fruiting structures regarding percentage of flower buds (Table 1). It is observed that brindle and spurs showed little difference for both cultivars, being numerically superior in spurs. This could cause a delay in the stopping of shoots growth.

Cultivar	Fruiting structures	
	Brindle	Spur
Gala	71.6	88.3
Fuji	75.0	78.3

Table 1. Flower buds percentage in Gala and Fuji apples under subtropical climate conditions. Caçador, Santa Catarina, Brazil, 2009.

Fruiting structures	'Gala'		'Daiane'	
	Leaf area (cm^2)*	Fruit diameter (mm)	Leaf area (cm^2)	Fruit diameter (mm)
1. Weak spur	34.3 c	56.4 c	43.6 c	51.0 c
2. Intermediate spur	60.4 c	62.1 b	58.4 c	57.3 b
3. Strong spur	107.2 b	66.2 a	103.1 b	65.9 a
4. Brindle	323.9 a	66.5 a	346.9 a	67.8 a

*Means followed by same letter do not differ by Duncan test ($p<0.05$)

Table 2. Leaf area and fruit diameter of 'Gala' and 'Daiane' apples in different fruiting structures. Caçador, Santa Catarina, Brazil, 2009.

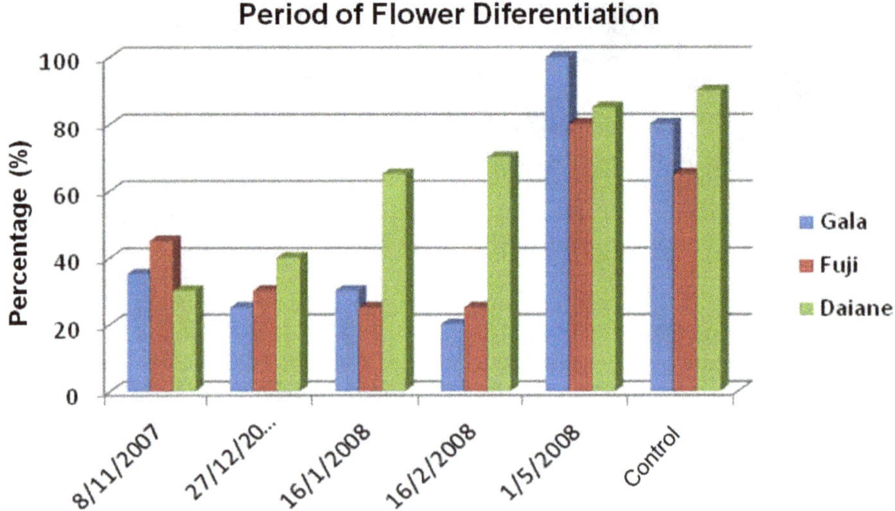

Fig. 2. Percentage of the return of bloom at brindles in different defoliation dates in the apple cvs. Gala, Fuji and Daiane. Caçador, SC, 2009.

Tree precocious defoliation affects negatively the following year's crop. At Fig. 2, it is possible to observe a reduction in the return bloom when defoliation was done up to 16th February in the cvs. Gala, Fuji and Daiane. The reduction was higher in fruiting spurs than in brindle (Table 2). Up to 16th February the reduction of the return bloom was higher in 'Gala' and 'Fuji'. From 16th February reduction was lesser accentuated in the brindles of Daiane (Table 2). This difference among cultivars could be attributed to the longer differentiation process seen in the cvs. Gala and Fuji. However, whether defoliation was done on 1st May the return of bloom was higher, pointing out that at this time the flower differentiation had already occurred. The results show that the apple flower induction process under warm climate conditions is prolonged for a period superior to 60 days after full bloom. Also, it was observed that the flower induction of axillary buds is associated to the growth period, mainly in the cultivar Gala what shows higher flowering intensity in the axillary buds whether compared to 'Fuji' and 'Daiane'.

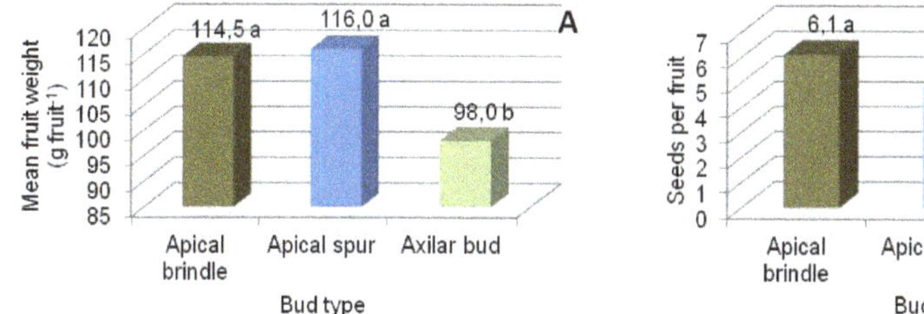

Fig. 3. Mean fruit weight (g fruit^{-1}) and seed per fruit in different bud types of 'Fuji' apples. Caçador, Santa Catarina, Brazil, 2009. Means followed by same letter do not differ by Tukey test ($p<0.05$).

Treatment	Fruit diameter						
	23/11*	05/12	20/12	03/01	20/01	11/04	C.V. (%)
1 fruit/cluster	21.4 abF	27.8 abE	34.3 abD	39.9 abC	45.4 aB	59.6 aA	14.9
2 fruit/cluster	21.3 abF	28.1 aE	35.4 aD	41.5 aC	46.5 aB	59.1 aA	13.0
3 fruit/cluster	22.6 aF	28.9 aE	35.4 aD	41.0 aC	45.5 aB	61.0 aA	8.64
4 fruit/cluster	19.9 bE	27.2 abD	34.3 abC	39.4 abB	43.6 abA	-	13.0
5 fruit/cluster	19.3 bE	25.7 bD	32.0 bC	37.3 bB	41.1 bA	-	13.4
C.V. (%)	17.8	12.8	12.1	11.4	11.0	14.1	

*Means followed by lowercase letters in the column, and means followed by uppercase letter in the line, do not differ by Duncan test ($p<0.05$)

Table 4. Fruit diameter according the number of fruits per cluster in Fuji apple trees. Caçador, Santa Catarina, Brazil, 2008.

Leaf area is variable, depending on fruiting structure. It was observed reduced leaf area in fruiting spurs, principally in weak spurs (Table 2). The largest leaf area is found in brindles where the largest fruits are produced. This increase in leaf area is not only determined by the size of leave, but also by the number of leaves, being smaller in weak spurs e higher in brindles (Table 2). The type and quality of fruiting organs also affect fruitset. Madail et al. (2010)

observed a significant increase of fruitset in longer brindles when compared to spurs. Larger leaf area structures also showed larger-diameter fruits, showing a direct relation between these two variables (Table 2). The lowest fruit fresh mass was observed in axillary buds, followed by spurs and brindles. The difference could be related to the number of seed per fruit which was superior in brindles whether compared to spurs and axillary buds (Fig. 3).

The results of mean diameter of fruits, growth index and fresh mass of fruits in relation to the number of fruits per inflorescence showed little differences. The presence of one or five fruits per inflorescence showed similar results (Table 4, 5 and 6). So, at fruit thinning and pruning period it must be given preference to fruit production in fruiting structure such as brindles and vigorous spurs.

Treatment	Fruit growth rate (%)					
	23/11-05/12* (12 days)	05/12-20/12 (15 days)	20/12-03/11 (13 days)	03/01-20/01 (17 days)	20/01-11/04 (81 days)	C.V. (%)
1 fruit/cluster	131.2 aA	123.4 aB	116.0 aC	113.9 aC	134.8 aA	5.8
2 fruit/cluster	136.4 aA	124.4 aB	116.3 aBC	112.5 abC	135.8 aA	8.0
3 fruit/cluster	129.3 aA	123.0 aB	116.6 aC	110.3 bD	122.8 aB	3.4
4 fruit/cluster	139.5 aA	126.2 aB	117.5 aC	112.4 abC	–	8.0
5 fruit/cluster	135.6 aA	124.7 aB	116.9 aC	112.8 abC	–	7.7
C.V. (%)	10.41	3.52	3.60	2.8	8.3	

*Means followed by lowercase letters in the column, and means followed by uppercase letter in the line, do not differ by Duncan test ($p<0.05$)

Table 5. Fruit growth rate according the number of fruits per cluster in Fuji apple trees. Caçador, Santa Catarina, Brazil, 2008.

Treatment	Harvest 14/11/2006		
	Fruit diameter (mm)*	Fruit length (mm)	Mean fruit weight (g)
1 fruit/cluster	59.6 a	47.6 a	104.3 a
2 fruit/cluster	59.1 a	49.6 a	112.9 a
3 fruit/cluster	61.0 a	54.4 a	147.0 a
C.V. (%)	14.1	14.9	40.7

*Means followed by same letter do not differ by Duncan test ($p<0.05$)

Table 6. Diameter, length and mean fruit weight of Fuji apples according the number of fruits per cluster in Fuji apple trees. Caçador, Santa Catarina, Brazil, 2008.

3. Apple phenological stages

The method used to determine the evolution of the phenological stages consists in the observation of the different stages of the buds, permitting to determine chronologically their development (Fig. 4). The result will provide information about the flower biology, will

Fig. 4. Apple phenological stages: A (dormant bud); B (silver tip); C-C3 (green tip); D-D2 (bud bursted); E (green leaf); E2 (pink); F-F2 (full bloom); G (beginning of petal fall); H (completely petal fall); I (fruit set); and J (fruit development).

Phenological Stages	Annual means (days)					Duration (days)		Mean dates*
	1995	**1996**	**1998**	**1999**	**2000**	**Mean**	**Accumu-lated**	
B e C	-	2	4	3	5	3.5	3.5	15/set.
C3	5	2	3	2	3	3.0	6.5	18/set.
D	3	1	2	2	2	2.0	8.5	21/set.
E	2	1	2	2	2	1.8	10.3	23/set.
E2	3	2	3	2	2	2.4	12.7	24/set.
F	2	3	4	5	3	3.4	16.1	26/set.
F2	2	2	3	3	3	2.6	18.7	29/set.
G	3	2	2	3	3	2.6	21.3	2/out.
H	3	2	2	3	4	2.8	24.1	4/out.
I	5	6	4	7	4	5.2	29.3	7/out.
J	3	6	5	7	-	5.2	34.5	12/out.
Σ	31	29	34	39	31			

*Referent to the mean of five years

Table 7. Duration of the different phenological stages of the cultivar Fuji during five years in Caçador, SC, Brazil.

make possible to compare the capacity of adaptation of the different cultivars in a same region and to verify the buds development and the influence of environmental factors, and moreover will contribute to some cultural practices such as fruit thinning, pollination and pest and diseases control.

Phenological Stages	Annual means (days)				Duration (days)		Mean dates*
	1995	**1998**	**1999**	**2000**	**Mean**	**Accumu-lated**	
C	-	-	2	4	3.0	3.0	20/set.
C3	4	3	3	2	3.0	6.0	21/set.
D	3	2	2	2	2.3	8.3	24/set.
E	2	2	2	2	2.0	10.3	26/set.
E2	2	4	3	4	3.3	13.6	28/set.
F	2	4	3	3	3.0	16.6	1/out.
F2	3	3	3	2	2.8	19.4	4/out.
G	2	2	3	3	2.5	21.9	6/out.
H	2	3	4	2	2.8	24.7	9/out.
I	5	4	5	4	4.5	29.2	1/out.
J	4	4	7	-	5.0	34.2	17/out.
Σ	29	35	37	28			

*Referent to the mean of four years

Table 8. Duration of the different phenological stages of the cultivar Golden Delicious during four years in Caçador, SC, Brazil.

Phenological Stages	Annual means (days)					Duration (days)		Mean dates*
	1995	**1996**	**1998**	**1999**	**2000**	**Mean**	**Accumu-lated**	
B e C	-	3	3	3	1	2.5	2.5	18/set.
C3	5	2	1	3	2	2.6	5.1	19/set.
D	4	2	2	3	2	2.6	7.7	21/set.
E	2	2	2	2	2	2.0	9.7	24/set.
E2	3	3	3	3	2	2.8	12.5	26/set.
F	2	3	5	5	3	3.6	16.1	29/set.
F2	2	2	4	3	3	2.8	18.9	2/out.
G	3	2	2	3	2	2.4	21.3	5/out.
H	3	4	3	3	3	3.2	24.5	8/out.
I	4	4	4	7	7	5.2	29.7	11/out.
J	3	6	5	9	-	5.8	35.5	16/out.
Σ	31	33	34	44	27			

*Referent to the mean of five years

Table 9. Duration of the different phenological stages of the cultivar Gala during five years in Caçador, SC, Brazil.

The duration of the stages varies according year and it is influenced by climatical factors. The mean duration of the phenological stages from silver tip (B) to fruit development (J) varied from 31 to 44 days in apples 'Gala'. The mean date of the occurrence of each phenological stage was estimated during five years of observation for the cvs. Fuji and Gala and four years for cv. Golden Delicious (Tables 7, 8 and 9). These variations were associated to the degree days (DD), the higher thermal amplitude the higher the sum of DD (see session 5).

In subtropical regions, the insufficient chilling accumulation required to the physiological necessities of the buds induces to a prolonged bloom period whether compared to temperate regions (Fig. 5). Erratic budbreak and flowering observed in warm regions is originated by the maximization of the heterogeneity of the buds regarding to chilling requirement and to the negative influence of the paradormancy promoted by the vigorous and advanced buds (Leite et al., 2004).

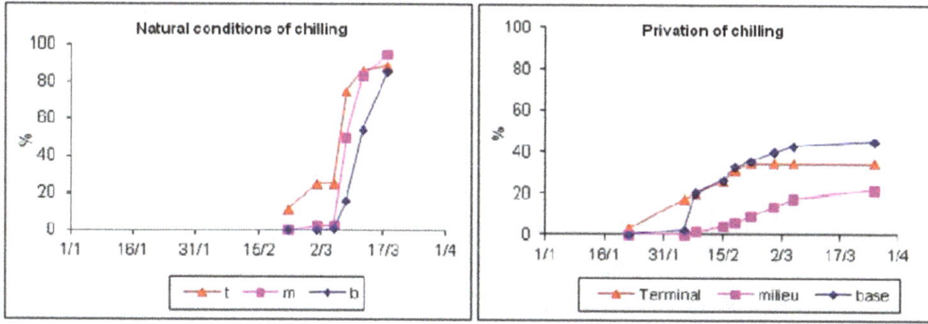

Fig. 5. Budbreak of the different buds along the brindle (t - terminal bud, m - median bud and b - basal bud) under natural conditions of chilling (left) and privation of chilling (right).

4. Effect of rootstocks and interstocks on apple phenology

In most economically explored temperate fruit species the use of rootstocks is an usual practice. The benefits of rootstocks are related to tree vigor control, precocious fruiting and high quality fruits (Jackson, 2003). The use of different rootstocks influences differently the vegetative and reproductive development of the trees and probably it would exert influence on the physiological process of the canopy dormancy (Erez, 2000).

Exposing six apple rootstocks to chilling privation conditions, with the previous exposition of the scion to chilling, Young & Werner (1985) verified deficient development of budbreak of the scion cultivar. Contradictory, Erez (2000) did not observed the same response in peaches under similar experimental conditions tested by Young & Werner (1985). In order to Finetto (2004) rootstocks can have significant contribution in apple budbreak when scion was submitted to conditions of insufficient winter chilling accumulation. The same author evaluated 'Golden Delicious' apples grown on different rootstocks in Italy and observed the rootstocks affect the chilling requirement of the scion cultivar when conditions were not sufficient. Under Brazilian conditions, Couvillon et al. (1984) verified that apples of high chilling requirements such as Rome Beauty can grow and develop according to the rootstock chilling requirement. According Hauagge & Cummins (2000), the chilling requirement of the apples rootstocks is an important factor that affects budbreak of 'Gala'. The use of the rootstocks 'MM111', 'M9' and 'M26' caused a delay on budbreak compared to 'MM106' and 'M7'; whereas the use of seedlings cultivar Anna tends to anticipate the budbreak.

The effect of rootstocks in the scion dormancy can result differences in the phenology of apples cultivars. The combination rootstock/scion alters the date of the beginning of flowering as well as the period of flowering (Fig. 6). 'Gala' raised on the rootstock 'Marubakaido' showed a delay in the beginning of flowering in comparison to 'Fuji' on the same rootstock. The duration of scion flowering on 'Marubakaido' was longer for 'Gala' than for 'Fuji', evidencing the scion cultivar dependence.

The use of 'M9' as interstock on the rootstocks 'MI.793', 'M.103' and 'Marubakaido' brings forward the beginning of flowering of the cultivar Fuji in comparison to the no use of interstock. Similar results were found by Scarpare Filho et al. (2000) who observed that the presence of plums interstock 'Januaria' anticipated flowering and budbreak of the peaches cultivars Tropical and Ouromel-2. Tomaz et al. (2010) also observed an influence of the length of the interstock on the duration of flowering; the longer the interstock the more precocious was flowering.

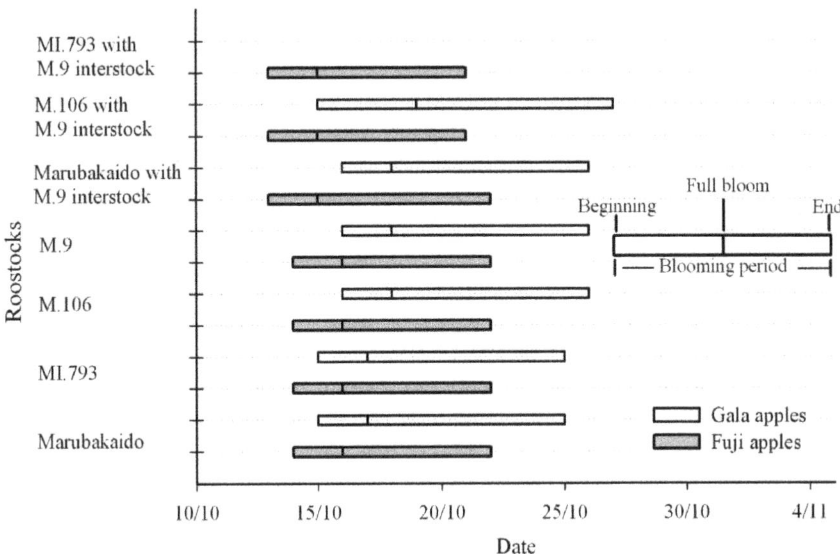

Fig. 6. Flowering period of 'Gala' and 'Fuji' apples grafted on different rootstocks and interstocks. Caçador, Santa Catarina state, Brazil, 2005.

5. Duration of flowering and degree-days

The physiological process and functions of the plants occur under thermal limits in their developing environmental. In order to complete every physiological subperiod of the life cycle some crops require an accumulation of certain heat amount. It is commonly expressed by degree-days index and represents the thermal sum above the minimum base temperature to the development. To complete each subperiod of the development plants need among other factors a thermal sum (Lozada & Angelocci, 1999).

The concept of degree-days admits that there is a base temperature which below plant growth and development are interrupted or extremely reduced. Besides, it is presupposed a linear relation between temperature and plant development since there is no limitation of other factors (Brunini et al., 1976). The sum of degree-days that plant requires to complete a subperiod or the whole cycle has been used to characterize the life cycle of the plants instead of the number of days. Also it is assumed as constant and independent from local and sowing time.

For all phenological phase it is necessary to calculate the index of heat units differently during the period, because the heat unit sum is one of the basic parameters that characterize the duration of the period from a phenological stage to another (Yazdanpanah et al., 2010).

Several models or expression have been suggested to the degree-days calculation or thermal units required to the plant to reach a certain stage of its development cycle (Brown, 1970; Arnold, 1959).

The heat unit index to each stage can be calculated using the following equation:

$$GD = \sum_{i=1}^{n}\left(\frac{T\max + T\min}{2} - Tb\right)$$

where: GD (DD) is the total accumulated degree-days; Tmax is the daily maximum air temperature (°C); Tmin is the daily minimum air temperature (°C); Tb is the base temperature (°C), normally used as 4.5°C according Richardson et al. (1975); n is the number of days of the flowering period.

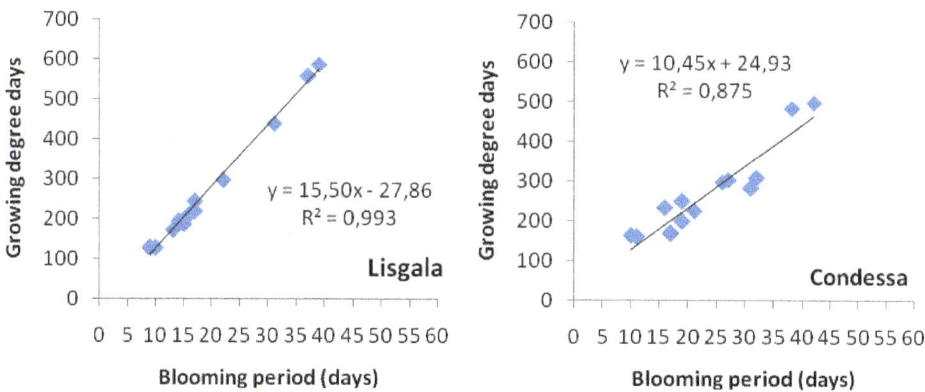

Fig. 7. Relation between duration in days and growing degree days of blooming period of 'Lisgala' and 'Condessa' apples during seven years of phenological observations. Caçador, Santa Catarina, Brazil.

Under subtropical conditions there is a high correlation between the duration of flowering period (in days) with the accumulated thermal sum. It can be observed in the Fig. 7 a high correlation between the duration of flowering and thermal sum in apples 'Lisgala' and 'Condessa' during seven years of phenological observation.

The relation between the duration of flowering and thermal sum is also observed among the subperiods of flowering. Considering the phenological observations of eight wild species (*Malus* spp.) in a period of eight years, it can be observed in the Fig. 8 a high correlation between the duration of full bloom to the end of flowering with the thermal sum, in the same magnitude of the period comprised between the begin and the end of flowering.

6. Phenology of wild apple species in subtropical climate conditions

The study of phenology has had great importance in the new apple planting system using high density. Wild apples have been used as pollinators for commercial cultivars due to their flowering intensity.

Aiming to define some pollinators Petri el al. (2008) studied the phenological behavior of trees belonging to 13 apple wild species (*M. atrosanguinea, M. baccata, M. eleyi, M. floribunda, M. hopa, M. platycarpa, M. robusta,* 'John Downil', 'Prof. Spengler', 'Milalew imuni', 'Profusion', 'Winter gold' and 'Yellow Siberian') and two commercial cultivars Gala and Fuji. The measurements were done from 2001 to 2007 in four trees per cultivar. It was recorded the dates of beginning of flowering, full bloom and end of flowering of each species. The beginning of flowering was considered when trees showed 5% of open flowers; full bloom when 80% of the flowers were opened and end of flowering with the last flowers open.

The period of flowering and the percentage of coincidence of flowering among the wild species with the cultivars 'Gala' and 'Fuji and their differences regarding the dates of full bloom were estimated for each year. At the end of the evaluation the mean behavior and the standard deviation for the percentage of coincidence among flowering and for differences among flowering date were estimated and the index of variability was obtained for each species. The index of variability is used as an estimative of regularity of the species or cultivars assessed throughout years. Thus, it could be identified the species/cultivars that showed great stability in relation to the coincidence of flowering (Table 10).

The wild species showed a great variability in flowering and duration during years (Fig. 8), indicating larger difference regarding chilling requirements. In 2003 it was observed a decrease in the period of flowering and large coincidence of flowering of the wild species with the cultivars Gala and Fuji. This could be related to the accumulation of chilling observed in this year (Tables 10, 11 and 12). 2006 was characterized by low chilling accumulation and frequent alternation between high and low temperatures during hibernal period. This condition determined the prolonged period of flowering in most evaluated species (with medium/high chilling requirement). Species with low chilling requirement anticipated flowering in relation to 'Gala' and 'Fuji'. According Petri et al. (1996) in warm winter conditions where chilling requirements are not fully satisfied to overcome dormancy it can occur variability in the date of flowering among years. Cultivars of low chilling requirement tend to flower earlier.

Among the species studied, *M. eleyi, M. floribunda, M. hopa, M. robusta,* 'Milalew Imuni' and 'Yellow Siberian' showed variation concerning flowering in the different years; and in some years, except *M. eleyi*, they showed low density of flowering indicating their low potential of use as pollinators (Fig. 8). The occurrence of abundant flowering is an indispensable condition for any cultivar or species that would be used as pollinator. Cultivars or species that show low flowering density or have tendency to alternate bearing should not be used as pollinators.

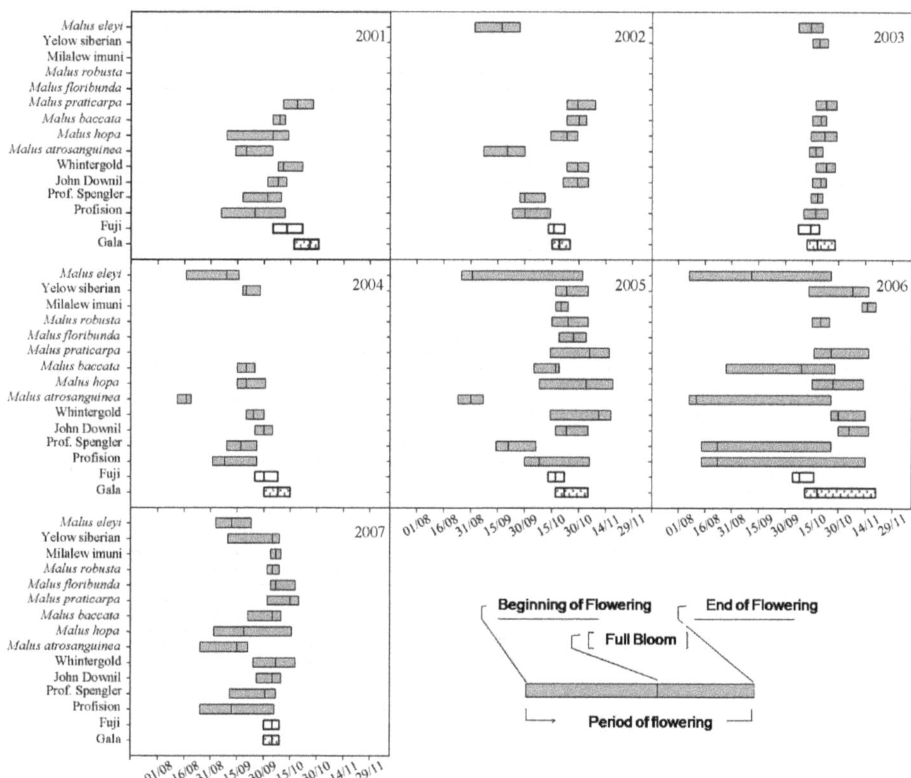

Fig. 8. Period of flowering of apple wild species and cultivars 'Gala' and 'Fuji' in seven years of phenological observation (2001-2007). Caçador, SC, Brazil, 2007. (Petri et al, 2008).

The occurrence of simultaneous flowering period and full bloom between pollinators and commercial cultivars increases the probability of having high fruitset index (Soltész, 2003). Any of the studied species showed perfect coincidence in the date of full bloom with the cultivars Gala and Fuji. Considering the species that flowered in all years 'John Downil', *M. platycarpa* and *M. baccata* had the date of full bloom closer to the observed for 'Gala' (Table 11). *M. baccata* and 'John Downil' differed 5 and 8.1 days (mean of years) in the date of full bloom in relation to cv. Fuji, respectively. However, these species showed great variability along the years. 'Gala' and 'Fuji' showed the lowest difference in the dates of full bloom and the lowest variability of this behavior along the evaluated years in comparison to the wild species. The percentage of coincidence of flowering of the wild species with 'Gala' and 'Fuji' varied among years and species (Table 11 and 12). The mean percentage of coincidence of the flowering period of the wild species with 'Gala' and 'Fuji' was 39.7 and 39.9%, respectively. In some years, some wild species had 100% of the flowering coinciding with the 'Gala' and 'Fuji', but in other years they did not coincide. This behavior highlights the high variability of some apple wild species regarding flowering under warm winter conditions.

Species/Cultivar	Difference of full bloom date in relation to 'Gala' apples	
	Mean ±standart error	Variability index
	Days*	%
Malus eleyi	29.3±15.6 ab	53.2
'Yellow Siberian'	7.8±9.8 d	125.4
'Milalew Imuni'	10.3±14.4 bcd	139.7
Malus robusta	1.3±1.2 d	86.6
Malus floribunda	3.5±2.1 d	60.6
Malus platycarpa	8.3±2.9 cd	35.3
Malus baccata	8.9±7.0 bcd	79.1
Malus hopa	11.9±6.7 bcd	56.7
Malus atrosanguinea	36.4±22.1 a	60.8
'Winter gold'	10.9±5.7 bcd	52.2
'John Downil'	8.0±7.5 cd	93.3
'Prof. Spengler'	22.0±18.2 abcd	82.8
'Profusion'	29.0±17.4 abc	60.2
Fuji	6.1±4.5 d	72.5
Specie/Cultivar	**Difference of full bloom date in relation to 'Fuji' apples**	
	Mean ±standart error	Variability index
	days	%
Malus eleyi	24.3±14.9 ab	61.0
'Yellow Siberian'	10.0±11.2 bc	112.0
'Milalew Imuni'	14.0±19.9 abc	142.3
Malus robusta	6.3±6.0 bc	95.2
Malus floribunda	6.0±5.7 bc	94.3
Malus platycarpa	11.0±5.7 bc	52.0
Malus baccata	5.0±5.4 c	107.7
Malus hopa	12.0±4.9 bc	40.5
Malus atrosanguinea	31.1±18.4 a	59.2
'Winter gold'	11.0±8.8 bc	79.6
'John Downil'	8.1±9.4 bc	115.5
'Prof. Spengler'	17.0±14.5 bc	85.1
'Profusion'	23.7±14.2 ab	59.9
Gala	6.1±4.5 bc	72.5

*Means followed by lowercase in the column do not differ by Tukey's test ($p<0.05$).

Table 10. Means and index of variability of the difference of full bloom date of apple wild species in relation to 'Gala' and 'Fuji' (*Malus domestica* Borkh.). Caçador/SC, 2007. (Petri et al, 2008).

Specie/cultivar	Year							Mean	Variability
	2001	2002	2003	2004	2005	2006	2007	±standart error	index (%)
Percentage of coincidence of the blooming period with 'Gala' apples (%)									
Malus eleyi	0.0*	0.0	56.3	0.0	82.4	38.5	0.0	25.3±34.0 ab	134.5
'Yellow Siberian'	0.0*	0.0*	56.3	0.0	100.0	84.6	44.4	40.8±42.2 ab	103.5
'Milalew Imuni'	0.0*	0.0*	0.0*	0.0*	41.2	20.5	55.6	16.7±23.2 b	138.7
Malus robusta	0.0*	10.0	0.0*	0.0*	100.0	25.6	77.8	30.5±41.4 ab	135.9
Malus floribunda	0.0*	0.0*	0.0*	0.0*	100.0	0.0*	55.6	22.2±40.1 ab	180.3
Malus platycarpa	78.6	20.0	68.8	0.0	100.0	76.9	77.8	60.3±36.1 ab	59.9
Malus baccata	0.0	20.0	50.0	0.0	11.8	43.6	100.0	32.2±35.8 ab	111.2
Malus hopa	0.0	100.0	87.5	6.7	100.0	82.1	100.0	68.0±44.8 a	65.8
Malus atrosanguinea	0.0	0.0	50.0	0.0	0.0	38.5	0.0	12.6±21.8 b	172.8
'Winter gold'	35.7	20.0	100.0	0.0	100.0	46.2	100.0	57.4±42.3 ab	73.7
'John Downil'	0.0	40.0	50.0	33.3	100.0	89.7	100.0	59.0±38.5 ab	65.2
'Prof. Spengler'	0.0	0.0	43.8	0.0	100.0	38.5	66.7	35.6±38.7 ab	108.9
'Profusion'	0.0	0.0	75.0	0.0	100.0	76.9	66.7	45.5±43.8 ab	96.1
Fuji	35.7	70.0	43.8	53.3	29.4	12.8	100.0	49.3±28.7 ab	58.3
Mean	10.7	20.0	48.7	6.7	76.1	48.2	67.5	39.7	107.5

*no blooming; ns no significant; Means followed by lowercase in the column do not differ by Tukey's test (p<0.05)

Table 11. Coincidence of blooming period of wild species in relation to 'Gala' apples in seven years of phenological observations. Caçador, Santa Catarina, Brazil, 2007. (Petri et al., 2008).

Petri (2006) reported that in areas with inconstant climate during winter apples have showed irregularity in flowering. These species not often coincide the flowering period with the main commercial cultivars grown in Southern Brazil, showing anticipation in the period more than 30 days.

M. atrosanguinea, 'Milalew imuni', *M. floribunda*, *M. eleyi*, *M. robusta*, *M. baccata*, 'Prof. Spengler' and 'Yellow Siberian' exhibit coincidence of flowering inferior to 41% in the mean of the years for 'Gala' (Table 11). 'Profusion', Fuji, 'Winter gold', 'John Downil', *M. platycarpa* and *M. hopa* were the species that showed the greatest coincidence of flowering with 'Gala', highlighting 'Profusion' and *M. hopa* by the flowering regularity and intensity in the several years. Some species showed no coincidence at some years observed.

'Milalew imuni', *M. floribunda*, *M. atrosanguinea*, *M. robusta*, 'Yellow Siberian', *M. platycarpa* and *M. eleyi* showed the lowest performance regarding to the coincidence of flowering with the cultivar Fuji (Table 12). Although *M. eleyi* does not show great coincidence of flowering with 'Gala' and 'Fuji' it has the advantage of having great density of flowering for a long period. 'Winter gold', 'John Downil', *M. baccata*, 'Prof. Spengler', 'Profusion', Gala and *M. hopa* were the species that better coincided the flowering with 'Fuji'. From these, 'Prof. Spengler', 'Profusion' and *M. hopa* showed high flowering intensity. Williams (1977) described the wild species as good producers of pollen, and were superior to commercial cultivars. He recommends two to four pollinators to make up for the difference of coincidence of flowering.

Specie/cultivar	Year							Mean ±standart error	Variability index (%)
	2001	2002	2003	2004	2005	2006	2007		
Percentage of coincidence of the blooming period with 'Fuji' apples (%)									
Malus eleyi	0.0*	0.0	100.0	0.0	100.0	100.0	0.0	42.9±53.5 ns	124.7
'Yellow Siberian'	0.0*	0.0*	33.3	23.1	55.6	25.0	44.4	25.9±20.9	80.8
'Milalew Imuni'	0.0*	0.0*	0.0*	0.0*	55.6	0.0	55.6	15.9±27.1	170.8
Malus robusta	0.0*	0.0	0.0*	0.0*	77.8	8.3	77.8	23.4±37.3	159.2
Malus floribunda	0.0*	0.0*	0.0*	0.0*	100.0	0.0*	55.6	22.2±40.1	180.3
Malus platycarpa	64.7	0.0	16.7	0.0	88.9	0.0	77.8	35.4±40.1	113.0
Malus baccata	41.2	0.0	33.3	0.0	66.7	100.0	100.0	48.7±42.1	86.3
Malus hopa	52.9	88.9	41.7	46.2	100.0	8.3	100.0	62.6±34.7	55.5
Malus atrosanguinea	5.9	0.0	50.0	0.0	0.0	100.0	0.0	22.3±38.9	174.5
'Winter gold'	82.4	0.0	16.7	38.5	88.9	0.0	100.0	46.6±43.3	92.8
'John Downil'	47.1	11.1	33.3	76.9	55.6	0.0	100.0	46.3±35.3	76.2
'Prof. Spengler'	29.4	0.0	41.7	7.7	100.0	100.0	66.7	49.3±40.9	83.0
'Profusion'	41.2	0.0	75.0	7.7	100.0	100.0	66.7	55.8±40.9	73.3
Gala	29.4	77.8	58.3	61.5	55.6	41.7	100.0	60.6±23.1	38.2
Mean	28.2	12.7	35.7	18.7	74.6	41.7	67.5	39.9	107.7

*no blooming; ns no significant; Means followed by lowercase in the column do not differ by Tukey's test ($p<0.05$)

Table 12. Coincidence of blooming period of wild species in relation to 'Fuji' apples in seven years of phenological observations. Caçador, Santa Catarina, Brazil, 2007. (Petri et al., 2008).

As exposed in the Table 13 the chilling accumulation during the hibernal period differed among years, having repercussion on the coincidence of flowering among cultivars (Table 11 and 12). Species that show lower variability along the years as for the coincidence of flowering with the cultivars Gala and Fuji, even when there was different chilling accumulation, show good ability as pollinators due their stable behavior at differentiated chilling regime during winter. The occurrence of low variability index in a determined pollinator may indicate that the pollinator responds similarly to the commercial cultivars under changeable chilling accumulation from one year to other. Considering the occurrence of winter with differenced chilling accumulation the use of pollinators that present similar behavior to their respective commercial cultivars is desirable in order to minimize the problems related to the non coincidence of pollination. Therefore, species showing the highest means of coincidence of flowering as well as the lowest index of variability are those that showed the best regularity regarding flowering period and are indicated to be used as pollinators whether considering the flowering period. It was observed that most cultivars showed index of variability superior to 100% which indicates that they have inconstant behavior along years and are greatly influenced by environmental conditions.

In general, it could be observed that no apple wild species had similar flowering phenological behavior to 'Gala' and 'Fuji'. Consequently, the use of a single specie as pollinator does not permit to embrace all the period of flowering of the commercial cultivars.

Soltész (2003) pointed out that it is necessary at least two cultivars to comprise the period of flowering of a particular cultivar of interest. Among those species 'Prof. Spengler', 'Profusion', 'Winter gold' and 'John Downil' are the apple wild species showing good potential of use because of the coincidence of flowering with 'Gala'. 'Profusion' and *M. hopa* were greater by the flowering regularity in the several years observed and flowering intensity in the year´s average. Some species showed no coincidence at some years observed.

Although *M. baccata* and *M. platycarpa* have showed good coincidence of flowering with 'Gala' and 'Fuji' they provided larger fruits which required to be eliminated after flowering to avoid alternating flowering. This condition makes these species inadequate.

Year	Chilling hours below 7.2ºC	Chilling units (Modified North Carolina model)
2001	418	782
2002	269	549
2003	500	824
2004	400	1056
2005	309	571
2006	363	939
2007	535	918

Table 13. Accumulation of chilling hours below 7.2ºC and chilling units according the model of Modified North Carolina from May to August in the years 2001 to 2007. Caçador. SC. Brazil. 2007.

7. Effect of budbreak promoters on apple phenology

The occurrence of simultaneous flowering between cultivars of interest pollinators is an essential condition to assure adequate fruit production for the apple crop. Problems related to pollination and fecundation can reduce apple production and fruit quality due to the low fruitset and number of seed formed per fruit (Brault & Oliveira, 1995; Keulemans et al., 1996).

The flowering period of the cultivars is affected by environmental conditions (Soltész, 2003). Apple cultivars that normally flower in the same time in regions of good adaptation, when grown under warm winter conditions where the chilling exigencies are not totally satisfied trees might present great variability in the period of flowering causing pollination problems and consequently, affecting fruit production. Considering that most of the apple grower regions of the subtropical climate chilling are insufficient to deal with the chilling requirements of the main apple cultivars used, the adoption of management practices aiming coincidence of flowering between cultivars is of a great importance. According Erez (2000) the enhancement of the coincidence of flowering between cultivars can be reached with the use of budbreak promoters which anticipate flowering of the late flowering cultivars. Budbreak promoter's substances can be used to reduce the chilling requirement of cultivars of low and medium exigencies and to modify the budbreak period, flowering and fruit maturation of temperate fruit species (George et al., 2002).

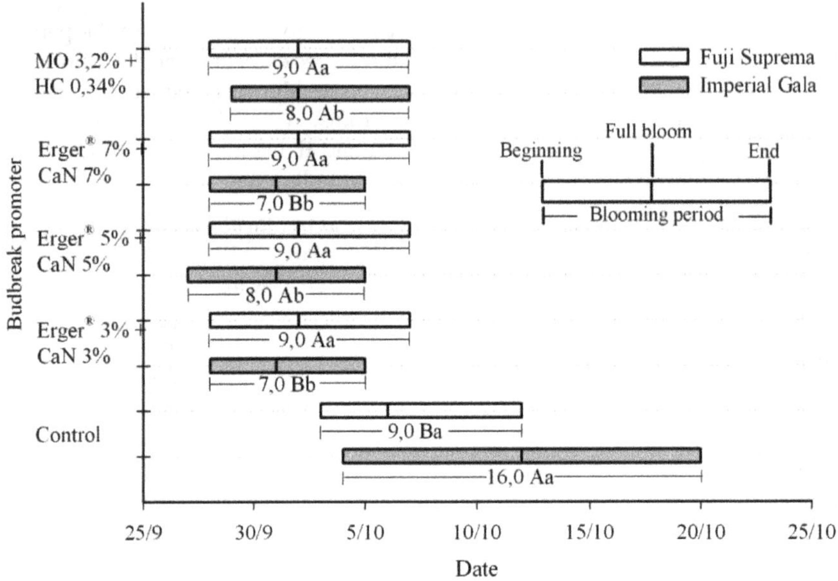

F (budbreak promoter x genotype) = 216,09** CV (%) = 4,96
** F signficant value at 1% of probability;
CaN = Calcium nitrate; MO = Mineral oil; HC = Hydrogen cyanamide;
Means followed by same lowercase letter in the same cultivar do not differ by Tukey test ($p<0.05$), and means followed by same uppercase letter in the same budbreak promoter do not differ by Tukey test ($p<0.05$).

Fig. 9. Flowering period of 'Imperial Gala' and 'Fuji Suprema' apples treated with different budbreak promoters. Caçador, SC, Brazil, 2010. (Hawerroth et al., 2010).

At Fig. 9 it can be observed that the duration of apple flowering differs between cultivars and among budbreak promoters. Results obtained by Hawerroth et al. (2010) under subtropical conditions show that when no budbreak promoter is applied the cultivars Imperial Gala and Fuji Suprema tend to increase the duration of flowering period. According Petri & Leite (2004) the prolongation of the budbreak and flowering phases in apple is a typical symptom of insufficiency of hibernal chilling. The longer the flowering period the more difficult will be the carrying out of some cultural practices such as fruit thinning and diseases control due to the occurrence of different phenological stages in the same tree. Under warm winter conditions, where the chilling exigencies are not completely satisfied, cultivars with distinct chilling requirements showed immense variability in the flowering period from year to year (Petri et al. 2008). The lack of synchrony of flowering between cultivars is clearly observed in the control treatment (Fig. 9) where the full bloom of 'Imperial Gala' coincides with the end of flowering of 'Fuji Suprema'.

The use of substances to induce budbreak can improve the synchrony of flowering between cultivars, evidenced by the greater coincidence in the dates of full bloom between cultivars treated with budbreak promoters (Fig. 10). The occurrence of simultaneous flowering of commercial cultivars and pollinators enhances the possibility to occur high indexes of fruitset (Soltész, 2003).

In relation to flowering period the application of budbreak promoters can anticipate the apple flowering, reducing the time between the application of the treatments and full bloom. The application of budbreak promoters such as mineral oil and hydrogen cyanamid were efficient in the anticipation of flowering of 'Imperial Gala' and 'Fuji Suprema', as observed by Hawerroth et al. (2009) (Table 14). The differenced behavior among the level of the factor budbreak promoters regarding flowering time proves the importance of the use of budbreak promoters as strategies of management to maximize pollination. This is due to the greater synchronization of flowering between cultivars of commercial importance and their pollinators.

Treatmet	Interval between application of treatments and beginning of flowering	
	Imperial Gala	Fuji Suprema
	Days	
Control	30.0Aa	26.0Ba
MO 3.2%	28.0Ab	25.0Bb
MO 3.2% + HC 0.20%	25.0Ac	23.0Bc
MO 3.2% + HC 0.39%	25.0Ac	23.0Bc
MO 3.2% + HC 0.59%	25.0Ac	23.0Bc
Mean	26.6	24.0
CV (%) = 1.12	F (budbreak promoters x genotype) = 29.90**	

** F value significant at 1% of probability. Means followed by uppercase letter in the line, and means followed by lowercase in the column do not differ by Tukey's test ($p<0.05$)

Table 14. Interval between application of treatments and beginning of bloom in 'Imperial Gala' and 'Fuji Suprema' cultivars treated with different concentrations of hydrogen cyanamide (CH) and mineral oil (MO). Caçador, SC, Brazil, 2008. (Hawerroth et al., 2009).

8. Conclusions

Under warm climate, such as the Brazilian conditions, the apple flower induction and differentiation tend to extend along growth cycle as there is vegetative growth. That means that in the meantime as buds flowers are formed the process of induction is still occurring. This particular situation might lead to an alteration in the tree physiology making necessary a differentiated management on tree training. So, cultural and climatic conditions alters positively or negatively the flower development to the next year.

In general, in fruit tree orchards, when trees are grown under unstable climate regions where cultural practices are altered, the timing of flowering and the phenology could have important impacts on fruit production because of possible indirect influences of phenology on pollination and fruit-set efficiency, and consequentially on bud differentiation.

9. References

Arnold, C.Y. (1959). The determination and significance of the base temperature in linear heat unit system. *Journal of the American Society for Horticultural Science*, Geneva, Vol.74, pp. 430-445.

Brault, A. & Oliveira, D. (1995). Seed number and an asymmetry index of 'McIntosh' apples. *HortScience*, Vol. 30, No. 1, pp. 44-46, ISSN 0018-5345.

Brown, D. M. *Fall Workdays in Ontario*. Ontario Agricultural Colege Facsheet: Guelph, 1970. 6p.

Brunini, O.; Lisbão, R. S.; Bernardi, J. B.; Fornasier, J. B.; Pedro Júnior, M. J. (1976). Temperatura-base para alface cultivar "White Boston" em um sistema de unidades térmicas. *Revista de Olericultura*, Lavras, Vol. 16, pp. 28-29.

Bulon, L. & Faust, M. (1995). New aspects of bud dormancy in apple trees. *Acta Hortculturae*, Vol. 395, pp.105-111, ISSN 0567-7572.

Chariani, K. & Stebbins, R.L. (1994). Chilling requirements of apple and pear cultivars. *Fruit Varieties Journal*, Vol. 48, No 4, pp. 215-222, ISSN 0091-3642.

Costes, E. (2003). Winter bud content according to position in 3-year-old branching system in "Granny Smith" apples. *Annals of Botany*, Vol. 92, pp.581-588, ISSN 0305-7364.

Couvillon, G. A., Finardi, N., Magnani, M. & Freire, C. (1984). Rootstock influences the chilling requirement of `Rome Beauty` apple in Brazil. *HortScience*, Vol. 19, No. 2, pp. 255-256, ISSN 0018-5345.

Erez, A. (2000). Bud dormancy: phenomenon, problems and solutions in the tropics and subtropics. In: *Temperate Fruit Crops in Warm Climates*, EREZ, A., pp. 17-48, Kluwer Academic Publishers, ISBN 041263290X, The Netherlands.

Finetto, G. A. (2004). The behavior of some apple rootstocks in relation to the chilling requirement. *Acta Horticulturae*, Vol. 662, pp. 245-251, ISSN 0567-7572.

George, A. P., Broadley, R. H., Nissen, R. J. & Ward, G. (2002) Effects of new rest-breaking chemicals on flowering, shoot production and yield of subtropical tree crops. *Acta Horticulturae*, Vol. 575, pp. 835-840, ISSN 0567-7572.

Hauagge, R. & Cummins, J. N. (2000). Pome fruit genetic pool for production in warm climates. In: *Temperate Fruit Crops in Warm Climates*, EREZ, A, pp. 267-303, Kluwer Academic Publishers, ISBN 041263290X, The Netherlands.

Hawerroth, F. J., Petri, J. L., Herter, F. G., Leite, G. B.; Leonetti, J. F., Marafon, A. C. & Simões, F. (2009). Fenologia, brotação de gemas e produção de frutos de macieira em resposta à aplicação de cianamida hidrogenada e óleo mineral. *Bragantia*, Vol.68, No.4, pp. 961-971, ISSN 0006-8705.

Hawerroth, F.J., Petri, J.L., Leite, G.B. & Herter, F.G. (2010) Brotação de gemas em macieiras 'Imperial Gala' e 'Fuji Suprema' pelo uso de Erger® e nitrato de cálcio. *Revista Brasileira de Fruticultura*, Vol. 32, No. 2, pp. 343-350, ISSN 0370-372X.

Iuchi, V.L., Iuchi, T., Brighenti, E. & Ditrich, R. (2002). Quebra de dormência da macieira (*Malus domestica* Borkh) em São Joaquim. *Revista Brasileira de Fruticultura*, Vol. 24, No. 1, pp. 168-174, ISSN 0370-372X.

Jackson, J. E. (2003). *Biology of apples and pears* (1), Cambridge University Press, ISBN 0521-38018-9, Cambridge.

Keulemans, J., Brusselle, A. & Eyssen, R. (1996). Fruit weight in apple as influenced by seed number and pollinizer. *Acta Horticulturae*, Vol. 42, pp. 201-210, ISSN 0567-7572.

Lauri, P., Maguylo, K. & Trottier, C. (2006). Architecture and size relations: an essay on the apple (*Malus domestica*, ROSACEAE) tree. *American Journal of Botany*, Vol. 93, No. 3, pp. 357-368, ISSN 0002-9122.

Lauri, P. & Trottier, C. (2004). Patterns of size and fate relationships of contiguous organs in the apple (*Malus domestica*) crown. *New Phytologist*, Vol. 163, No. 3, pp. 533-546, ISSN 1469-8137.

Leite, G.B., Bonhomme, M., Lacointe, A., Rageau, R., Sakr, S., Guilliot, A., Maurel, K., Petel, G. & Couto-Rodriguez, A. (2004). Influence of lack of chilling on bud-break patterns and evolution of sugar contents in buds and stem tissues along the one-year-old shoot of the peach trees. *Acta Horticulturae*, Vol. 662, PP. 61-71, ISSN 0567-7572.

Lozada, B.I.; Angelocci, L.R. (1999) Determinação da temperatura-base e de graus-dia para a estimativa do subperíodo da semeadura à floração de um híbrido de milho (*Zea mays*). *Revista Brasileira de Agrometeorologia*, Vol. 7, No.1, pp. 31-36.

Madail, R., Herter, F. G, Leite, G. B, & Petri, J. L. (2010). Influence of Flower Structure in the Flower Production and Fruit Set in Some Apple Cultivars. *Acta Horticulturae*, Vol. 872, pp. 309-318, ISSN 0567-7572.

MOTA, F. S. da. (1957) Os invernos de Pelotas-RS, em relação às exigências das árvores frutíferas de folhas caducas.*Boletim Técnico*. Instituto Agronômico do Sul, Pelotas, 38 p.

Petri, J. L. (2002). Formação de flores, polinização e fertilização. In. *A cultura da macieira*, Epagri, Florianópolis, SC, pp. 229 - 260.

Petri, J.L. (2006) Formação de flores, polinização e fertilização. In: EPAGRI: *A cultura da macieira*. Florianópolis, pp.229-260.

Petri, J. L., Hawerroth, F. J. & Leite, G. B. (2008). Fenologia de espécies silvestres de macieira como polinizadoras das cultivares Gala e Fuji. *Revista Brasileira de Fruticultura*, Vol. 30, No. 4, pp. 868-874, ISSN 0370-372X.

Petri, J. L. & Leite, G. B. (2004). Consequences of Insufficient Winter Chilling on Apple Tree Bud-break. *Acta Horticulturae*, Vol. 662, pp. 53-60, ISSN 0567-7572.

Petri, J. L., Souza, A. & Leite, G.B. (2006). Análise de gemas de macieira como subsídio para orientação da poda. *Agropecuária Catarinense*, Vol. 19, pp. 53-56, ISSN 0103-0779.

Petri, J.L., Palladini, L. A., Schuck, E., Ducroquet, J.P., Matos, C. S. & Pola, A. C. (1996). Dormência e indução da brotação de fruteiras de clima temperado. *Epagri Boletim Técnico*, No. 75, PP. 1-110, ISSN 0100-7416.

Richardson, E.A., Seeley, S.D., Walker, D.R., Anderson, J.L.M., Ashcroft, G.L. (1975). Pheno-climatography of spring peach bud development. *HortScience*, Alexandria, Vol. 10, No. 3, pp. 236-237.

Rodriguez, R. & Rodriguez, M. (1997). Efecto de la posición y del tipo de estructuras fructíferas sobre la calidad del fruto de manzanos cv. Fuji em Rio Negro y Neuquén Argentina. *Fruticultura Profesional*, No. 90, pp. 10-12, ISSN 1131-5660.

Saraiva, I. (1973) *Fenologia das pomóideas*. Frutos 72/73: pp. 25-44.

Saure, M.C. (1985). Dormancy release in deciduous fruit trees. *Horticultural Review*, No. 7, pp. 239-299, ISSN 0163-7851.

Scarpare Filho, J.A., Kluge, R.A., Victória Filho, R., Tessarioli Neto, J. & Jacomino, A.P. (2000). Comportamento de duas cultivares de pessegueiro com interenxerto da ameixeira 'Januária'. *Pesquisa Agropecuária Brasileira*, Vol. 35, pp. 757-765, ISSN 0100-204X.

Skinner, J.E. (1964). Delayed foliation. *Deciduous Fruit Grower*, Vol. 14, No. 7, pp. 195-197, ISSN 0302-7074.

Soltész, M. (2003). Apple, In: *Floral Biology, Pollination and Fertilization Zone Fruit Species and Grape*. Kozna, P., Nyéki, J., Soltész, M. & Szabo, Z., Akadémia Kiadó, p.237-316, Budapest, ISBN 963-05-7816-6.

Tomaz, Z.F.P., Lima, C.S.M., Gonçalves, M.A., Rufato, L. & Rufato, A.D. (2010). Crescimento vegetativo, floração e frutificação efetiva do pessegueiro 'Jubileu' submetido a diferentes comprimentos de interenxertos. *Pesquisa Agropecuária Brasileira*, Vol. 45, No. 9, pp. 973-979, ISSN 0100-204X.

Webster, D. (1976) Factors affecting shape of McIntosh apple fruit. *Canadian Journal of Plant Science*, Vol. 56, pp. 95-105, ISSN 0008-4220.

Williams, R.R.; Maier, M. (1997). Pseudo-compatibility after self-pollination of apple Cox's Orange Pippin. *Journal of Horticultural Science*, Ashford, Vol.52, pp.475-483.

Yazdanpanah, H., Ohadi, D., Tabar, M.S. (2010) Forecasting Different Phenological Phases of Apple Using Artificial Neural Network. *Journal of Research in Agricultural Science* Vol. 6, No. 2, pp. 97 - 106.

Young, E. & Werner, D. J. (1985). Chilling unit and growing hour requirements for vegetative bud break in six apple rootstocks. *Journal of the American for Horticultural Science*, Vol. 110, No. 3, pp. 411-413, ISSN 0003-1062.

12

Mixed Cacti Orchards – A Horticultural Alternative for Mexican Semiarid Tropics

Rodolfo López Gómez[1*], Heriberto Martínez López[2] and Luis María Suárez Rodríguez[1]

[1]Universidad Michoacana de San Nicolás de Hidalgo, Instituto de Investigaciones Químico-Biológicas, Edificio B-1. Francisco J Mujica S/N Col Felicitas del Rio,

[2]Centro Interdisciplinario de Investigaciones para el Desarrollo Integral Regional, Unidad Oaxaca, México

1. Introduction

One of the most determinant phenomenon in horticulture is the phenology of cultivated species, a factor that is poorly known for cacti. Sexual reproduction in the majority of species belonging to the deciduous tropical forest communities (including some cacti) takes place during the middle or the end of the dry season (Rzedowski, 1978). There is an increasing interest in the fruits of some columnar cacti with economic value in the national and international markets. Such is the case in the state of Oaxaca, México of fruits of *Stenocereus griseus* (Haw) Buxbaum, locally known as "pitaya de mayo," *S. stellatus* (Pfeiffer) Riccobono, "tunillo" or "pitaya de aguas" and *Escontria chiotilla* (Weber) Rose, "jiotilla," species which produce fruits of economic relevance for several marginal communities in the state. The fruits of these columnar cacti are consumed fresh or used for making ice cream. The diversity in pulp color is an additional attractive of the fruits of the "pitayas" (*Stenocereus* spp.), which can be red, yellow, orange, purple or colorless, while those of "jiotilla" are invariably of red color. Several works have been made about the nutritional characteristics and diversity of the fruits of these cacti, mainly in their size and color (Ayala & Beltrán, 2007; Benito et al., 1992; Piña, 1977; Beltran et al., 2005; Casas, 2005; Tenango, 2005). The objective of the present work was to observe the phenological behavior of these species in their natural ecosystem, in order to obtain information about differences in their reproductive phenology that could be used for the establishment of mixed cacti orchards in the semiarid tropical lands of Mexico.

2. Materials and methods

A sampling area for phenological data was delimited for each one of the three studied species (Figure 1): 32 plants of "pitaya de mayo" were marked in a rural orchard at Joluxtla,

* Corresponding Author

in the municipality of Cosoltepec, district of Huajuapan de León, Oaxaca, at 1640 m.a.s.l. (Figure 1A); 30 plants of "tunillo" were marked in a rural orchard at Agua del Espino, municipality of La Compañía, district of Ejutla, Oaxaca, at 1420 m.a.s.l. (Figure 1B); and for "jiotilla," an area of approximately 0.1 ha was delimited located 7 Km S of San Pedro Totolápam, municipality of San Pedro Totolápam, district of Tlacolula, Oaxaca, at 950 m.a.s.l. (Figure 1C) 45 plants were marked.

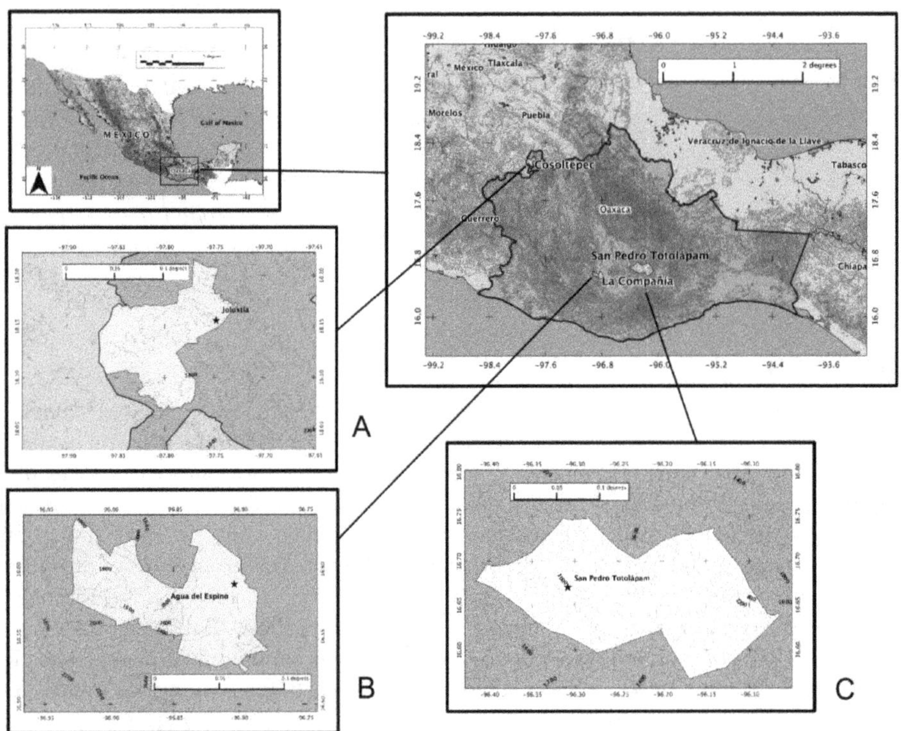

Fig. 1. Localization of study area and sampling sites for phenological data. A. Joluxtla: *Stenocereus griseus,* "pitaya de mayo;" B. Agua del Espino: *Stenocereus stellatus,* "tunillo;" C. San Pedro Totolápam: *Escontria chiotilla,* "jiotilla."

The sampling areas were visited every 30 days throughout the 2005 year and phenological data were recorded. The phenological stages were defined as following:

Fl_1.- Buds before anthesis
Fl_2.- Flowers after anthesis
Fr_1.- Immature fruits
Fr_2.- Mature fruits
Veg.- Vegetative

For the three phenological data sampling areas, ombrothermic diagrams were constructed using the climatic data for the last ten years from the nearest meteorological stations.

3. Results and discussion

The climatic conditions for the three study areas are represent in the ombrothermic diagrams shown in Figures 2A, 2B and 2C. These areas present similar climatic conditions with average temperature between 25-30 °C , a midsummer drought and an average annual precipitation of 600-800 mm.

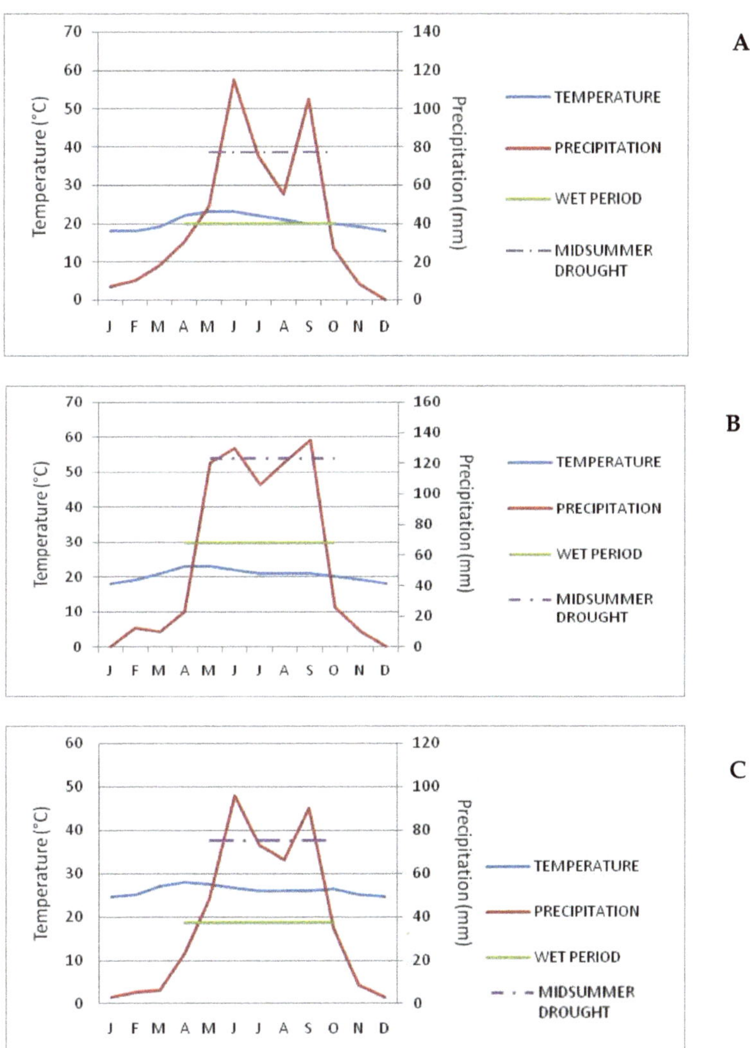

Fig. 2. Ombrothermic diagrams of sampling sites of phenological data of fruit producing columnar cacti; A) Chazumba for "pitaya" B) Ejutla for "tunillo" and C) Totolapan for "jiotilla."

In the "pitaya" area, 76.6% of the plants were observed to have reproductive structures during the months from January to June (Figure 3) with a maximal of fruit production during May (hence the local name of "pitaya de mayo," alluding to the month of May). 90% of the plants in reproductive phase produced mature fruits. *Stenocereus griseus* has also been reported to be of economic importance in Colombia, with two fruit production periods associated to rain seasons (Villalobos et al., 2007).

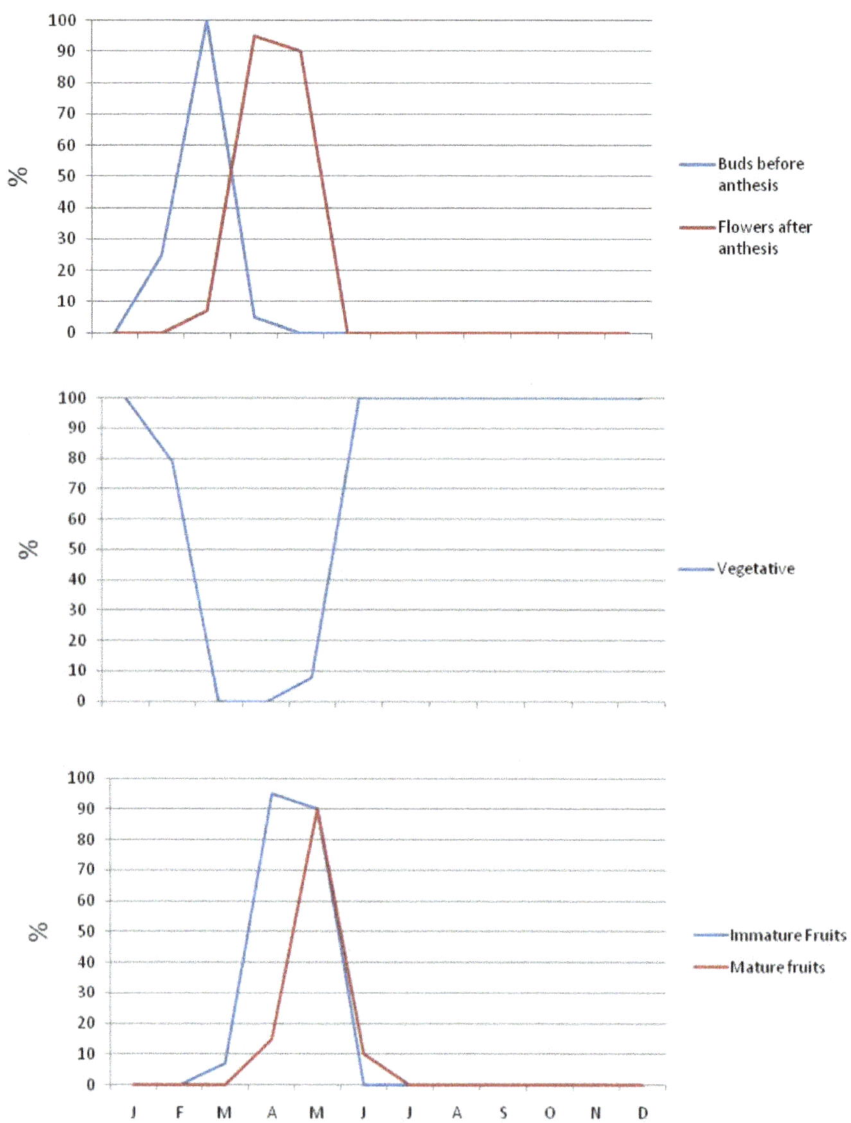

Fig. 3. Phenological diagram for "pitaya" (*Stenocereus griseus*).

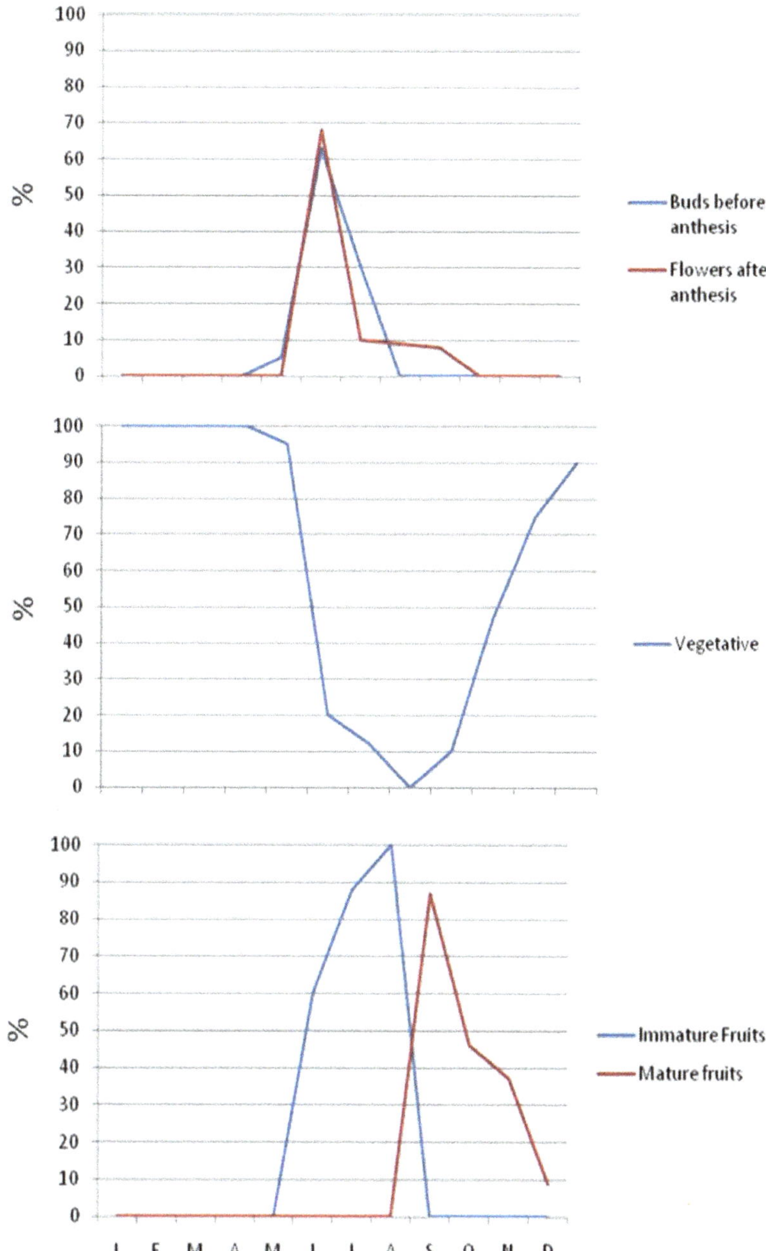

Fig. 4. Phenological diagram for "tunillo" (*Stenocereus stellatus*)

The phenological behavior of "tunillo" (*Stenocereus stellatus*) is represented in Figure 4. The reproductive period of this species is from May to October and mature fruits are present during the time period from June to September, i.e., during the rainy season. A relationship between precipitation and fruit growth has been reported for this species (García-Suárez et al., 2007). The phenology of *Escontria chiotilla* differs from that of *Stenocereus queretaroensis,* which is phenologically more similar to *S. griseus* ("pitaya de mayo") (Pimienta et al., 1994). The majority (80%) of the monitored plants of tunillo presented fruits. The fruits are only produced in the apical part of the stems. Some observations in *Stenocereus stellaus* shown that cultivated populations have more branches and fruit production that wild populations (Casas et al., 1999)

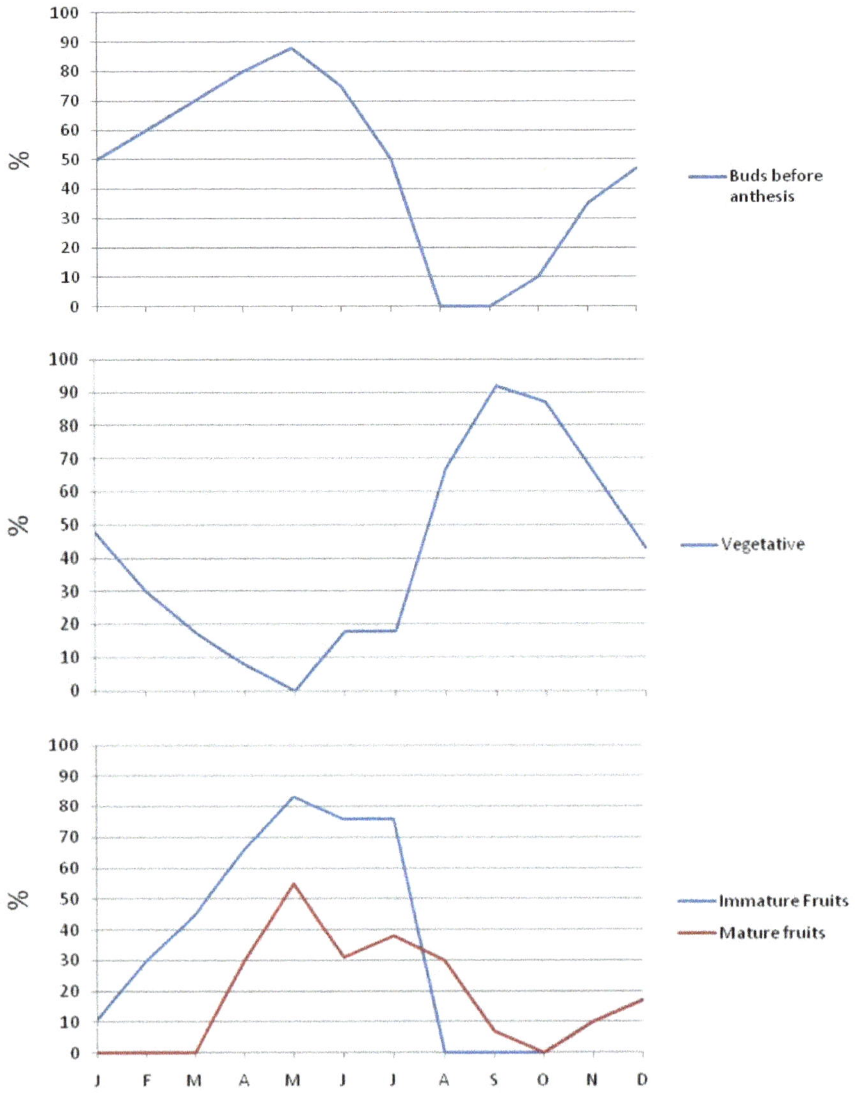

Fig. 5. Phenological diagram for "jiotilla" (*Escontria chiotilla*).

The fruits of "jiotilla" (*Escontria chiotilla*) were produced between January and September and during the time of monitoring only 25.8% of the plants presented reproductive structures, of which 80% produced mature fruits (Figure 5). However, we observed that the majority of marked plants (74.2%) did not bear reproductive structures. This could be the consequence of the presence of many immature plants in the monitored wild population.

Our results suggest that "pitaya" and "jiotilla" have a reproductive period that is similar to the majority of the plants of the tropical deciduous forest, i.e. in the middle or at the end of the dry season, which corresponds to the short day (SD) period. The difference in the reproductive period of the tunillo could be the consequence of a response to either the rainy season or to long days (LD) occurring during the rainy season. Due to the morphological characteristics of cacti flowers, their pollination is carried out by insects or by bats (Osborn, 1988). In the case of *Stenocereus stellatus*, the pollination is principally accomplished by bats and flower anthesis is nocturnal (Casas et al., 1999). These characteristics led us to believe that reproduction of these species could be related to the existence or high activity period of their pollinators. "Jiotilla's" flowers also undergo anthesis during the night.

Our observations show that the studied species respond to climatic conditions in different ways, given that despite growing in similar climatic conditions as seen in the similarity of the ombrothermic diagrams (Fig. 2), their phenological responses are different. Bullock & Solis (1990) state that most elements of the tropical deciduous forest reproduce during the wet season, and that there is a relation between phenology and water availability as is suggested by the flowering of some species during atypical rainfall events in December or January.

Assuming that humidity may be determinant for fruit development in cacti, the pitaya and the jiotilla appear to store water in their stems (Gibson & Novel, 1986) during the rainy season to be used for filling of fruits after the end of the dry season. A bimodal reproductive behavior has been reported for *Stenocereus griseus* (pitaya) in Colombia (Villalobos et al., 2007; Bustamante & Búrquez 2005), suggesting the effect of latitude in the phenology of that specie.

In the case of tunillo (*Stenocereus stellatus*), we observed that water availability determines the change to a reproductive phenological state. García-Suárez et al., (2007) demonstrated a correlation exists between fruit development and the wet season. In arid ecosystems most plants respond to precipitation. The pulse-reserve model addresses the response of individual plants to precipitation and predicts that there are "biologically important" rain events that stimulate plant growth and reproduction (Ogle & Reynolds, 2004).

These two phenological behaviors suggest differences in the flowering-fructification genetic programs of these cacti in response to light (Valverde et al., 2004)

Adequate fruit development depends on a successful pollination. Although the flowers of cacti are hermaphrodite, most species display genetic self-incompatibility so fruit setting is largely dependent on pollinating insects or bats (Bustamante & Búrquez, 2005). In the case of the studied species, with flowers having nocturnal anthesis, bats have been reported to be the main pollinators (Casas et al., 1999; Nassar et al., 1997). The presence of common pollinators for the three studied species strengthens our proposal.

Although the environmental factors determining the phenology of the studied species may not be established based on the data herein presented, and that the phenology of cacti may respond to light too, day length in particular, or to a combination of environmental factors (De la Barrera et al., 2009), it is possible to use phenological behavior for the establishment of orchards with elements of the native plant communities that allow for the sustainable development of marginal rural communities inhabiting the semiarid tropical regions of our country. From a commercial point of view and based on the results of the present work, we propose the establishment of "mixed orchards" using the three studied cacti species taking advantage of their phenological behavior and climatic conditions. Actually, some regions have been proposed for the use of only one of these species in rural orchards (Piña, 1997; Pimienta et al., 1994; Sánchez-Cortéz, 2011). Figure 6 represents the F_2 phenophase for the three studied species; and as can be seen in the graph, marketable fruits were found to be present throughout an 8 month period. If we add that the physicochemical behavior of the fruits of the studied species is known to be similar (Benito et al., 1992), evidence is found supporting that, by establishing mixed orchards of these columnar cacti, local producers in the study area could harvest a sufficient amount of fruit to sustain a diversified agroindustry for the production of processed commodities such as jams, juices, ice cream and colorants, in addition to the marketing of fresh fruit.

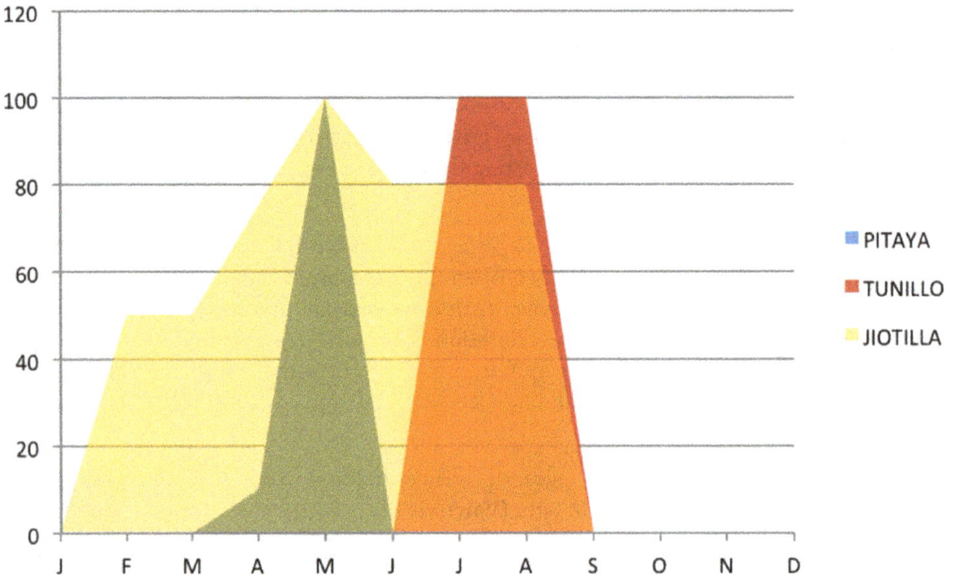

Fig. 6. Annual fruit production periods of "pitaya," "tunillo," and "jiotilla."

4. References

Ayala K., Beltrán M.C. 2007. Caracterización de frutos de Pitaya *Stenocereus griseus* H. IX Congreso de Ciencia de los Alimentos y V Foro de Ciencia y Tecnologia de Alimentos. Memorias 244-251.

Beltran O.M.C., Oliva C.T.G., Gallardo V.T. 2005. La pitaya *Stenocereus stellatus,* como alimento funcional. www.alfa-editores.com/alimentaria/Noviembre

Benito B.P., López G.R., Chávez S.L. 1992. Biologia y Aprovechamiento de los frutos de tres cacataceas *Stenocereus griseus* (Haw, Buxbaum), *Stenocereus stellatus* (Pfeiffer, Riccobono), *Escontria chiotilla* (Weber, Rose). Cuadernos de Investigacion. CIIDIR-Oaxaca. Instituto Politecnico Nacional.

Bullock S.H. & Solis-Magallanes A. 1990. Phenology of canopy trees of a tropical deciduous forest in México. Biotropica 22(1): 22-35.

Bustamante E. & Búrquez A. 2005. Fenología y biología reproductiva de las cactáceas columnares. Cactáceas y Suculentas Mexicanas 50(3): 68-88

Casas A., Valiente-Banuet A., Rojas-Martínez A., Dávila P. 1999. Reproductive biology and the process of domestication of the columnar cactus *Stenocererus stellatus* in central mexico. American Journal of Botany 86(4): 534-542.

Casas A. 2005. El manejo tradicional de una especie puede incrementar la diversidad biologica: el caso xoconochtly. Biodiversitas No 60: 2-6

De la Barrera E., Pimienta-Barrios E., Schondube J.E. 2009. Reproductive Ecophysiology In: Perspectives in Biophysical Plant Ecophysiology: A Tribute to Park S. Nobel, pp. 301-335 Universidad Nacional Autónoma de México. México

García-Suárez F., Carreto-Montoya L., Cárdenas-Navarro R., Díaz-Pérez J.C., López-Gómez R. 2007. Pitaya (*Stenocereus stellatus*) fruit growth is associated to wet season in Mexican dry tropic. Pyton 76: 19-26.

Gibson A.C. & Novel P.S. 1986. The cactus primer. Harvard University Press. Cambridge Massachusetts. USA.

Nassar J.M., Ramírez N., Linares O. 1997. Comparative pollination biology of Venezuelan columnar cacti and the role of nectar-feeding bats in their sexual reproduction. American Journal of Botany 84(8): 918-927.

Ogle K. & Reynolds J.F. 2004. Plant responses to precipitation in desert ecosystems: integrating functional types, pulses, thresholds, and delays. Oecologia 141: 282-294.

Pimienta-Barrios E. & Nobel P.S. 1994. Pitaya (*Stenocereus* spp., Cactaceae): An ancient and modern fruit crop of Mexico. Economic Botany 48(1): 76-83.

Piña L.I. 1977. Pitayas y otras Cáctaceas afines del estado de Oaxaca. Cácatceas y Suculentas Mexicanas XXII: 3-14.

Sánchez-Cortés H. 2011. El cultivo de la pitaya de agosto (*Stenocereus stellatus*) en los municipios de Tlalixtaquilla y Xochihuehetlán, Gro. Desarrollo Sostenible de Zonas Indígenas. Maestría Tecnológica. www.biblio.colpos.mx:8080/jspui/handle/10521/291

Villalobos S., Vargas O., Melo S. 2007. Uso, manejo y conservación de "yosú" *Stenocereus griseus* (Cactaceae), en la alta guajira colombiana. Acta Biol. Colombiana 12(1): 99-112.

Tenango C.E. 2005. Germinación de escontria chiotilla (Rose) Cactaceae. Tesis profesional. Universidad Autonoma Metropolitana Unidad Ixtapalapa.

13

Influence of the Environmental Conditions, the Variety, and Different Cultural Practices on the Phenology of Peach in the Central Area of Santa Fe (Argentina)

Norberto Gariglio, Marcela Weber,
Damián Castro and Norma Micheloud
Facultad de Ciencias Agrarias, Universidad Nacional del Litoral, Argentina

1. Introduction

Phenology is the study of the rhythm of repetitive biological events, the biotic and abiotic causes of these events, and the relationship between phenophases (i.e. periods during which specific phenological events occur) for individual or different species (Lieth, 1974). The knowledge of the phenological patterns of fruit tree crops is essential to estimate their behavior in a new growing area by extrapolation of experimental results from one region to another, or to predict changes in the tree phenology in a given area caused by the variation of the environmental factors between years (Broadhead et al., 2003). Thus, phenological studies can provide criteria for selection of suitable genotypes, helping to improve fruit yield and quality, and reduce environmental risk (Petri et al., 2008).

The time of flowering strongly influences the reproductive traits of animal-pollinated plants (Kudo, 2006), being one of the most important phenological events of the temperate fruit tree crops. The occurrence of late frosts during blooming or at fruit set can severely affect the annual fruit tree production, mainly in early flowering plants such as stone fruits. Harvesting time also has a great agronomic interest in temperate fruit trees, since each variety of peach, plum and nectarine has its own specific harvest time which lasts for about 10 days. Furthermore, these fruits have a short shelf life (Gorny et al., 1999); therefore, cultivation of a large number of crop varieties with a harvesting period in succession in time is required so that a farm or an entire production area can achieve a wide harvest period and a long presence in the regional market.

Fruit tree production in Argentina is over 7.5 million T, mainly including citrus (32.5%), grape (31.5%), pomes fruits (26.5%), stone fruits (5.3%), tropical fruits (3%), and other minor fruits (1.3%) such as walnuts, cherries and figs (CFI, 2011). The Santa Fe province contributes a small proportion of the whole fruit tree production of Argentina (INDEC, 2002), but there is an important encouragement for the expansion of fruit tree crops, due to the propitious ecological conditions, the abundant availability of high quality water, and the possibility of obtaining early fruit production (Gariglio et al., 2009). Nearly 50 new small-scale enterprises over 50 hectares of different crops such as low chilling peach and apple, raspberry, fig and citrus appeared in the last four years in the central area of the province, allowing the diversification of the production system characterized by the cultivation of annual horticultural crops.

One of the most widespread fruit tree crops in Argentina is peach (*Prunus persica* L. Batsch), which is grown on 29,660 hectares mainly located in the Mendoza and Buenos Aires provinces (INDEC, 2002). Both area of production are characterized by the use of medium and high chilling varieties (Carrá de Toloza, 2001; Valentini, 2002). In contrast, in the central-east area of Santa Fe, early maturing peach varieties are mainly used (Gariglio et al., 2009) because their low chilling requirements are compatible with the accumulation of cold in the region (≈ 300 chilling hours), which is lower than half in comparison with the traditional peach production areas of Argentina (Gariglio et al., 2006a).

The quantity of chilling hours accumulated in a certain area is a very important ecological requirement for the adaptation of temperate fruit trees (Erez et al., 1998; Faust, 1989; Lang, 1996; Squire, 1990) due to the influence of chilling on the release of dormancy (Dennis, 2002; Faust et al., 1997). Currently there is no valid method for the quantification of chilling accumulation in the region. Consequently, it is important to test the agronomic behavior of a wide range of varieties, mainly those with low chilling requirements, which are not widely grown and known in Argentina (Alayón Luaces et al., 2004). The introduction and evaluation of low chilling fruit trees in Santa Fe occurred from 1996, and now at least 10 varieties have been recommended for cultivation in view of their adequate phenological and reproductive traits (Gariglio et al., 1999, 2006b, 2009; Ortiz de Zárate et al., 2007). However, these studies are still continuing due to the rapid varietal replacement and the interaction between environment and genotype, which requires many years of research.

The phenology and the agronomic performance of peach trees may be greatly affected by some cultural practices. Autumn defoliation, for example, can modify the time of spring bloom. However, the results have been contradictory; in some cases it was observed that autumn defoliation delayed spring bloom of peach trees (Couvillon & Lloyd, 1978), whereas in others it was reported that warm field temperatures delayed defoliation extending the rest period (Walser et al., 1981). Moreover, removal of leaves of low chilling apple at the onset of winter released dormancy and advanced flowering (Mohamed, 2008).

The study of dormancy is difficult under field conditions where solar radiation, diurnal fluctuations in temperature and other factors cannot be controlled (Dennis, 2003). Consequently, it is considered that experiments using isolated buds as one-node cuttings (Balandier et al., 1993) or excised shoots (Dennis, 2003) are appropriate methods for studying dormancy. This was the methodology adopted in our experiments.

Pruning also greatly affects the performance of the fruit trees. The traditional pruning of peach trees is done during winter (Marini, 2003), and is a time-consuming cultural practice that demands highly skilled labor. In addition, low chilling peach trees in our region have a short period of winter rest which limits the time for winter pruning to only 30–40 days. As a consequence, this agronomical practice is difficult to implement and usually fails to be completed. However, low chilling peach has a long growing season after the harvest period. Thus, pruning the fruit-bearing shoots after harvesting is an alternative cultural practice that may be used to extend the time for pruning and to encourage the growth of the new desirable shoots (Weber et al., 2011). Summer pruning is generally used as a complementary method to winter pruning, but it has not been sufficiently evaluated as an alternative to traditional winter pruning for reproductive shoots renovation. Thus, the effect of summer pruning on the phenology and on a range of vegetative and reproductive parameters of peach trees may be evaluated before recommending this cultural practice.

The aim of this work was to study the phenological behavior and the reproductive traits of different peach and nectarine varieties in the central area of Santa Fe, Argentina. Extreme changes in the phenology of peach varieties in a nine-year period were analyzed taking into account temperature, chilling requirements of the varieties, and the influence of cultural practices such as artificial defoliation and summer pruning.

2. Material and methods

This research was carried out in Esperanza city (31° 26' S; 60° 56' W.; 40 m above sea level), in Santa Fe, Argentina. Table 1 summarizes temperature and rainfall conditions in the study area.

	Average	Maximum	Minimum
Annual temperature (°C)	18.95	19.55	18.08
Maximum annual temperature (°C)	25.61	26.85	24.55
Minimum annual temperature (°C)	13.24	13.63	12.67
Warmest month temperature (January)	25.94	27.20	25.10
Coldest month temperature (July)	12.07	15.50	9.00
Chilling accumulation (CH H)	301	493	183
Annual rainfall	1004	1426	571

Table 1. Main temperature and rainfall parameters in the central-east area of Santa Fe, Argentina. Data are the means of nine years (2002-2010).

Four-year-old peach trees (*Prunus persica* L. Batsch) were used, planted 5 x 3 m apart in a silty-loamy soil and grafted onto 'Cuaresmillo' seedling rootstock, with complementary drip irrigation and trained to the standard open vase system. Fifteen peach and five nectarine (n) cultivars with different chilling requirements were evaluated: 'Spring Lady', (650 chilling hour (CH) requirement), 'Maycrest' (600 CH), 'June gold' (520 CH), 'Flavorcrest' (509 CH), 'Flordaking' (450 CH), 'Hermosillo' (350 CH), 'Don Agustín' (Fla. 81-12; 300 CH), 'San Pedro 1633' (297 CH), 'Early grande' (260 CH), 'Tropic snow' (250 CH), 'Flordastar' (250 CH), 'Flordaprince' (150 CH) (Valentini, 2002), 'Fla. 91-8c' (100 CH, according to its behavior in Santa Fe, Argentina), 'Fla 1-8', 'Fla 1-8 bis', 'Lara' (n) (350 CH), 'Carolina' (n) (325 CH) (INIA Salto Grande, 2003), 'Nectarina 22', 'Nectarina 24', and 'Nectarina 29'. Where cultivars' chilling requirements are not presented, it is because they are unknown, but according to their vegetative and reproductive traits, using the methodology of excised shoots exposed to artificial chilling (Dennis, 2003), and observations of their behavior in the experimental orchards in the central area of Santa Fe (unpublished data), their need is lower than 350 CH.

2.1 Phenology and reproductive traits of peach and nectarine

Trees of each variety were chosen by their uniformity in size and vigor. During the winter, 10 homogeneous current season shoots per plant were randomly selected at 1.8 m above the ground level and their length was measured. The phenological phases of leafing, beginning of flowering, full flowering, and fruit set, were observed weekly from the end of the winter rest to the end of harvest time in these twigs by using the Biologische Bundesanstalt, Bundessortenamt and Chemical industry (BBCH) scale for stone fruit (Meier et al., 1994). Furthermore, the number of new vegetative shoots, flowers and fruits were measured weekly on the one-year-old selected shoots from the release of dormancy to the stage of pit hardening. Data were expressed as the mean number of flowers per meter of shoot length, percentage of vegetative bud break, and percentage of fruit set. The percentage of fruit set was calculated as the ratio between the number of fruit at pit hardening and the maximum number of flowers measured at full bloom in each selected twig. The experiment was conducted over a nine-year period (2002–2010), with the exception of 'Fla 1-8', 'Fla 1-8 bis', 'Nectarina 22', 'Nectarina 24', and 'Nectarina 29', in which the study was conducted over three years (2008–2010).

A randomized complete-block design with one-tree plots and eight replications per cultivar was used. Analysis of regression and variance was performed on the data, and means were compared by Tukey's test ($p \leq 0.05$). Extreme date of the occurrence of each phenological stage in the nine-year period were recorded and expressed in days. Extreme variation of the medium month temperature was also registered and expressed as a percentage of the medium value. The relationships between the variability of the phenological stage with the chilling requirement of the variety, and with climatic data were performed using the Infostat software.

2.2 Autumn defoliation and peach phenology

This experiment was conducted over three consecutive years (2005–2007) using trees of the cv. 'Flordaking'. Two treatments were performed; at the beginning of leaf fall plants were manually defoliated (D) whereas another group of plants remains as control (C) in which the physiological process of autumn leaf fall occurred naturally. Ten twigs per tree were randomly and periodically collected from both treatments (20, 50, 65, and 90 days after leaf fall) from leaf fall to the end of July, to measure the evolution of dormancy during the rest period. Twigs were cut into segments 15 cm long obtaining 80 stem cuttings for each treatment. Only one node with three axillary buds (two floral buds and one central leaf bud) per stem cutting was kept, removing the remaining buds.

Excised shoots were placed with their basal tip in water and forced in a phytotron with an 8-h photoperiod (22.5 mmol m^{-2} s^{-1}) (Citadin et al., 2001), at 20.0 ± 1.0 °C. The basal ends of the shoots were cut weekly and water was replaced daily (Balandier et al., 1993; Citadin et al., 1998). Floral and leaf budbreak were observed three times a week. The number of buds that reached the balloon or green tip stage was recorded (Citadin et al., 2001). Results were expressed as the percentage of excised shoots that reached floral and vegetative budbreak, and as the mean time of budbreak (MTB). MTB was expressed in days (arithmetic mean of each eight groups of 10 excised shoots) (Balandier et al., 1993). This trial was conducted in a complete randomized design with eight replicates of 10 cuttings per treatment. In the field, phenological phases and reproductive traits of peach trees were observed on selected shoots as was described previously (see item 2.1). A randomized complete-block design with one-tree plots of eight replications each was used.

2.3 Summer pruning and peach phenology

Two experiments were conducted over seven consecutive years (2002–2008), using six-year-old peach trees [*Prunus persica* (L.) Batsch], cvs. 'Flordaking' and 'Early grande' grafted onto 'Cuaresmillo' seedling rootstocks.

In the first experiment (2002–2004), representative trees of the cv. 'Flordaking' selected for their uniformity in size and trunk girths received different treatments after harvesting, during the first week of November. One group of plants remained unpruned (NP) as control trees, whereas a second group of plants was pruned each year (P). Summer pruning consisted of heading back the recently harvested bearing shoots to basal current season shoots which were at least 10 cm long; bearing shoots were thinned out when a new current season shoot (10 cm long) protruded directly from the branch framework. In order to avoid sunburn, care was taken not to prune branches that protected the scaffold from direct sunlight. In both treatments, vigorous watersprouts were removed after harvesting. NP trees received traditional dormant pruning 15–25 days before the beginning of their bloom; in P trees, a light pruning was also performed during the dormant period to remove branches that were previously left to avoid sunburn of the tree scaffold.

In the second experiment, using the cv. 'Early grande', pruning was performed at different times from harvest to 15–20 days before the next flowering. Six times of pruning were evaluated: November, December, February, March, April, and June. The combination of summer pruning with traditional winter pruning (WP) was also evaluated, being performed on 11 treatments: (Nov-P, Nov-P+WP, Dec-P, Dec-P+WP, Feb-P, Feb-P+WP, Mar-P, Mar-P+WP, Apr-P, Apr-P+WP and Jun-P). The last treatment, in which pruning was performed 15–25 days before the beginning of blooming, was considered as control treatment. In trees that received summer and winter pruning, the last were performed to remove branches that were previously left to avoid sunburn of the tree scaffold and to adjust the number of bearing shoots.

In both experiments, phenological phases and reproductive traits of the treatments were evaluated on selected shoots as was mentioned in section 2.1. Moreover, the number and distribution of fruits per plant were evaluated at harvest taking into account their distribution in three levels of plant height; L1: lower, up to 1.5 m height; L2: intermediate, from 1.5 to 2.5 m height; and L3: upper, from 2.5 m and higher.

A randomized complete block design with a single tree plot of five replications was used. The data was tested for normal distribution and variance homogeneity and means were compared via an LSD test using the software Statgraphics plus for Windows 3.1 (Statistical Graphics Corp). A 5% probability level was used to indicate significant differences between treatments.

3. Result and discussion

3.1 Field behavior of peach and nectarine

3.1.1 Time of occurrence of the different phenological stages

The most precocious peach variety began to sprout the first week of July ('Fla. 91-8c'), but 80% of the low chilling cultivars sprouted during 10 days from July 11th. High chilling requirement genotypes (> 500 CH) showed sprouting during the second and third weeks of September, two months later than the low chilling group. With the exception of 'Maycrest' and 'Spring Lady', all cultivars reached over 60% of budbreak, despite its chilling requirement. These data did not show significant changes from data reported previously for a minor number of varieties and for a shorter period of analysis (Gariglio et al., 2009).

Full bloom occurred from the middle of July to the first week of August for the low chilling varieties; 'Carolina', 'Lara', 'Flordaprince', and 'Flordaking' being the later cultivars of this group (Fig. 1). On the other hand, full blooming was observed from September 10th to 20th for high chilling varieties. Most low chilling cultivars reached fruit set from August 10th to 23rd, whereas those varieties with higher chilling requirements reached fruit set from the end of September, around two months later than the most precocious genotypes (Fig. 1).

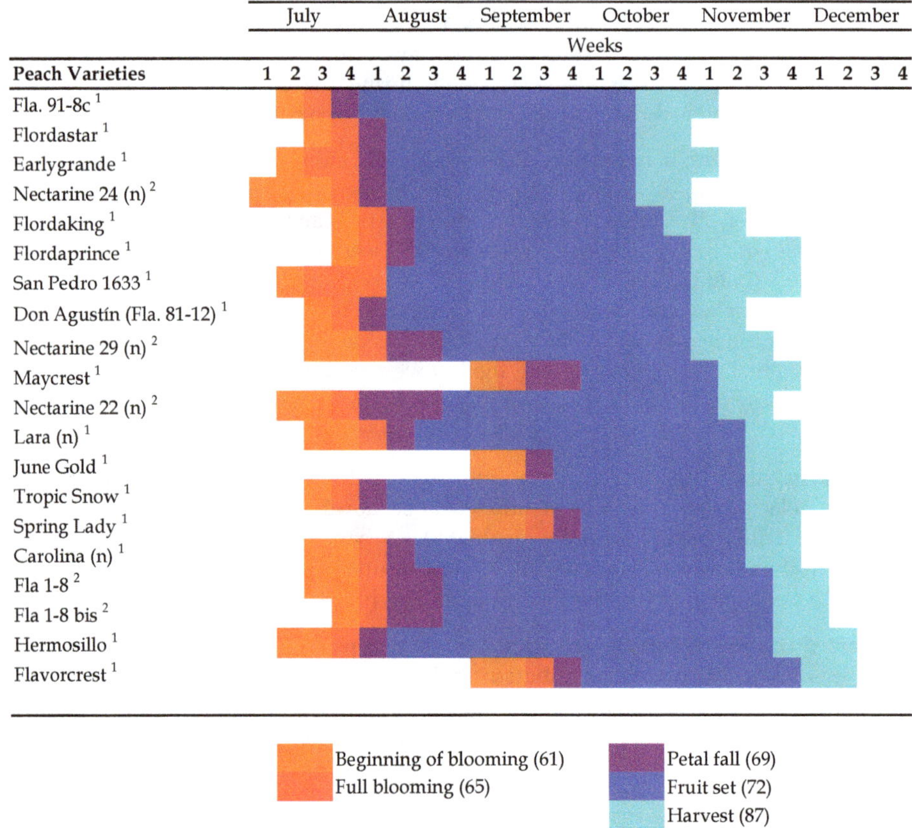

Fig. 1. Main phenological stages for different peach and nectarine (n) cultivars grown in the central-east area of Santa Fe, Argentina. Data are the means of nine-year (2002–2010) (1) or three-year (2008-2010) (2) periods. Numbers in the color legend indicate the phenological stages according to the BBCH scale (Meier et al., 1994).

Harvesting began on October 20th with the cv. 'Flordastar' (Fig. 1), which was around 10 days before the harvesting date of the most important peach production areas of the central region of Argentina (Valentini & Arroyo, 2000), and finished in the middle of December with the cv. 'Flavorcrest'. The period of harvesting was reduced in 10 days when only low chilling varieties were taken into account, with 'Hermosillo' as the latest harvest variety.

The harvest showed continuity but the incorporation of new low chilling varieties should be evaluated to prolong the harvesting period. The use of varieties more precocious than 'Flordastar' does not seem to be appropriate because of the risk of late frost occurrence and the difficulty in obtaining a proper fruit size. However, it is possible to extend the whole harvest period, and improve fruit size and quality, with the use of varieties with a later harvest period than 'Hermosillo'. In this way, new low chilling varieties recently incorporated in our experimental orchard are being tested with promising results; therefore, in the future, the harvest period would be extended, reaching nearly three months of duration.

3.1.2 Influence of the chilling requirements of the variety and the climate on the tree phenology

The extreme variation in the occurrence of the phenological stages of 15 varieties with different chilling requirements during nine years of study showed that the phenological variability diminished with the advance of the growing cycle (Fig. 2). The extreme variation at the beginning of flowering was nearly 35 days, whereas it reached 21 days at fruit set and only 11 days at harvest (Fig. 2). However, it was observed that the variability at the beginning of blooming reached 55 days for the cv. 'Flordastar', and 48 and 49 days for the cvs. 'Tropic snow' and 'Hermosillo', respectively. On the other hand, the extreme variation at this stage reached only 18 days in 'June gold' and 23 days in 'Maycrest', indicating a reduction in the phenological variability with the increase of the chilling requirement of the variety (Table 2). The model that expresses the relationship between the chilling requirement (x) and the extreme variation (in days) of each variety at a given phenological stage (y) was significant ($P < 0.05$) from the beginning of flowering to the beginning of fruit set, despite the medium value reached by the coefficient of regression. In contrast, the variation of the phenological stages between extreme years did not show any relationship with the chilling requirement of the variety from the stages of fruit set to harvesting (Table 2).

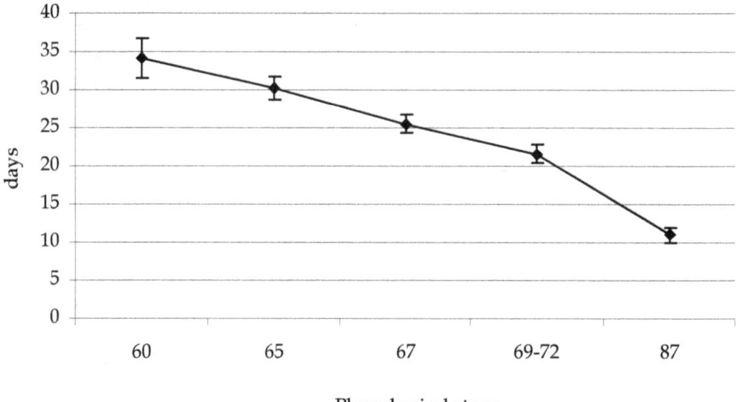

Fig. 2. Maximum variation (in days) in the occurrence of different phenological stages of peach and nectarine trees in the central-east area of Santa Fe, Argentina. Data are the means of 15 varieties with different chilling requirements (150-650 CH) during nine years. Phenological stages according to the BBCH scale (Meiers et al., 1994): 60 = beginning of flowering; 65 = full flowering; 67 = beginning of fruit set; 69-72 = fruit set; 87 = harvesting.

Phenological stage	Model	r^2	p-value
Beginning of flowering	y = −0,0344x + 46,65	0.23	0.046
Full flowering	y = −0,0280x + 40,39	0.32	0.045
Beginning of fruit set	y = −0,0239x + 34,21	0.38	0.025
Fruit set	y = −0,0023x + 22,43	0.003	0.850
Harvesting	y = 0,0078x + 8,184	0.071	0.380

Table 2. Equation, coefficient of regression (r^2) and p-value of the model that represents the relationships between the extreme variations (*y*, in days) of different phenological stage occurrence during the nine-year period (2002–2010) with the chilling requirement of the variety (*x*).

The higher phenological variability observed in low chilling peach trees can be explained by taking into account the physiology of dormancy of peach and the temperature conditions in the central area of Santa Fe. The depth of dormancy (measured by the MTB value) declines with the accumulation of chilling in all peach varieties (Balandier et al., 1993; Citadin et al., 2001; Gariglio et al., 2006b). This means that the need of heat, expressed as growing degree day (GDD) accumulation over 7°C to reach flowering decreases with chilling (Balandier, 1993; Citadin et al., 2001). However, low chilling peach trees have a very low depth of dormancy that allows them to reach more than 50% of floral budbreak in less than 30 days (MTB value) when they were forced at 25°C and without any need of chilling (Gariglio et al., 2006b). On the other hand, peach varieties with higher chilling requirements (> 500 CH) were unable to reach budbreak when they were forced at 25°C due to their major depth of dormancy (major MTB value), and they need a certain accumulation of chilling to diminish their heat requirement for blooming. However, during May and June, only 120 chilling hours were accumulated as an average of the last nine years in the central area of Santa Fe (Gariglio et al., 2006a), which is insufficient to cause an important modification of the depth of dormancy of these varieties.

Furthermore, the major variability of the average medium temperature and that of the monthly heat accumulation observed in the central area of Santa Fe occurs from May to August, being highest during July (Fig. 3), coinciding with the period of dormancy and flowering of low chilling peach. As a consequence, the variability of the temperatures between years that occurs during the first period of dormancy, and that modifies chilling and GDD accumulation, mainly affects the phenology of the next growing cycle of low chilling varieties, explaining the major variation of the phenology of these groups in comparison with that of high chilling varieties. This last group reaches blooming when the climatic variability decline and the GDD accumulation increases nearly 80% in comparison with that of the time of flowering of low chilling peach. Furthermore, the high GDD accumulation allows compensation for the differences in the phenology between years of high chilling varieties in a minor lapse of days, reducing its variability. In the same way, high GDD accumulation and the lower climatic variability can explain why the difference in the occurrence of the phenological events between years diminished with the advance of the growing cycle (Fig. 2).

3.1.3 Reproductive traits of peach and nectarine

Most of the studied varieties showed a flower density (FD) between 20 and 30 flowers per meter of mixed shoots (FMMS). Nectarines reached the higher FD, around 50 FMMS, with the exception of 'Nectarine 24' (Fig. 4A). 'Tropic snow' and 'Hermosillo' were the only peach cultivars that were able to reach an FD like nectarines, whereas 'Flordastar', 'Fla 1-8 bis', and 'Nectarine 24' showed more than 30 FMMS. The remaining cultivars presented lower values of FD, the lowest one being 'Flavorcrest' (Fig. 4A). Flower density showed relatively low values in comparison with other researches. 'Springlady' peach cultivated under Mediterranean climatic conditions showed an FD 2.2-fold higher (51.1 flowers m^{-1}) (González Rossia et al., 2007), whereas the average FD of 33 peach germplasm was 67 flowers m^{-1} in the highlands of central Mexico, with only two peach varieties that showed less than 30 flowers m^{-1} (Pérez, 2004).

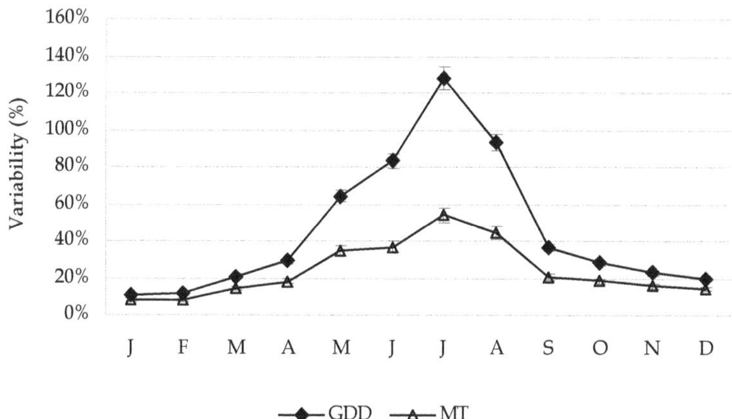

Fig. 3. Maximum variability of the monthly growing degree day (GDD) accumulation and the medium month temperature (MT) during a nine-year period (2002–2010) in the central area of Santa Fe, Argentina. GDD was calculated using a base temperature of 7°C. Standard error is represented by the vertical bar.

Fruit set was around 30% in several varieties, with only three genotypes that reached over 40% of fruit set ('Flordastar', 'Flordaprince' and 'Flordaking') (Fig. 4B). 'Don Agustín', 'Lara', 'Nectarine 22' and 'Nectarine 29' were the low chilling varieties that showed the lowest fruit set, similar to that observed in high chilling ones (6.3% to 15.7%) (Fig. 4B).

Fruit size showed great differences among genotypes. 'Hermosillo' and 'Fla 1-8 bis' had the biggest fruit size, whereas six other varieties reached a fruit size near 120 g $fruit^{-1}$ ('Early grande', 'Fla 1-8', 'Flordaking', 'June gold', 'Nectarine 22', and 'Tropic snow') (Fig. 4C). 'Flordastar' and nectarine 'Carolina' were the low chilling varieties that had the lower fruit size. High chilling varieties also showed low fruit size with the exception of 'June gold'. 'Flavorcrest' was not able to develop normal fruit, being small with a high percentage of aborted embryos. Fruit size was lower in this experiment than that observed in Valencia, Spain (Badenes et al., 1998), but these differences varied among varieties; it was slightly different for 'Tropic snow' but significant for 'Flordaking', 'Flordastar', and 'Flavorcrest'.

As a consequence of their better reproductive behavior, 'Hermosillo' and 'Tropic snow' had the highest fruit yield (Fig. 5). The other low chilling varieties (≤ 450 CH) had lower fruit yield than these varieties. In 'Flordastar' and 'Carolina', fruit yield was mainly limited by fruit size, whereas in 'Don Agustín', 'Lara', 'Nectarine 22', 'Nectarine 24' and 'Nectarine 29', fruit yield was limited by fruit set. Fruit yield of high chilling requirement varieties (≥ 500 CH) was insignificant due to its lower fruit set (Fig. 4B), and late fruit drop.

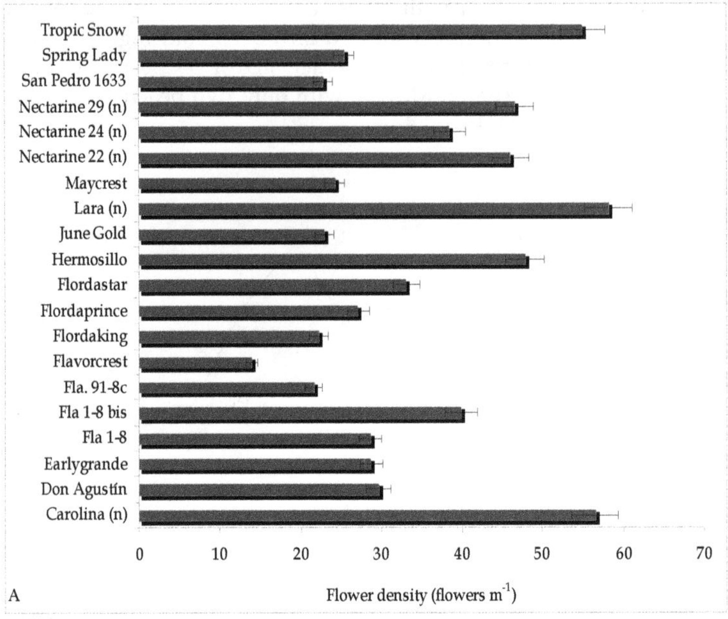

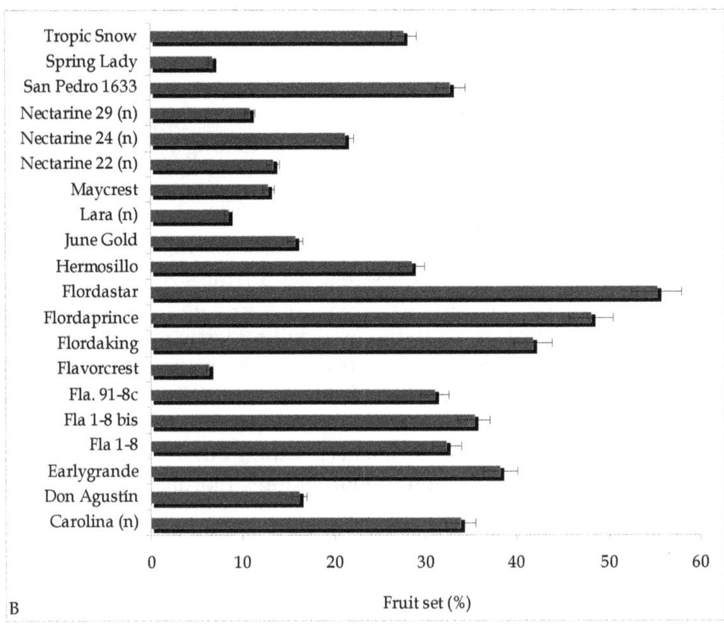

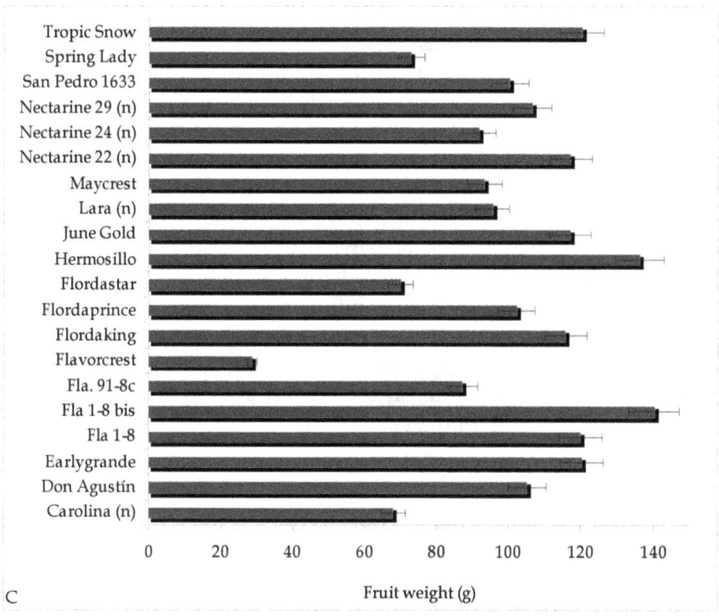

Fig. 4. Flower density (A), fruit set (B), and fruit weight (C) of different peach and nectarine (n) varieties grown in the central-east area of Santa Fe, Argentina.

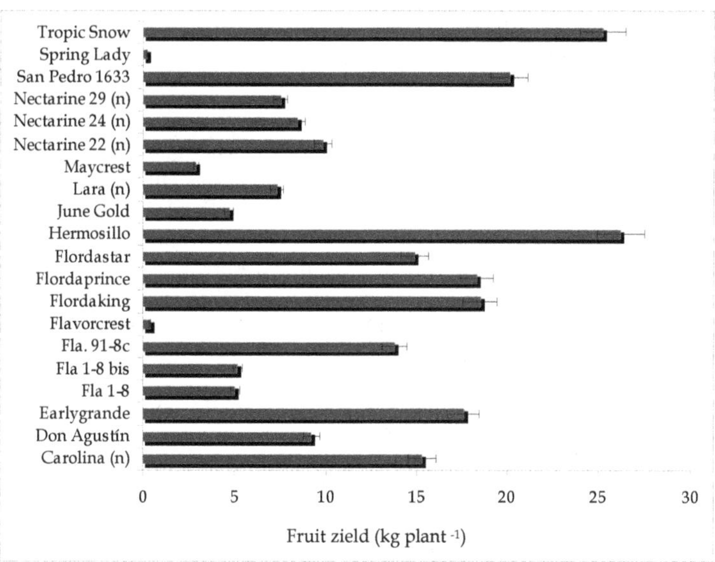

Fig. 5. Fruit yield of different peach and nectarine (n) varieties grown in the central-east area of Santa Fe, Argentina. Data are the means of a nine-year period (2002–2010) beginning with four-year-old plants, with the exception of 'Fla 1-8', 'Fla 1-8 bis', 'Nect. 22', 'Nect. 24', and 'Nect. 29', which are the means of a three-year period.

In the temperate fruit tree crops, the most critical phenological period affected by frost is from full bloom until fruit set (stages 71 and 72 in the BBCH scale for stone fruits) (Fideghelli, 1987). September 5th is the average date for last frost occurrence in the central area of the province of Santa Fe (Panigatti, 1980). Consequently, in our experiment, blooming and fruit set took place during the period of frequent late frost occurrence for low chilling but not for high chilling varieties. However, although high chilling varieties can avoid spring frost, they showed a very low fruit yield because they were unable to cover its chilling requirement in the central area of Santa Fe (Gariglio et al., 2006a), explaining the negative correlation between flowering density, fruit set and crop load with the chilling requirement of the cultivar that was observed for peaches and nectarines grown in the region (Gariglio et al., 2009). In contrast, low chilling varieties showed a better adaptation in spite of the high risk of late winter frost, because these cultivars exhibited an extended blooming period and had a higher opportunity for fruit set after the occurrence of low temperature (Pérez, 2004). Furthermore, in the central area of Santa Fe, the Paraná River tempers the intensity and duration of frosts.

Low yield of the group of high chilling varieties can also be attributed to the reduction of the pollination and fertilization caused by relatively warm temperature occurrence at the time of flowering and fruit set as was mentioned for peach accessions with late blooming in the Mexican subtropics (Pérez, 2004).

3.2 Modification of peach phenology caused by the time of autumn defoliation

Artificial defoliation of 'Flordaking' peach performed at the beginning of natural leaf fall advanced sprouting and beginning of blooming of the next growing cycle by 15 days, in comparison with the control. This difference between treatments decreased to 8 days at full flowering and fruit set, and disappeared at harvest time. Furthermore, autumn defoliation did not affect the percentage of budbreak, but reduced fruit set (−40%) and crop load at harvest (−35%).

The depth of dormancy (MTB value) of floral buds from shoots excised during winter rest was highest at leaf fall for both treatments, and decreased during 65 days, remaining constant until the end of dormancy (Table 3). However, artificial defoliation reduced to half the depth of dormancy at leaf fall in comparison with the control. Differences between treatments diminished one month later, and it disappeared from 50 days after leaf fall to the end of dormancy (Table 3). The MTB value of vegetative buds was also affected by defoliation but the reduction was only 12% at leaf fall and the differences between treatments disappeared 20 days later.

Defoliation only modified MTB on shoots excised from the trees during the first 40-45 days after leaf fall because the endodormancy of 'Flordaking' peach is released at this time; the depth of dormancy (MTB value) of its buds did not decrease further over time, i.e. MTB does not show further decreases with chilling, indicating that buds are under eco-dormancy (Balandier et al., 1993; Dennis, 2003). Defoliation also reduced fruit set with regard to control trees, as in apple (Mohamed, 2008), perhaps for the major probability of lower temperature occurrence and lower activity of pollinators during flowering.

The role of the presence of leaves during autumn in the onset and progression of dormancy is attributed to the perception of the photoperiodic stimulus via phytochrome-mediated signaling that triggers the onset of dormancy of deciduous trees under a shortening photoperiod (Rinne & van der Schoot, 2004), mainly by the stimulation of the synthesis of ABA and other growth inhibitors (Tanino, 2004). Thus, defoliation near the period of natural leaf fall interferes with the physiological process of the onset of dormancy, explaining the great reduction of the depth of dormancy (Table 3) and the advances of blooming in the next spring that was observed in this work.

Time of shoot extraction	MTB (days)		Significance
	Control	Defoliation	
May 3rd	47.0	22.5	*
May 24th	27.1	20.0	*
June 22nd	11.5	11.0	ns
July 7th	7.2	8.5	ns
July 29th	6.5	8.8	ns

* = significant; ns = not significant at $P \leq 0.05$

Table 3. Effect of autumn defoliation on mean time to budbreak (MTB) of floral buds from shoots of 'Flordaking' peach excised at different times of winter rest period and forced at a constant temperature (20°C).

3.3 Modification of peach behavior caused by summer pruning

3.3.1 Summer pruning and autumn defoliation

Summer pruning showed a great effect on the time of autumn leaf fall, and blooming in the next spring. Natural leaf fall of 'Flordaking' peach began at the end of March and pruning performed after harvest delayed it 11 days (Fig. 6). Summer pruning after harvest also changed the patterns of defoliation, which showed significant differences between treatments until late April. However, complete defoliation was reached during the first fortnight of May in both treatments (Fig. 6).

When pruning was performed at a different time of the growing cycle using the cv. 'Early grande', it was observed that autumn defoliation was also delayed by summer pruning after harvest (November), regardless of whether the trees were pruned again during winter or were only summer pruned (Fig. 7). On the other hand, autumn defoliation occurred earlier when peach trees were pruned at the beginning of February or February and June; these trees reached 50% of defoliation in the middle of February, one month earlier than trees pruned during April or June, and nearly two months earlier in comparison with trees pruned during November (Fig. 7). Pruning performed at the beginning of April did not affect the evolution of leaf drop in comparison with trees that were pruned only during the winter period (Control = Jun-P, Fig. 7).

In Japan, summer pruning after harvest also delayed leaf fall by two months in comparison with non-summer pruned trees (Hossain et al., 2005). Leaf senescence is an active process programmed by genetic information and modulated by several plant hormones and environmental factors, but basically it is an age dependent process (Srivastava, 2002). Consequently, the delay of leaf fall caused by summer pruning can be explained by the presence of younger leaves at the time of leaf fall in trees that were summer-pruned after harvest (November) in comparison with trees that were not summer-pruned or were pruned in February. Summer pruning after harvest but not summer pruning at the beginning of

February induces the appearance of new axes which are able to continue growing almost to the end of the vegetative period (Médiène et al., 2002; Weber et al., 2011; Weibel et al., 2003).

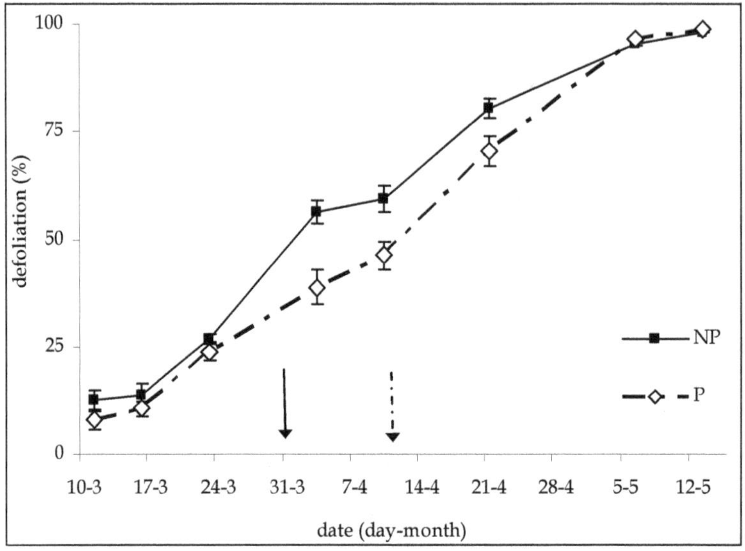

Fig. 6. Pattern of leaf senescence (percent of leaf defoliation) of peach trees, cv. 'Flordaking', pruned (P) or not pruned (NP) after harvest. Vertical bars represent the standard error for each sample. Arrows indicate the time at which each treatment reached 50% of defoliation.

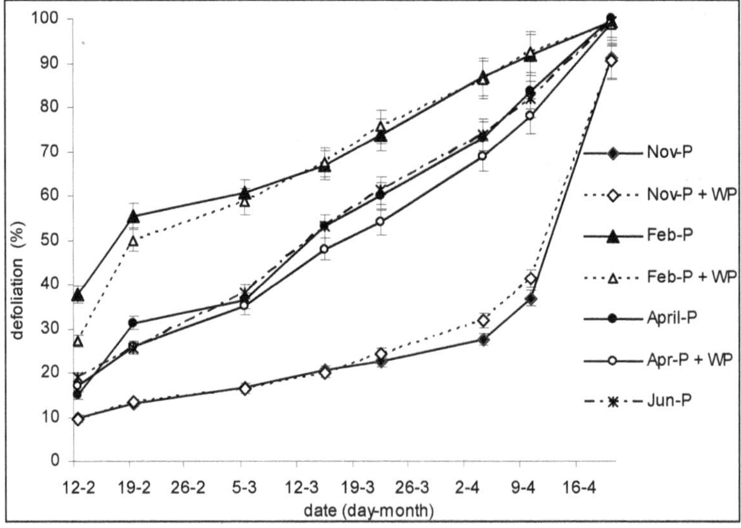

Fig. 7. Pattern of autumn defoliation evolution in 'Early grande' peach trees pruned only in summer (Nov-P; Feb-P; Apr-P), both in summer and winter periods (Nov-P + WP; Feb-P + WP; Apr-P + WP), or only during winter (Jun-P, control treatment). Data correspond to 2007.

3.3.2 Summer pruning and time of blooming

As was discussed previously for the autumn defoliation experiment, the time of leaf fall can also affect dormancy induction and release. Thus, the delay of leaf senescence caused by summer pruning after harvest can explain why this treatment delayed the occurrence of the different phenological stages of flower buds of 'Flordaking' peach in the next spring. Non-summer pruned plants reached the beginning of flowering eight days before and were harvested three days earlier in comparison with summer-pruned plants (Table 4).

Phenological growth stage	BBCH[1]	Date		Difference
		NP	P	(Days)
Full flowering	65	August 28th	September 05th	+8
Fruit set	69–72	September 11th	September 14th	+3
Harvest	87	November 10th	November 13th	+3
First leaves emerging	10	August 15th	August 18th	+3

(1)Phenological growth stage according to BBCH (Meiers et al., 1994)

Table 4. Main phenological events of peach trees, cv. 'Flordaking', pruned (P) or not pruned (NP) after harvest.

However, the modification of the tree phenology by summer pruning in relation with autumn defoliation was easier to observe in the experiment in which 'Early grande' peach was pruned at a different time of the growing cycle. The treatments of pruning that delayed autumn defoliation (Nov-P) also delayed spring bloom by 7 to 10 days in comparison with the control; on the other hand, the treatments that advanced autumn defoliation (Feb-P) also advanced spring bloom by 6 to 8 days (Table 5). This effect was explained by the modification of the depth of dormancy caused by summer pruning; pruning during February (late summer) showed the lowest depth of dormancy at leaf fall in comparison with the other treatments, and this effect was attributed to the early defoliation caused by pruning (Lloyd & Firth, 1993; Weber et al., 2010). Trees defoliated closer to the time of natural defoliation advanced the blooming date over naturally defoliated controls; however, this effect was variable when defoliation occurred early (Lloyd & Firth, 1990).

Treatment	Full flowering deviation (days)
Nov-P	+10
Nov-P + WP	+7
Feb-P	−8
Feb-P + WP	−6
April-P	−1
April-P + WP	0

Table 5. Phenology variation (days) regarding control (winter pruning; Jun-P) at the phenological stage of full flowering in peach trees, cv. 'Early grande', pruned only during summer (Nov-P; Feb-P; Apr-P), or both in summer and winter period (Nov-P + WP; Feb-P + WP; Apr-P + WP).

3.3.3 Relationships between summer pruning, autumn defoliation and time of blooming

Summer pruning treatments performed at a different time of the growing cycle modified the time of autumn defoliation and the time of spring blooming with a strong and direct relationship between both variables (Fig. 8). Treatments of pruning that advance autumn leaf fall also advanced spring bloom whereas treatment that delayed autumn defoliation also delayed flowering in the next spring. According to the slope value of the fit line (Fig. 8), the time of spring flowering was modified in one day when the time of autumn defoliation was changed in three days. However, this effect was clearly observed in only one of three years of experimentation.

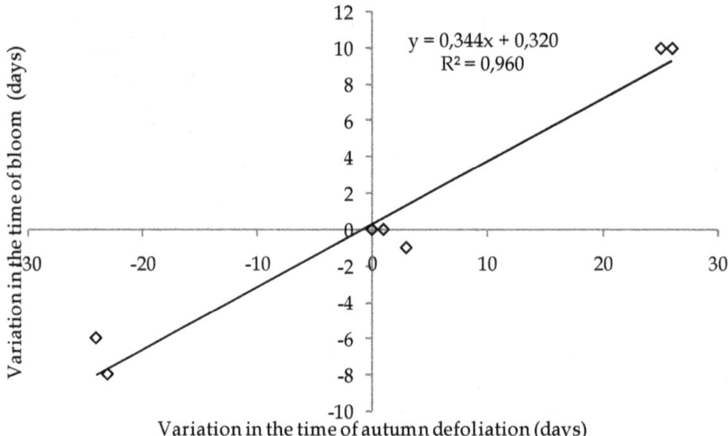

Fig. 8. Relationships between the modification of the time of autumn defoliation and the consequent variation in the time of blooming in the next spring caused by summer pruning performed at a different time of the growing cycle. Data were expressed in comparison with trees that were pruned during winter rest, in peach trees, cv. 'Early grande'.

These results showed the relationships between summer pruning and artificial defoliation on the time of leaf fall, and provide further evidence of the importance of the presence of leaf during autumn and the time of leaf abscission on the phenology of low chilling peach varieties in the next spring. Any cause (abiotic, biotic, cultural practices) that modifies the time of leaf fall occurrence changes the period of tree exposition to the short day inductive condition, and may alter the physiological process of dormancy in temperate fruit trees (Heide, 2008). As low chilling peach also had a low depth of dormancy (Citadin et al., 2001; Gariglio et al., 2006b), any modification, however small, in the depth of dormancy seems to have a great effect on the tree phenology in the next spring. These observations can also explain different results obtained with the defoliation technique on the time of blooming in the next spring (Citadin et al., 2001; Egea et al., 2003; Ganji Moghadam & Mokhtarian, 2006). In addition, it is expected that the treatments that reduce the depth of dormancy can increase the variability of peach phenology as was previously discussed when the phenological variability of low and high chilling requirement varieties were mentioned (see Fig. 2). Otherwise, the time of sprouting was less affected by summer pruning in relation with blooming (Table 4).

It is important to mention that summer pruning performed during February caused an important density of flowering during the autumn period. In the middle of April, Feb-P trees showed open flowers in 21% of its mixed shoots, causing a budbreak that reached up to 35% of the flower buds of these branches. No vegetative budbreak during the autumn period was stimulated by this pruning. Clearly, the occurrence of autumn flowering was due to the reduction on the depth of dormancy caused by summer pruning performed at the beginning of February. Furthermore, as flower bud had lower depth of dormancy than leaf bud (Gariglio et al., 2006b), blooming but not sprouting was stimulated by summer pruning at the beginning of February.

3.3.4 Summer pruning on vegetative and reproductive traits of peach trees

The pattern of the percentage of sprouting evolution was also modified by summer pruning after harvest (Fig. 9); vegetative budbreak increased faster in non-summer pruned trees but not in pruned ones. Non-summer pruned trees reached 50% of sprouting before August 25th whereas summer pruned plants arrived at this threshold value during the second week of September. The final percentage of vegetative budbreak was slightly above 60% in both treatments (Fig. 9). This behavior of vegetative budbreak observed in summer pruned trees can be a consequence of the delays on the time of sprouting and blooming of the next growing cycle, and on the increases of the depth of dormancy of vegetative and reproductive buds caused by summer pruning after harvest, mentioned and discussed previously.

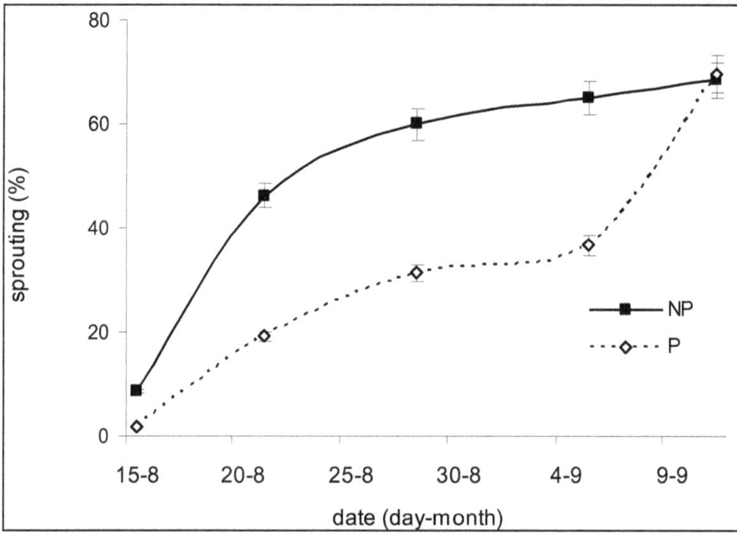

Fig. 9. Pattern of the vegetative budbreak evolution at the beginning of the growing cycle in 'Flordaking' peach trees that were pruned (P) or not pruned (NP) after harvest.

Summer pruning after harvest significantly reduced fruit size at the beginning of the fruit growth period (−23.7%), but this difference diminished from the last 15 days of fruit growth, becoming insignificant at harvest.

In relation to the reproductive traits of peach trees, summer pruning after harvest did not affect flower density and fruit set of 'Flordaking' peach. Floral induction of low and medium chilling peach varieties occurred after harvest (González-Rossia et al., 2007), and summer pruning performed at this time stimulated vegetative growth rate (Weber et al., 2011). Therefore, as vegetative growth and floral induction are a competitive process (Reig et al., 2006), it is expected that summer pruning after harvest caused a reduction on the flower density of pruned trees. However, summer pruning stimulates vegetative growth, but also improves light penetration into the canopy (Marini, 2003; Marini & Barden, 1982; Weber et al., 2011) which is required for floral induction and floral bud differentiation (Marini, 2003). Thus, it is possible that both effects may be neutralized and flower intensity would not be affected.

Although pruning treatment did not significantly affect fruit load over three years of experimentation (NP, 231 and P, 259 fruits per plant in average), it did significantly affect the distribution of the fruits in the plant height, increasing the proportion of those located in the intermediate stratum of the plant height (+10% to +40%), and decreasing those of the upper one (−45% to −50%). This effect is related to the major proportion of reproductive shoots observed in the medium stratum for summer pruned trees in comparison with non-pruned ones (Weber et al., 2011). This effect of pruning is explained by an improvement of the internal distribution of light (Stebbins, 1997; Weber et al., 2011) and a change in the dominant position of vegetative growth which is transferred from the apical to the basal zone of the canopy after pruning (Li et al., 1994a).

Unlike the previous experiment, summer pruning performed at a different time of the growing cycle using 'Early grande' peach caused a significant modification on fruit set, and consequently on the number of fruits per plant at harvest. Both variables (fruit set and number of fruit per plant) were significantly reduced by winter pruning during the years characterized by light late frost occurrence (Table 6). In this experiment, trees that were only pruned during the summer period showed a fruit set from 1.25- to 3.2-fold higher and the number of fruit per plant was 1.17- to 5.5-fold higher in comparison to trees that were pruned during winter. This effect was not observed during years without late frost occurrence.

Treatment	Fruit set (%)	Fruits/plant
Jun-P	3.8 c	77 d
Nov-P	16.0 a	293 a
Nov-P + WP	4.1 c	87 d
Feb-P	12.2 a	206 b
Feb-P + WP	4.9 c	54 e
Apr-P	6.2 b	102 c
Apr-P + WP	5.0 c	55 e

Table 6. Fruit set and total fruits per plant in 'Early grande' peach trees that were pruned in summer (Nov-P, Feb-P and Apr-P), both in summer and winter periods (Nov-P + WP, Feb-P + WP and Apr-P + WP) or only during winter (Jun-P, control). Data are the means of two years (2007 and 2008), characterized by light spring frost occurrence. Means with different letters in the same column differ significantly ($P \leq 0.05$).

This effect of winter pruning on fruit set is difficult to explain because the treatments showed different times of blooming (Table 5) and the sensitivity to frost increases with the development of the flowers (Racskó et al., 2007); however, comparing trees that received pruning only during the summer period with trees that were pruned twice (summer and winter time) and that reached blooming at the same time (Nov-P and Nov-P + WP; Feb-P and Feb-P + WP; Apr-P, Apr-P + WP, and Jun-P), showed that treatments that received winter pruning had a lower fruit set (Table 6). It is also possible to demonstrate that this effect is due to the winter pruning and not by the number of prunings that received each treatment because the control (Jun-P) received only one pruning during winter and showed the lowest fruit set in comparison with trees that received one pruning treatment during the growing cycle or two pruning treatments (summer and winter pruning).

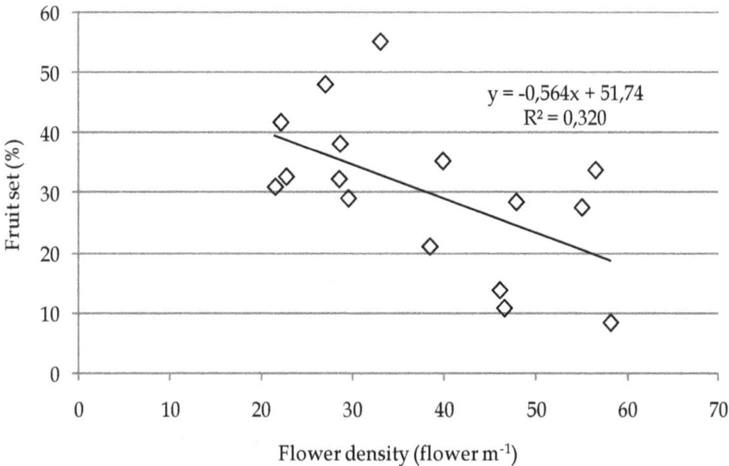

Fig. 10. Relationship between flower density and fruit set of different low chilling peach cultivars growing in the central-east area of Santa Fe, Argentina. Data are the means of a nine-year period (2002–2010).

This negative effect of winter pruning on fruit set was previously observed when long pruning and short pruning techniques were evaluated in peach trees growing in China (Li et al., 1994b; Singh & Daulta, 1985). They observed that long pruning increased resistance of flowers to late frost in spring because the high quality buds that have a higher resistance to frost were generally situated at the mid region of the shoot, and these buds were eliminated when short pruning was practiced (Li et al., 1994b). However, in our experiments long pruning was practiced during winter and consequently the importance of bud quality in the mid region of the shoot cannot explain our results. The competence between growing organs caused a great diminution of fruit set in citrus (Agustí, 2003) and other subtropical crops such us mango (Ramírez & Davenport, 2010). Fruit set of stone fruits seems to be unaffected by flower intensity (González-Rossia et al., 2007), but despite this we found a negative tendency of fruit set of peach with the flower density of the cultivars (Fig. 10); fruit set decreased by half (40% to 20%) when the flower density increased from 20 to 60 flowers per meters of mixed shoots. Furthermore, we did not observe changes of flower density caused by pruning treatments. Instead, there is good evidence that winter pruning stimulates vegetative growth (Li et al., 1994a; Marini, 2003) of deciduous trees, and that competition for carbohydrates from shoot growth accentuates early fruit abscission, even at a relatively low crop load (Racskó et al., 2007).

In low chilling peach a great proportion of floral bud abortion was observed when excised shoots received excessive chilling that stimulated vegetative growth when they were forced at 25°C (Gariglio et al., 2006b). Similar effects were induced by the use of hydrogen cyanamide to stimulate budbreak under conditions of low chilling accumulation (Lloyd & Firth, 1993). Thus, it is possible that winter pruning diminishes fruit set under suboptimal low temperature conditions by increasing competition between vegetative and reproductive growth.

4. Conclusion

Low chilling peach and nectarine varieties (≤ 450 CH) grown in the central area of Santa Fe, Argentina, showed high flower density, fruit set, number of fruit per plant, and consequently, higher fruit yield with regard to peach varieties with higher chilling requirement (> 500 CH). The phenology of high chilling requirement varieties had the advantage that blooming and fruit set occurred after the period of risk of late frost occurrence; however, they did not satisfy their chilling requirement, showing an inadequate vegetative and reproductive behavior. Phenology variability between years decreased with the advance of the growing cycle and with the chilling requirement of the variety, and this behavior is explained by the low depth of dormancy of low chilling peach, and the high variability of the temperature during the period of dormancy and blooming of low chilling peach.

Artificial defoliation at the beginning of leaf fall and summer pruning greatly affect phenology of low chilling peach. Summer pruning delayed autumn leaf fall when it was performed after harvest (November), but advanced leaf senescence when it was done at the beginning of February. The time of blooming and sprouting in the next spring was directly related to the time of autumn leaf fall. This effect can be explained by the role of the presence of leaf on the onset and progression of dormancy; thus, the advances of leaf fall (artificial autumn defoliation and summer pruning during February) reduce the depth of dormancy, advancing blooming and sprouting, while the delay in leaf fall (summer pruning during November) increases the depth of dormancy of low chilling peach, delaying blooming and sprouting in the next spring. As a consequence, by changing the time of autumn leaf fall through the use of different cultural practices such as artificial defoliation or summer pruning, it is possible to modify by nearly 20 days the time of flowering in the next spring, allowing growers to improve precocity or to avoid late frost occurrence, among other possible objectives.

5. References

Agustí, M. (2003). *Citricultura.* Second Edition. Mundi Prensa, ISBN 9788484761587, Madrid, Spain

Alayón Luaces, P., Rodríguez, V.A. & Bertuzzi, S.M. (2004). El cultivo del manzano (*Malus domestica* Borkh.) en Corrientes. Primer año de investigación. *Comunicación científica y tecnológica A-029,* pp. 1-4. Accessed November 2007, Available from: http://www.unne.edu.ar/Web/cyt/com2004/5-Agrarias/A-029.pdf

Badenes, M.L., Martínez-Calvo, J. & Llacer, G. (1998). Estudio comparativo de la calidad de los frutos de 26 cultivares de melocotonero de origen norteamericano y dos variedades-población de origen español. *Investigación Agraria: Producción y Protección Vegetal*, Vol.13, No.1-2, (May 1997), pp. 57-70, ISSN 0213-5000

Balandier, P., Bonhomme, M., Rageau, R., Capitan, F. & Parisot, E. (1993). Leaf bud endodormancy release in peach trees: evaluation of temperature models in temperate and tropical climate. *Agricultural and Forest Meteorology*, Vol.67, No.1-2, (December 1993), pp. 95-113, ISSN 0168-1923

Broadhead, J.S., Ong, C.K. & Black, C.R. (2003). Tree phenology and water availability in semi-arid agroforestry systems. *Forest Ecology and Management*, Vol.180, No.1, (July 2003), pp. 61–73, ISSN 0378-1127

Carra de Toloza, M.S. (2001). Selección de variedades de duraznos de mesa y nectarinos promisorios y recomendados para su cultivo en la provincia de Mendoza. *Horticultura Argentina*, Vol.20, No.48, (October 2001), pp. 44, ISSN 0327-3431

CFI. Consejo Federal de Inversiones. (2011). Sector frutícola argentino. Accessed June 2011, Available from: http://www.cfired.org.ar/Default.aspx?nId=328

Citadin, I., Raseira, M.C.B. & Quezada, A.C. (1998). Substrato para conservacao de ramos destacados de pessegueiro *Prunus persica* L. (Bastch). *Agropecuária Clima Temperado*, Vol.1, No.1, (February 1998), pp. 55-59, ISSN 1415-6822

Citadin, I., Raseira, M.C.B., Verter, F.G. & Baptista da Silva, J. (2001). Heat requirement for blooming and leafing in peach. *HortScience*, Vol.36, No.2, (May 2001), pp. 305-307, ISSN 0018-5345

Couvillon, G. & Lloyd, D. (1978). Summer defoliation effects on spring bud development. *HortScience*, Vol.13, pp. 53-54, ISSN 0018-5345

Dennis Jr., F.G. (2002). Dormancy: Manifestations and causes. In: *Handbook of plant and crop physiology*, Pessarakli, M. (Ed.), Chapter 8, pp. 161-179, Marcel Dekker, ISBN 9780824705466, New York

Dennis Jr., F.G. (2003). Problems in standarizing methods for evaluating the chilling requirements for the breaking dormancy in buds of woody plants. In: Standarizing methods for evaluating the chilling requirements to break dormancy in sedes and buds (Including Geophytes). *HortScience*, Vol.38, No.3, (June 2003), pp. 347–350, ISSN 0018-5345

Egea, J., Orgega, E., Martínez-Gómez, P. & Dicenta, F. (2003). Chilling and heat requirement of almond cultivars for flowering. *Environmental and Experimental Botany*, Vol.50, No.1, (August 2003), pp. 79-85, ISSN 0098-8472

Erez, A., Faust, M. & Line, M. (1998). Changes in water status in peach buds on induction, development and release from dormancy. *Scientia Horticulturae*, Vol.73, No.2-3, (March 1998), pp. 111-123. ISSN 0304-4238

Faust, M. (1989). *Physiology of temperate zone fruit trees*. John Wiley & Sons, ISBN 0471817813, New York

Faust, M., Erez, A., Rowland, L.J., Wang, S.Y. & Norman, H.A. (1997). Bud dormancy in perennial fruit trees: Physiological basis for dormancy induction, maintenance, and release. *HortScience*, Vol.32, No.4, (July 1997), pp. 623-629, ISSN 0018-5345

Fideghelli, C. (1987). *El Melocotonero*. Ediciones Mundi Prensa, ISBN 84-7114-166-3, Madrid

Ganji Moghadam, A. & Mokhtarian, A. (2006). Delaying apricot (cv Shahroudi) flower induction by growth regulators application. *Journal Applied Sciences*, Vol.6, No.2, pp. 266-269, ISSN 1812-5654

Gariglio, N.F., Baldi, B. & Enrique, A. (1999). Comportamiento del duraznero en la región central de Santa Fe. *VII Jornadas de Jovens Pesquisadores do Grupo Montevideo*. Ciência para a Paz. Libro de resúmenes, p. 21

Gariglio, N., Dovis, V.L., Leva, P., García, M.S. & Bouzo, C.A. (2006a). Acumulación de Horas de Frío en la zona centro-oeste de Santa Fe (Argentina) para frutales caducifolios. *Horticultura Argentina*, Vol.25, No.58, (June 2006), pp. 26-32, ISSN 0327-3431

Gariglio, N.F., González Rossia, D.E., Mendow, M., Reig, C. & Agustí, M. (2006b). Effect of artificial chilling on the depth of endodormancy and leaf and flower budbreak of peach and nectarine cultivars using excised shoots. *Scientia Horticulturae*, Vol.108, No.4, (May 2006), pp. 371-377, ISSN 0304-4238

Gariglio, N., Mendow, M., Weber, M., Favaro, M.A., González-Rossia, D. & Pilatti, R.A. (2009). Phenology and reproductive traits of peaches and nectarines in central-east Argentina. *Revista Scientia Agrícola*, Vol.66, No.6, (November 2009), pp. 757-763, ISSN 0103-9016

González Rossia, D., Reig, C., Juan, M. & Agustí, M. (2007). Horticultural factors regulating effectiveness of GA3 inhibiting flowering in peaches and nectarines (*Prunus persica* L. Batsch). *Scientia Horticulturae*, Vol.111, No.4, (February 2007), pp. 352-357, ISSN 0304-4238

Gorny, J.R., Hess-Pierce, B. & Kader, A.A. (1999). Quality changes in fresh-cut peach and nectarine slices as affected by cultivar, storage atmosphere and chemical treatments. *Journal of Food Science*, Vol.64, No.3, (May 1999), pp. 429-432, ISSN 1750-3841

Heide, O.M. (2008). Interaction of photoperiod and temperature in the control of growth and dormancy of *Prunus* species. *Scientia Horticulturae*, Vol. 115, No. 3, (February 2008), pp. 309-314, ISSN 0304-4238

Hossain, A.B.M.S., Mizutani, F., Onguso, J.M. & Yamada, H. (2005). Effect of summer and winter pruning of peach as slender spindle bush type on growth, yield and quality of fruit. Journal of Applied Horticulture, Vol.7, No.1, pp. 11-15. ISSN 0972-1045

INDEC. (2002). Censo Nacional Agropecuario. Accessed November 2005. Available from: http://www.indec.mecon.gov.ar/agropecuario/cna_principal.asp

INIA Salto Grande. (2003). Reunión anual de frutales de carozo. Salto Grande. Accessed February 2006. Available from: http://www.inia.org.uy/publicaciones/documentos/sa/ad/2003/ad_334.pdf

Kudo, G. (2006). Flowering phenologies of animal-pollinated plants: reproductive strategies and agents of selection. In: *Ecology and evolution of flowers*, Harder, L.D. & Barrett, S.C.H., eds, Chapter 8, pp. 139-158, Oxford University Press, ISBN 978-0-19-857085-1, New York, USA

Lang, G.A. (1996). *Plant Dormancy: Physiology, Biochemistry and Molecular Biology*, CAB International, ISBN 0851989780, Wallingford, UK

Li, S., Zhang, X., Meng, Z. & Wang, X. (1994a). Responses of peach trees to modified pruning. I. Vegetative growth. *New Zealand Journal of Crop and Horticultural Science*, Vol.22, No.4, (May 1994), pp. 401-409, ISSN 0114-0671

Li, S., Zhang, X., Meng, Z., Wang, X. & Liu, G. (1994b). Responses of peach trees to modified pruning. 2. Cropping and fruit quality. *New Zealand Journal of Crop and Horticultural Science*, Vol.22, No.4, (May 1994), pp. 411-417, ISSN 0114-0671

Lieth, H. (1974). *Phenology and Seasonality Modelling.* (Ecological Studies-Analysis and Synthesis Series, Vol 8). Springer, ISBN 0387065245, Berlin, Germany

Loyd, J. & Firth, D. (1990). Effect of defoliation time on depth of dormancy and bloom time for low-chill peaches. *HortScience,* Vol.25, No.12, (December 1990), pp. 1575-1578, ISSN 0018-5345

Lloyd, J., Firth, D. (1993). Effect of hydrogen cyanamide and promalin on flowering, fruit set and harvest time of 'Flordaprince' peach (*Prunus persica* (L.) Batsch) in subtropical Australia. *Journal of Horticultural Science,* Vol.68, No.2, pp. 177-183

Marini, R.P. (2003). Peach fruit weight, yield, and crop value are affected by number of fruiting shoots per tree. *HortScience,* Vol.38, Nro.4, (July 2003), pp. 512-514, ISSN 0018-5345

Marini, R.P. & Barden, J.A. (1982). Growth and flowering of vigorous apple trees as affected by summer or dormant pruning. *J. Amer. Soc. Hort. Sci.,* Vol.107, No.1, (January 1982), pp. 34-39, ISSN 0003-1062

Médiène, S., Jordan, M.O., Pages, L., Le Bot, J. & Adamowicz, S. (2002). The influence of severe shoot pruning on growth, carbon and nitrogen status in young peach trees (*Prunus persica*). *Tree Physiology,* Vol.22, No.18, (May 2002), pp. 1289-1296, ISSN 0829-318X

Meier, U., Graf, H., Hack, H., Hess, M., Kennel, W., Klose, R., Mappes, D., Seipp, D., Stauss, R., Streif, J. & Van Den Boom, T. (1994). Phänologische entwick-lungsstadien des kernobstes (*Malus domestica* Borkh. und *Pyrus communis* L.), des steinobtes (*Prunus-Arten*), der johannisbeere (*Ribes-Arten*) und der erdbeere (*Fragaria × ananassa* Dutch.). *Nachrichtenbl Deutsch Pflazenschutzd,* Vol.46, No.7, pp. 141-153

Mohamed, A.K.A. (2008). The effect of chilling, defoliation and hydrogen cyanamide on dormancy release, bud break and fruiting of Anna apple cultivar. *Scientia Horticulturae,* Vol.118, No.1-2, (September 2008), pp. 25-32, ISSN, 0304-4238

Ortiz de Zárate, V.G., Güemes, D.R., Piagentini, A.M., Gariglio, N.F. & Pirovani, M.E. (2007). Comparación de la calidad de duraznos de diferentes variedades cultivadas en la región Centro-Este de la provincia de Santa Fe. *Revista FAVE-Ciencias Agrarias,* Vol.5/6, No.1-2, (November 2006), pp. 27-33, ISSN 1666-7719

Panigatti, J.L. (1980). Consideraciones sobre el clima del centro-oeste de la provincia de Santa Fe. INTA, Estación Experimental Regional Agropecuaria Rafaela. *Boletín interno de divulgación* No.15

Pérez, S. (2004). Yield stability of peach germplasm differing in dormancy and blooming season in the Mexican subtropics. *Scientia Horticulturae,* Vol.100, No.1-4, (March 2004), pp. 15-21, ISSN, 0304-4238

Petri, J.L., Hawerroth, F.J. & Leite, G.B. (2008). Fenologia de espécies silvestres de macieira como polinizadoras das cultivares Gala e Fuji. *Revista Brasileira de Fruticultura,* Vol.30, No.4, (December 2008), pp. 868-874, ISSN 0100-2945

Ramírez, F. & Davenport, T.L. (2010). Mango (*Mangifera indica* L.) flowering physiology. *Scientia Horticulturae,* Vol.126, No.2, (September 2010), pp. 65-72, ISSN, 0304-4238

Racskó, J., Leite, G.B., Petri, J.L., Zhongfu, S., Wang, Y., Szabó, Z., Soltész, M. & Nyéki, J. (2007). Fruit drop: The role of inner agents and environmental factors in the drop of flowers and fruits. *International Journal of Horticultural Science,* Vol.13, Nro.3, pp. 13–23. ISSN 1585-0404

Reig, C., González-Rossia, D., Juan, M. & Agustí, M. (2006). Effects of fruit load on flower bud initiation and development in peach. *Journal of Horticultural Science & Biotechnology*, Vol.81, No.6, (November 2006), pp. 1079-1085, ISSN 1462-0316

Rinne, P.L.H. & van der Schoot, C. (2004). Cell-Cell communication as a key factor in dormancy cicling, In: *Adaptations and responses of woody plants to environmental stresses*, Arora, R. ed., pp. 113-156, The Haworth Press Inc., ISBN 81-8189-107-4, Binghamton, NY

Singh, D. & Daulta, B.S. (1985). Studies on pruning severity in peach (*Prunus persica* Batsch) cv. Sharbati. I. Effect on sprouting, flowering, fruit set and fruit drop. *Haryana journal of horticultural sciences*, Vol.14, pp. 33-36, ISSN 0970-2873

Squire, G.R. (1990). *The Physiology of Tropical Crop Production*. CAB International, ISBN 0851986773, Wallingford, UK

Srivastava, L.M. (2002). *Plant growth and development: Hormones and the environment*, Academic Press, ISBN 0-12-660570-X, San Diego, USA

Stebbins, R.L. (1997). Training and Pruning commercial peach orchards. *Oregon State University Extension Service*. Accessed December 2003. Available from: htpp://www.eesc.orst.edu/agcomwebfile/edmat/EC1236.pdf

Tanino, K.K. (2004). Hormones and endodormancy induction in Woody plants, In: *Adaptations and responses of woody plants to environmental stresses*. Arora, R. ed., pp. 157-199, The Haworth Press, ISBN 81-8189-107-4, Binghamton, NY

Valentini, G. (2002). Variedades de duraznero y nectarina para el NE de la provincia de Buenos Aires. Accessed March 2004, Available from: <http://www.inta.gov.ar/sanpedro/info/doc/fru/gv_003.htm>

Valentini, G. & Arroyo, L. (2000). Variedades y Portainjertos. Jornada de Actualización en diferentes aspectos del cultivo de duraznero. INTA, Estación Experimental Regional Agropecuaria San Pedro, Libro de conferencias, pp. 2-8

Walser, R., Walker, D. & Seeley, S. (1981). Effect of temperature, fall defoliation and gibberelic acid on the rest period of peach buds. *J. Am. Soc. Hortic. Sci.*, Vol.106, No.1, (January 1981), pp. 91-94, ISSN 0003-1062

Weber, M.E., Pilatti, R., Sordo, M., Castro, D., Rista, L. & Gariglio, N. (2010). La fecha de poda modifica la profundidad de la dormición y la intensidad de su ruptura en yemas vegetativas y reproductivas del duraznero, cv. \`Early grande'. *Proceedings of XXXIII Congreso Argentino de Horticultura y I Simposio Internacional de la Frutilla*, p. 214. Rosario, Argentina, September 2010

Weber, M.E., Pilatti, R., Sordo, M.H., García, M.S., Castro, D. & Gariglio, N.F. (2011). Changes in the vegetative growth of the low-chill peach tree in response to reproductive shoot pruning after harvesting. *New Zealand Journal of Crop & Horticultural Science*, Vol.39, No.3 (September 2011), pp. 153-160, ISSN 1175-8783, Available from: http://www.tandfonline.com/doi/abs/10.1080/01140671.2011.559255

Weibel, A., Johnson, R.S. & DeJong, T.M. (2003). Comparative vegetative growth responses of two peach cultivars grown on size-controlling versus standard rootstocks. *Journal of the American Society for Horticultural Science*, Vol.128, No.4, (April, 2003), pp. 463-471, ISSN 0003-1062

Part 4

Remote Sensing Phenology

14

NDVI Time Series for Mapping Phenological Variability of Forests Across the Cerrado Biome in Minas Gerais, Brazil

Thomaz Oliveira, Luis Carvalho, Luciano Oliveira, Wilian Lacerda and Fausto Acerbi Junior
UFLA, Brazil

1.Introduction

The constant threat that natural forests of the world have suffered over the years led to strategies that intend to prevent these from further losses in the near future. The tropical rain forest that exists in northern Brazil is the main concern of preservation, but other areas such as the "Mata Atlântica" have had losses over the years and almost 75 % of its remnants are under threat of deforestation. The main activities that lead to deforestation in the region are the expansion of agricultural frontiers, the extraction of coal and other minerals, timber exploitation, and other anthropogenic activities (Oliveira, 2004). Another area in Brazil that is under threat is the Cerrado Biome in Minas Gerais. The Cerrado biome of tropical South America covers about two million squared kilometres, representing almost 22% of the Brazilian territory. The biome was named due to its predominant vegetation type, a fairly dense woody savannah composed by shrubs and small trees. The term "cerrado" (Portuguese for closed or dense) was probably applied to this vegetation because of the difficulty of traversing it on horseback (Oliveira-Filho et al. 2002). The constant threat to the Brazilian Cerrado has led to the necessity of developing strategies and measures to promote the monitoring and mapping of this biome. The Cerrado has a rich biodiversity but its fragmentation throughout the years caused losses of a number of species from this biome (Oliveira 2004). The Cerrado has many types of phytophysiognomies, these include Semi-Deciduous Forests, Deciduous Forests, Cerradão (dense Savanna), Cerrado (Savanna), Floresta Ombrófila and other more specific types of vegetation. In order to prevent further deforestation of this biome, monitoring by means of remote sensing is regarded as an efficient tool. This technique can provide accurate mapping of the occurrence of each phytophysiognomy of the Cerrado Biome and promote efficient monitoring in order to detect changes so legal actions can be taken in due time.

Mapping land cover using remotely sensed data has been an area of growing research interest throughout the past decades. Its complexity, peculiarities and state of the art concerning computational aids and processing routines differ a lot from past conventional cartographic tools. Developments in computer science have aided a better information extraction from remotely sensed images, as well as an effective use of geographical information systems to store, analyse and present all sorts of land cover information (Carvalho 2001).

The Statewide Vegetation Monitoring and Mapping of Minas Gerais was conducted by Carvalho (2007) with the use of multitemporal Landsat images. This project includes a dataset of images acquired in different seasons of a year so that the dry and wet seasons could be captured in order to characterize the phenological cycle of deciduous and semideciduous forests (Figure 1).

Fig. 1. Visual Difference of phytophysiognomies (a) Deciduous forest, (b) semi-deciduous forest Source Oliveira (2004). Semi-deciduous forests loose less leaves in the dry season.

Deciduous forests lose 70% of its leafs in the dry season, the semi-deciduous forests loose less of its leafs in this season, with less variation in greenness.

According to Jensen (2000), temporal signatures are very important when mapping different vegetation types or extracting vegetation biophysical information (e.g. biomass) from remotely sensed data. Multi-temporal imagery are currently of standard use when studying large areas at regional and global scales (Carrão et al. 2007).

However, errors occurred in the Official state map due to the characteristics of one specific phytophysiognmy within this biome - the Deciduous Forests. The deciduousness of these forests is determined by an alternating cycle of dry and wet seasons, where more than 70 % of the leaves are off (Figure 2) in the dry season (Oliveira-Filho, 2006). The period of dryness occurs from mid-April till September. The wet season starts in October and goes up to March. The variation of greenness of the semi-deciduous forest - another phytophysiognomy of the Cerrado Biome - is not as intense as the deciduous counterpart, due to its occurrence in regions of intensified humidity (Oliveira 2004), like riparian areas.

Some objects on the Earth´s surface reflect the electro-magnetic energy in the same way when sensed with a multi-spectral scanner. In the present case, it is difficult to differentiate deciduous and semi-deciduous forests when leaves are on using single date remote sensing.

Nevertheless, objects' reflectance may vary according to growth stage, phenology, humidity, atmospheric transparency, illumination conditions etc. These characteristics led to a search for alternative features to enable the discrimination of land cover classes with similar reflectance behaviour (Carvalho et al. 2004). Phenological timing is an example of such features, which have the potential to reduce mapping errors.

Fig. 2. Two picture of the same location of a Deciduous forrest of the Cerrado Biome, illustrating the difference of greenness: Wet season (a) and dry season (b). Oliveira-Filho (2011).

Considering the peculiarities of this vegetation type, there is a need to develop specific image processing approaches when mapping land cover in the Cerrado Biome due to this temporal variation in greenness. Since land cover features that have similar spectral reflectance are difficult to be differentiated when using single-date remote sensing imagery, the analysis of time series of images that capture vegetation seasonality may provide improved results. These include the quantification of the seasonal profile of Deciduous and Semi-deciduous forests. Errors among these classes were commonly found in the Official state map, which does not capture deciduous forests fragments present in the region of the Triângulo Mineiro, western Minas Gerais. Research however, suggests that remotely sensed time series data could possibly improve misclassification and the accuracy of mapping deciduous forests (Oliveira, 2004).

These errors may be associated with a time-shift in the phenological patterns of these forests that occur in different areas. One forest remnant in the eastern part of the state may have a lag of time on its cycle that may lead to mapping errors. Thus, temporal images should be acquired in different dates in order to properly characterize leaf-on and leaf-off periods across the State.

Sano et al. (2008) have used NDVI time series for studying the Cerrado, in order to produce a semi detailed land cover mapping of the area considering the following classes: croplands, planted pasturelands, reforestations, urban settlements and mining areas. In another study conducted by Carvalho Júnior et al.(2006), the characterization of Deciduous Forests time profile was performed with the Minimum Noise Fraction (MNF) transformation in order to remove outliers from the time profile. This study was conducted in the states of Goiás and Tocantins, Brazil. The authors have also considered different vegetation classes when compared to the present study: deciduous forests, savannah, dense cerrado, and "Campo Sujo". Their methodology did not consider the discrimination between semi-deciduous and deciduous forests, nor timing of phonological cycles. In the work carried out by Silveira et al. (2008) in Minas Gerais, vegetation was classified using smoothed MODIS time series with excellent results. Nevertheless, the study did not consider different geographic locations within the State. No previous work was carried out to specifically characterize the deciduous forests of Minas Gerais with the use of Multi-temporal imagery and to quantify its phenological cycle.

The present work is concerned with the quantification of phenological time shifts among deciduous forests that are geographically distant. Preprocessing techniques were applied to the MODIS time series in in order reduce noise. The outcomes have the potential to provide less errors in future maps that could be benefited by the acquisition of images that take this time shift into account.

This study was motivated by the following questions surrounding deciduous forests:

1. Do Geographically distant deciduous forests have different timing in their phenological cycle during the year, leading to an annual shift (figure 3) in their cycle?
2. Can MODIS filtered NDVI (Normalized Difference Vegetation Index) time series be used to generate better maps of deciduous forests in different regions?
3. Which of two filtering techniques, viz. HANTS Fourier Analysis and Wavelet Filtering, produces the best smoothed time series for mapping this phytophysiognomy.

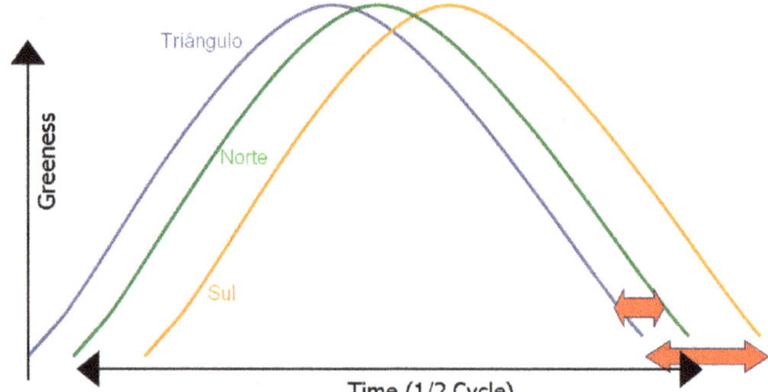

Fig. 3. Time shift analysis of phenology. The arrows indicate the time delay that was calculated later in this work. The plot indicates the variation of greenness in time in different regions of deciduous forests of the Cerrado for ½ year period.

2 Methods

2.1 Vegetation indices

Temporal information used in this study comprised time series of vegetation indices, viz. the Normalized Difference Vegetation Index (NDVI). Since the 1960's, scientists have extracted and modelled vegetation biophysical variables using remotely sensed data. Much of the effort has gone into the development of vegetation indices - defined as dimensionless radiometric measures that function as indicators of relevant abundance and activity of green vegetation, often including leaf-area-index (LAI), percentage green cover, chlorophyll content, green biomass, and absorbed photosynthetically active radiation (APAR). There are more than 20 vegetation indices in use. A vegetation index should maximize sensitivity to plant biophysical parameters; normalize or model external effects such as sun angle, viewing angle, and the atmosphere for consistent spatial and temporal comparisons; normalize internal effects such as canopy background variations. A vegetation index may preferably couple with a measurable biophysical parameter such as biomass, LAI, or APAR (Jensen et al., 2000).

Vegetation dynamics indicate important short and long-term ecological process. Continuous temporal observations of land surface parameters using satellite reveal seasonal and inter-annual developments. Vegetation indices have been extensively applied to characterize the state and dynamics of vegetation, in particular multiple NDVI datasets of the Advanced Very High Resolution Radiometer (AVHRR) instrument used during the last 25 years (Coldiz et al., 2007; Jensen, 2000)

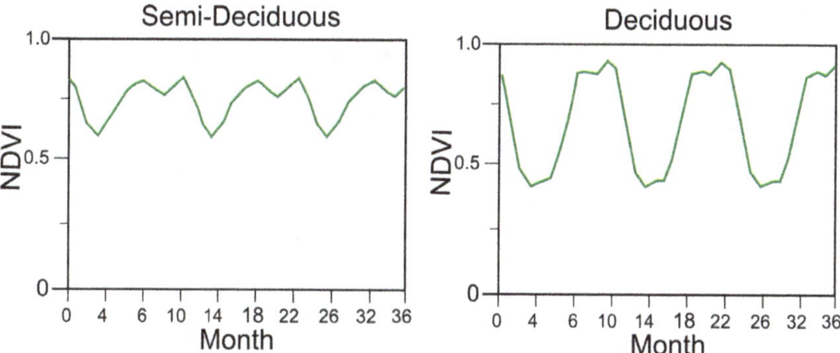

Fig. 4. Denoised MODIS NDVI time series for Deciduos and Semi-deciduous Forests. The difference in amplitude is noticeable in this plot for these two phytophysiognomies, indicating the different time signature among them.

Different vegetation types exhibit distinctive seasonal patterns on NDVI variation (Yu et al., 2004). Vegetation profiles of deciduous and semi-deciduous forests are illustrated in figure 4. In most cases, different types of vegetation have different phenolgical patterns. For example, evergreen plants will have a more steady temporal dynamics throughout the year when compared to tropical plants that lose their leaves (Bruce et al., 2006).

Spatial and temporal variability in vegetation indices arise from several vegetation related properties, including LAI, canopy structure/architecture, species composition, land cover type, leaf optics, canopy crown cover, understory vegetation, and green leaf biomass (Huete et al., 2002).

2.2 MODIS NDVI time series

In the present study, NDVI time series from the Moderate-resolution Imaging Spectroradiometer (MODIS) were used. MODIS data products offer a great opportunity for phenology-based land-cover and land use change studies by combining characteristics of both AVHRR and Landsat, including: moderate resolution, frequent observations, enhanced spectral resolution, and improved atmospheric calibration (Galford et al., 2007). The AVHRR sensor was originally designed for meteorological applications, and has only two spectral bands (red and near-infrared) that can be used to generate spectral indices of vegetation. The new generation MODIS sensor has a number of advantages over AVHRR, including more spectral bands that can be used for vegetation analysis (Yu et al., 2004).

MODIS vegetation indices are appropriate for vegetation dynamics studies and characterization. They are found to be sensitive to multi-temporal (seasonal) vegetation variations and to be correlated with LAI across a range of canopy structure, species composition, lifeforms, and land cover types. The MODIS-NVDI demonstrates a good dynamic range and sensitivity for monitoring and assessing spatial and temporal variations in vegetation amount and condition. The seasonal profiles provided by the MODIS-NDVI outperform in sensitivity and fidelity the equivalent AVHRR-NDVI profiles, particularly when the atmosphere has a relatively high content of water vapor (Huete et al., 2002).

2.3 Dataset and study site

Due to the widespread occurrence of deciduous forests in the state of Minas Gerais and its large extent, 586.528 km^2, four different areas of interest were chosen so that time signatures of geographically separated forests could be compared. These locations were primarily chosen because of known occurrences of Deciduous forests according to the Treeatlan data base (Oliveira-Filho, 2009) (Figure 5), and the Official State Map by Carvalho (2007). A set of temporal images from the Landsat TM sensor were used as auxiliary data. The Landsat images from each location were acquired in the dry and wet seasons in order to identify fragments of deciduous forests by visual interpretation.

The NDVI time series were derived from the MOD13 product, which has a spatial resolution of 250m, and 16-day compositing period.

The original images were pre-processed using the MODIS Reprojection Tool (MRT). The data set was sampled to 23 values per year, approximately two images per month. This dataset included the years of 2007, 2008, and 2009. The representation of this time series is pictured in figure 6.

2.4 MODIS compositing methods

Several factors such as cloud contamination, atmospheric variability, and bi-directional reflectance, affect the stability of the satellite derived NDVI. Thus compositing methods have been developed to eliminate these effects. The compositing method for the AVHRR NDVI data source is the MVC (Maximum value composite), which selects the maximum NDVI value on a per pixel basis over a set of compositing period (Wang et al., 2004).

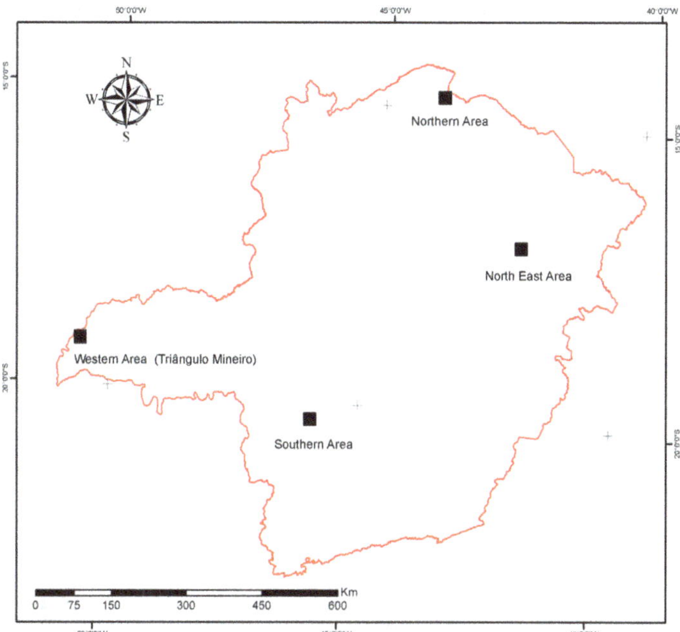

Fig. 5. Locations of Deciduous Forests of four different geographical areas chosen to calculate the timing difference among each location. Each area was chosen accordingly to the occurrence of Deciduous Forests from the TreeAtlan database and the Official State Map.

Fig. 6. MODIS NDVI time series dataset. The multi-temporal remote sensing MOD13 data is layered as described in this figure in order to acquire a time series for each pixel.

The MODIS compositing method operates on a per-pixel basis and relies on multiple observations over a 16-day period to generate a composite Vegetation Indice. Due to sensor orbit overlap and multiple observations, a maximum of 64 observations may be collected in a 16 days compositing period. Once all the 16 days of observation are collected, the MODIS VI algorithm applies a filter to the data based on quality, cloud, and viewing geometry. Only the higher quality cloud free, filtered data are retained for compositing (Huete et al., 2002).

At regional and global scales, variations in community composition, micro and regional climate, soils, and land management result in complex spatio-temporal variation in phenology. Furthermore, some vegetation types exhibit multiple modes of growth and senescence within a single annual cycle. Therefore compositing methods need to be sufficiently flexible to allow for this type of variability (Zhang et al., 2003).

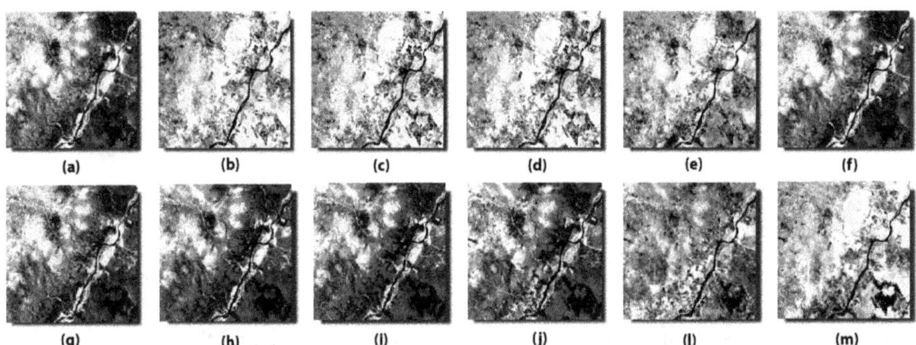

Fig. 7. Original MODIS MOD13 NDVI time series with compositing procedures. From (a) to (m) this time series corresponds to one full year cycle of the northern studied area

Figure 7 exhibits 1 full year of the NDVI original time series before the noise reduction algorithms.

2.5 Signal denoising

In order to extract pertinent features from time signatures for potential target applications, the signals must first be denoised. The main noise from remote sensing time series comes from pixels that were cloud contaminated, these appear as low NDVI values in the time signatures, as observed in figure 9, in the datasets prior to denoising procedures. Authors have investigated automated methods for denoising, including straightforward methods such as median filters and moving-average filtering, as well as more advanced methods such as wavelet denoising (Bruce et al., 2006).

Curve fitting parameterization using logistic functions have also succeeded in generating time signatures of MODIS (Zahng et al. 2003). Other methods for phenology curve fitting and noise reduction are included in the Timesat software (Jönsson & Lars 2004) and its enhanced version by Tan et al. (2010) which uses traditional least square fit for polynomial, Gaussian and Sigmoidal functions, they are also useful for computing derivates which correspond to annual senescence and growth and other important phonological metrics.

A study by Couto-Júnior (2011) concluded that MODIS is an adequate tool for monitoring and extracting metrics for the Cerrado biome due to its high temporal resolution and availability. Although one of its drawbacks is its high signal-to-noise ratio leading to the use of advanced signal processing methods.

2.6 Fourier transform

The Fourier Transform has been traditionally used to solve differential and partial equations in Mathematics and Physics. Its main objective is to approximate a function in the time domain by a linear combination of harmonics (sinusoids) (Morettin, 2006). The most basic property of the sinusoids that makes them suitable for the analysis of time series is their simple behaviour under a change in time scale (Bloomfield, 1976).

Fourier analysis have been used for denoising and curve fitting in MODIS vegetation index data sets (Colditz et al., 2007; Bruce et al., 2006; Yu et al., 2004; Wang et al., 2004). If the original time series is discrete rather than continuous, the Discrete Fourier transform (DFT), which requires regular spacing on samples within the temporal domains, should be applied (Wang et al., 2004). Eq 1 depict the DFT: the original signal, x[n], has N samples. Two vectors containing N/2 values where ReX is the Real part vector and ImX is the imaginary part vector of the transformation, k is the index of these transformation vectors. Eq 2 depicts the inverse Fourier transform or synthesis equation in discrete time, where the original signal [x] can be completely resynthesized from the ImX and the ReX vectors.

The algorithm chosen to implement the Discrete Fourier Transform was the HANTS algorithm (Harmonic Analysis of Time Series) (Verhuef, 1996; Roerink et al., 2000). The algorithm was developed to deal with time series of irregularly spaced observations and to identify and remove cloud contaminated observations. Since the NDVI time series of this study were acquired thorough compositing, the pixels have different acquiring dates that lead unequal time spacing.

$$\mathrm{ReX}[k] = \sum_{i=0}^{N-1} x\,[i\,]\cos(2\pi k\, i\,/N\,) \tag{1}$$

$$\mathrm{ImX}[k] = \sum_{i=0}^{N-1} x\,[i\,]\sin(2\pi k\, i\,/N\,) \tag{2}$$

$$x[i] = \sum_{k=0}^{N/2} \mathrm{Re}\,\overline{\mathrm{X}}\,[k\,]\cos(2\pi k\, i\,/N\,) + \sum_{k=0}^{N/2} \mathrm{Im}\,\overline{\mathrm{X}}\,[k\,]\sin(2\pi k\, i\,/N\,) \tag{3}$$

Equations 1, 2 and 3 - Fourrier analysis equation in discrete time, Equation (5) sysnthesis equation in frequency domain - Source Smith (1998)

HANTS considers only the most significant frequencies expected to be present in the time profiles (determined, for instance, from a preceding FFT analysis), and applies a least squares curve fitting procedure based on harmonic components (sines and cosines) (Verhoef et al., 1996; Roerink, et al. 2000). For each frequency, the amplitude and phase of the cosine function is determined during an iterative procedure. Input data points that have a large positive or negative deviation from the current curve are removed by assigning a weight of zero to them. After recalculation of the coefficients on the basis of the remaining points, the procedure is repeated until a predefined maximum error is achieved or the number of remaining points has become too small. (Roerink, et. al. 2000).

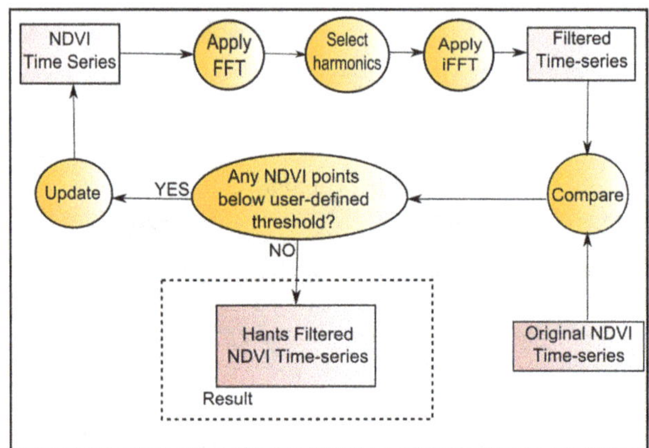

Fig. 8. Flowchart of the FFT based HANTS algorithm, it is a cyclic algorithm that removes the pixels that are below a user chosen threshold - Source (De Wit, 2005)

The algorithm starts in the upper left block with the raw NDVI time series. These are used as input in the FFT and the relevant frequencies (usually mean, annual and half-year signal) are selected from the Fourier spectrum. The inverse FFT (iFFT) then transforms the spectrum back into a filtered NDVI time-series. Next, a comparison is made between the filtered NDVI time-series and the original NDVI time-series (Figure 8). The difference is calculated between the filtered and the original NDVI time-series. Any point in the original NDVI time-series that are below a user-defined threshold are considered 'cloudy' and are replaced with the value of the filtered NDVI time-series. However, by replacing values in the NDVI time-series, the average of the entire profile becomes larger. Therefore, a next iteration is needed and the NDVI time-series is searched again for possible cloud contaminated NDVI observations. This process continues until no new points are found. Many different phenological indicators have been defined in various satellite-based studies. The advantage of the HANTS algorithm is that the output consists of a completely smoothed NDVI profile which is convenient for calculating derivatives. (De Wit, 2005) The calculations of derivates are important to estimate the start of growing season and senescence (Sakamoto et. al 2005).

The version of HANTS used was implemented in IDL by De Wit (2005) and is under the GNU General Public License.

Among the resulting files, the algorithm outputs a FFT file which has a complex number pair, this pair is the Fourier transform of each pixel location regarding its NDVI time series.

2.7 Calculation of amplitude and phase

Harmonic analysis can be used aiming at reducing the dimensionality of the data. Another advantage is that each pixel is treated individually, being independent from the rest of the image. It is also possible to choose the period of analysis relating to the frequency of the studied phenomenon, thus this technique serves well to deal with noise originated from cloud contamination in the time series and from noise resulting from pre-processing that is not periodic. The magnitude and phase of a waveform can be calculated from the complex number resulting from the FFT. The magnitude corresponds to half of wave's peak value, and the phase corresponds to the shift from the origin to the wave's peak value from 0 to Π (eq. 6 and 7) (Lacruz, 2006). The amplitude/phase vector corresponds to the polar form of the DFT (Smith, 1998). The output of the HANTS algorithm contain the complex form for each harmonic (eq. 3, 4 and5) and the mean value of the time series (this value was also used for discussion and comparison among the different areas).

Since the content in the original FFT prior to the transformation to the polar form is not intuitive (Smith, 1998), equations 6 and 7 were used to compute the amplitude and phase of each harmonic. Coldiz et al. (2007) suggest that only the amplitude and phase of the first three harmonics depict biophysical parameters. Some authors, such as Yu (2004), state that forest classification can be carried out in the amplitude/phase space. Previous work by Oliveira et al. (2009) suggests, however, that information is lost in the dimensionally reduction and efficient forest classification is not possible in a rich and complex environment such as the Cerrado.

$$Cj = \sqrt{Aj^2 + Bj^2} \tag{4}$$

$$\phi = \tan^{-1}\frac{B_j}{A_j} \tag{5}$$

Equations 4 and 5 - Calculation of amplitude (equation 1) and phase (equation 2) for to the polar form of the DFT

2.8 Phase statistics of NDVI of deciduous forests

The exact location of the deciduous forests for this study was determined by the use of the official state map by Carvalho (2007) , auxiliary Landsat images, and using information from Oliveira-Filho and Ratter (2002) which describes deciduous forest fragments which were not captured in the official state map. With this information, it was possible to calculate the amplitude-phase statics of each forest fragment within the four chosen areas.

After the calculation of amplitude and phase of each harmonic of the original NDVI images of different locations, a subset of the original images was generated. This subset contains only the pixels corresponding to the occurrences of Deciduous Forests. Statistics were computed to the phase of the first harmonic so that annual shifts of deciduous forests of different geographic locations could be quantified. Each phase value ranges from $-\pi$ to π, where 2π corresponds to a full year cycle. The phase values were multiplied by 182.5, which correspond to half year, in order to calculate annual shifts in days.

2.9 Wavelet transform

Fourier series are ideal for analysing periodic signals, since harmonics modes used in the expansions are themselves periodic. In contrast, the Fourier integral transform is a far less natural tool because it uses periodic functions to expand nonperiodic signals. Two possible

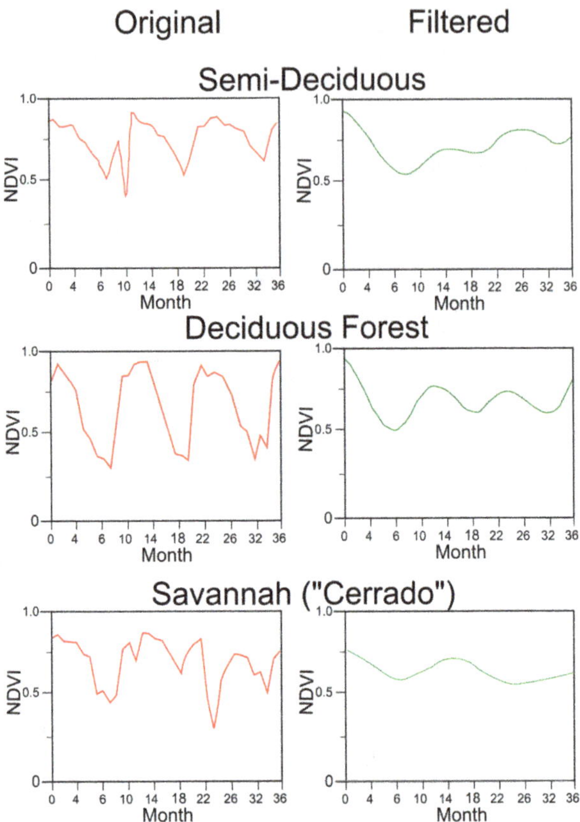

Fig. 9. Denoised phenology time series by wavelet transform. It is clear that before denoising, the data set has many outliers most of them being cloud contaminated pixels. Denoising procedures remove these pixels smoothing the time signatures.

substitutes are the windowed Fourier transform (WFT) and the wavelet transform. The windowed Fourier transform can, however, be an inefficient tool to analyse regular time behaviour that is either very rapid or very slow relative to the size of the analysing window. The Wavelet transform solves both of these problems by replacing modulation with scaling to achieve frequency localization. The WFT might also be an inefficient tool when very short time intervals are of interest. On the other hand, a similar situation occurs when very long and smooth features of the signal are to be reproduced by the WFT. (Kaiser, 1994).

Different from the infinite sinusoidal waves of the Fourier transform, a wavelet is a small wave localized in time or space. Since a wavelet has compact support, which means that its value becomes 0 outside a certain interval of time, the time components of time-series can be maintained during the wavelet transformation (Sakamoto et al. 2005).

Previous work reveal that the wavelet transform is a powerful tool for denoising data sets and for curve fitting procedures in NDVI time series (Sakamoto et al., 2005; Galford et al., 2007; Bruce et al., 2006).

For the present work, we used the methodology proposed by (Carvalho, 2001). In remote sensing, outliers caused by clouds and shadows (noise) appear as peaks with narrow bandwidth in the temporal spectrum. They appear similarly in the spatial domain, but with variable bandwidth. If we consider the presence of clouds and shadows as signal response against a "noisy" background, a framework for their detection can be based on noise modeling in transformed space. The discrete wavelet transform was implemented with the 'à trous' algorithm with a linear spline as the wavelet prototype. It produces a vector of wavelet coefficients d at each scale j, with j=0,...,J. The original function f(t) was then expressed as the sum of all wavelets scales and the smoothed version aj. The input signal was decomposed using one scale, two scales and three scales. Figure 9 shows smoothed time series of NDVI data by denoising via wavelet transform. The resulting different data sets were used as inputs for image classification, described as followed.

2.10 Image classification

The main objective of this work is to compare different filtering techniques and their output vegetation signature for time series of NDVI. One way to accomplish this is to use smoothed time series as input vectors to automated image classification.

For Moreira (2003) automatic image identification and classification can be understood as the analyses and the manipulation of images through computational techniques, with the goal of extracting information regarding an object of the real world.

2.11 Artificial neural networks

Humans and other animals process information with neural networks. These are formed from trillions of neurons (nerve cells) exchanging brief electrical pulses called action potentials. Computer algorithms that mimic these biological structures are formally called artificial neural networks to distinguish them from the squishy things inside of animals (Smith, 1998). These biological inspired models are extremely efficient when the pattern of classification is not a simple and trivial one. Theses networks have shown to be helpful in the resolution of problems of practical scope. Problems such as voice recognition, optical character recognition, medical diagnosis and other practical scope problems are by no means complex problems to the human brain and sensor as they are for a computer to resolve.

Even though, some researchers do not recognize the artificial neural networks as being the general natural solution surrounding the problems of recognizing patterns on processed signals, it can be noticed that a well trained network is capable of classifying highly complex data. The use of artificial neural networks in pattern recognition and classification has grown in the last years in the field of remote sensing (Kanellopolous, 1997).

This work proceeded with 2 filtered data sets per region, these data sets included one HANTS filtered time series and one Wavelets filtered time series. The samples were separated in 3 datasets. These contained samples of each phytophysiognomy of the four different locations. Each location was processed separately, with 2 different datasets. One dataset was used for training the network, and the other for validation. These datasets were input into a neural network with the following characteristics: Sigmoidal activation function, 0.01 learning rate, momentum factor of 0.5, sigmoid constant of 1.0, 14 hidden layers, with 69 neurons per layer. For training the network, 10000 iterations were used, with RMS error of 0.0001. These parameters were extracted from literature based on standard applications of Neural networks to remote sensing image classification.

3. Results

Results in table 1 explicit differences in the phase of the annual frequency of the NDVI value of the deciduous forests of Minas Gerais. These results can be very useful for future vegetation classification and to quantify the geographic differences among apparently similar fragments of this phytophysiognomy.

The largest time shift was observed for the Triângulo Mineiro Region (Western Minas Gerais) which is on average 13.45 days ahead in the annual cycle when compared to the Northern region. This difference could be explained by the fact that the Deciduous Vegetation in this region is mixed with other phytophysiognomies such as the Savanna resembled "Cerradão" and other formations. The mean value of this phytophysiognomy´s time series does not differ substantially from the others (Table 1). These similarities in the mean NVDI time series value confirm that deciduous forests do not have discrepancies in their amplitude value suggesting that the analysed forest fragments are not mixed with other vegetation that have higher mean NDVI value such as the semi-deciduous forests.

This explains the errors occurred in the Official state map in the Triângulo Mineiro region. This indicates that this time lag should be taken into account when mapping Deciduous Forests of this region.

Region	Fundamental harmonic's phase mean value	Phase value expressed in days shifted value	Time series mean value
North East	1.04325	60.6	0.687018
North	1.035954	60,2	0.641661
South	1.141607	66.2	0.735416
West (Triângulo Mineiro)	1.267896	73.65	0.685943

Table 1. Statistics of harmonic analysis of the four different study areas.

Previous work from Sakamoto (2005) rely on the use of derivates and wavelet transforms to obtain the days of harvest and plantation of paddy rice in Japan with the use of MODIS NDVI images. Changes in cropping system, management, and climate make the times-series collected over agricultural areas closer to non-stationary signals, which are better handled by the wavelet transform. In the case of native forests that exhibit a stationary behaviour, our FFT approach is most suitable.

The proximity in results regarding mean value of the phase in the North and the North West areas (table 1) can be partly explained by the geographical proximity of the areas, thus reinforcing that there is a shift in the annual cycle of the deciduous forests due to geographical differences. The Southern area also has a 6 days shift in the average phase value of the annual NDVI frequency and thus also reinforces our hypothesis.

3.1 Classification results

Classification results as shown in table 2 confirm that no time series filtering technique necessarily produces a more accurate classified map. In some cases the classified maps produced from HANTS filtered time series generated more accurate results. In others cases it produced less accurate results. The kappa coefficient for the classification results can be either classified as substantial or almost perfect (Landis, 1977). Different from our findings, previous work carried out by Burce et al. (2006), which have also used filtered time series from Wavelet and Fourier transforms for image classification, showed that the former produced more accurate results. This can be partly explained by the fact that the HANTS algorithm have some enhancements over traditional Fourier based algorithms which was present in the cited work.

Region	Kappa Coefficient (wavelet filtering)	Kappa Coefficient (HANTS filtering)
North East	0.9480	0.8257
North	0.8476	0.8333
South	0.6355	0.7412
West region (Triângulo Mineiro)	0.8729	0.9051

Table 2. Classification results.

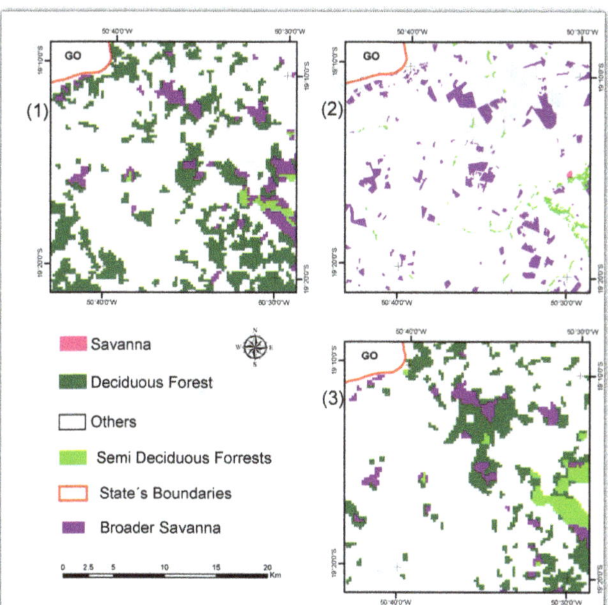

Fig. 10. Triangulo Mineiro Region Classification Results with Neural Networks: (1) Wavelet transform filtered time series, (2) Official State Map, (3) HANTS filtered time series.

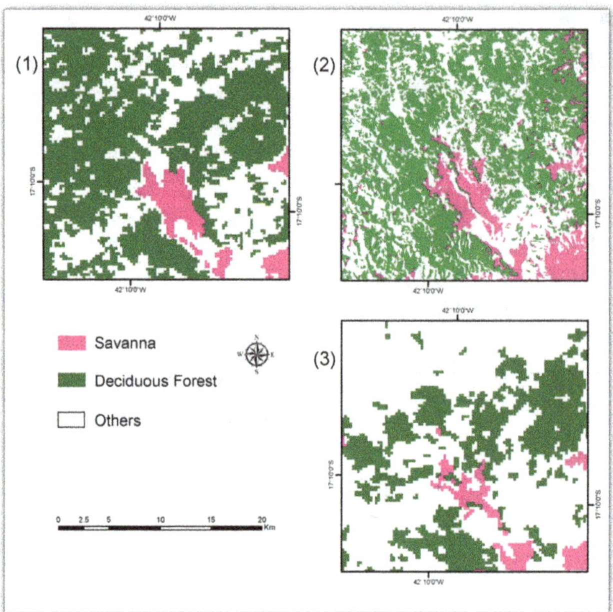

Fig. 11. NothEast Area Classification Results with Neural Networks: (1) Wavelet transform filtered time series, (2) Official State Map, (3) HANTS filtered time series.

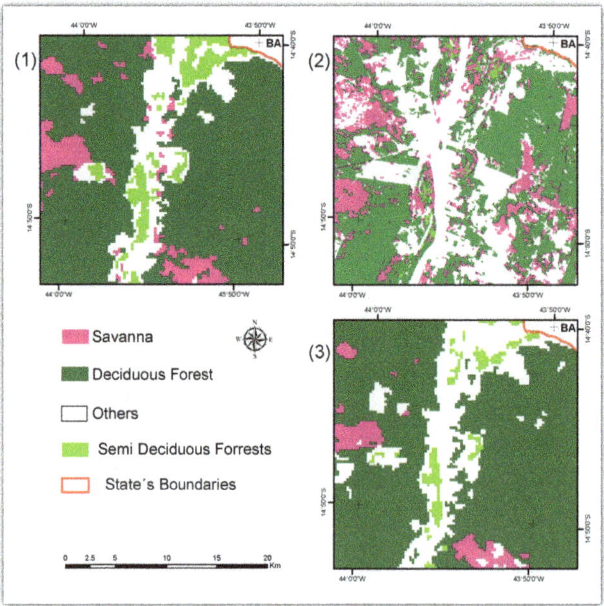

Fig. 12. Northern Area Classification Results with Neural Networks: (1) Wavelet transform filtered time series, (2) Official State Map, (3) HANTS filtered time series.

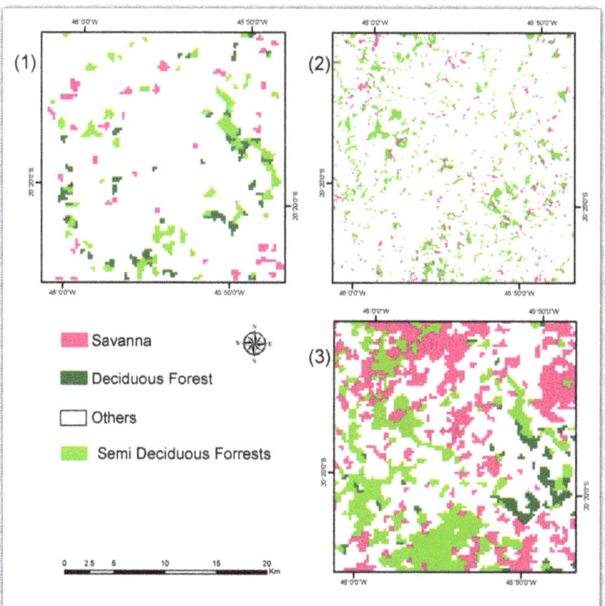

Fig. 13. Southern Area Classification Results with Neural Networks: (1) Wavelet transform filtered time series, (2) Official State Map, (3) HANTS filtered time series.

The rows of figures 10 to 13 show the classification results in the four study areas. The results of Wavelet filtering time series, used as input for classification are in the right hand column. In the left hand column, the HANTS filtered time series as input to the neural network are illustrated. In the middle column we have the official forest map of Minas Gerais, carried out by Carvalho (2008), with a 30m spatial resolution. The proposed methods captured the general characteristics of vegetation of each area. In some cases such as the North East area, the classification results resemble the general "shape" of forest fragments. Both Northern area data sets have similar patterns when compared to the official state map. The other areas, however, do not show these similarities with the general shape of forest fragments from the official map. Note that the spatial resolution has important implications in the map comparisons. The Northern areas have a more accurate "shaping" of vegetation classification.

4. Conclusions

This research suggests that there is an annual shift in the phenological curve of the Deciduous forests of Minas Gerais that are geographically distant, however these differences are not great in value. Different regions demonstrate different annual shifts in the time profile of their deciduous forests of about half month. However, the spatial resolution of the MODIS sensor limit its application resulting in few pixels to calculate statistics on small forest fragments such as the West region of Triângulo Mineiro. The combined use of MODIS NDVI time series and higher spatial resolution sensors, ground truth data and Geostatistics might improve the discrimination of deciduous forests in Minas

Gerais. However this work could be further improved by investigating how large the impact of the time delay of the Deciduous Forests is on mapping.

This work developed an efficient methodology to map the deciduous forests present in the Cerrado Biome by using MODIS temporal attributes and artificial intelligence neural networks algorithm for classification. It is concluded that MODIS filtered NDVI (Normalized Difference Vegetation Index) time series might generate accurate mapping of deciduous forests in different regions.

- The maps generated from both HANTS and wavelet transformation curve smoothing procedures showed very similar and high accuracy indicating that any of these procedures can be used to denoise similar data sets.
- In the Northern areas, the maps generated from temporal features resemble the general "shape" of forest fragments, having similar patterns when compared to the official state map
- This methodology was capable of detecting fragments of deciduous forests in the Triângulo Mineiro region where the official state map of forest did not.

5. References

Bloomfield, P. (1976). *Fourier Analysis of Time Series*. New York : John Wiley & Sons Inc. 257p.

Bruce, L. M.; et al. (2006). *Denoising and Wavelet-Based Feature Extraction of MODIS Multi-Temporal Vegetation Signatures*, GIScience & Remote Sensing, 43,. pp.170 - 180

Carrão, H.; et al. (2007) *Contribution of multiespectral and multitemporal information from MODIS images to land cover classification*. Remote Sensing of Environment pp. 123-140

Carvalho Júnior et al. (2006). *Identificação regional da Floresta Estacional Decidual na bacia do Rio Paraná a partir da análise multitemporal de imagens MODIS*, Rev. Bras. Geof. [online]., vol.24, n.3, pp. 319-332. ISSN 0102-261X.

Carvalho, L. M. T. (2001). *Mapping and monitoring forest remnants: a multi-scale analysis of spatio-temporal data*, 140 p. Thesis (Doctor) - Wageningen University, Wageningen

Carvalho, L.M.T. et al. (2004). *Selection of imagery data and classifiers for mapping Brazilian semideciduous Atlantic forests*. Internacional Journal of Applied Earth Observation and Geoinformation 5 173 - 186

Cerrado utilizando séries temporais do sensor MODIS, Revista Árvore, Viçosa-MG, Vol.35, No.3, Edição Especial, pp.699-705

Colditz, R.R. et al. (2007). *Dynamics of MODIS time series for ecological applications in southern África. In Fourth International Workshop on the Analysis of Multitemporal Remote Sensing Images*, Leuven, Belgium, (Multitemp 2007), 18. - 20.

Couto-Junior, A. F., et al. (2011). *Tratamento de ruídos e caracterização de fisionomias do*

Galford, G. L. et al. (2007). *Wavelet analysis of MODIS time series to detect expansion and intensification of row-crop agriculture in Brazil*. Remote Sensing of the Environment doi 10.10016/j.rse 2007-05-017

Huete, A. et al. (2002). *Overview of the radiometric and biophysical performance of the MODIS vegetation indices*. Remote Sensing of Environment, Vol 83, No. 1, November. pp. 195-213(19)

Jensen, J. R. (2000). *Remote Sensing of the enviroment: An Earth resource perspective,* Pretince Hall Series in geographic information science, Upper SaddleRiver, New Jersey 07458,

Jönsson,Per., Lars, P. (2004). *TIMESAT – a program for analyzing time-series of satellite sensor data,* Computers & Geosciences No.30 833–845

Kanellopoulos G.G. (1997). Wilkinson F.Roli, J.Austin, *Neuro-computation in Remote Sensing Data Analysis*

Kohonen, T.(1999). *The Self-Organizing Map*. Proceedings of the IEEE, Vol. 78: No. 1464-80.

Lacruz, M. S, P; Santos, J. R. (2006). *Monitoriamento da paisagem de unidades de Conservação* in: Rudorff, B. F. T.; Shimabukuro, Y. E.; Ceballos, J. C. (Org.), *O sensor MODIS e suas aplicações no Brasil,* São José dos Campos: Editora Parêntese, 2007, cap. 13, pp.173-183

Mather, P. M. (199). *Computer Processing Of Remotely-Sensed Images: An Introduction.* 2. ed. Nottingham, UK: Johb Wiley, 292 p.

Moreira, M. A.(2003). *Fundamentos de Sensoriamento Remoto e Metodologias de Aplicação,* 2ª Edição Revista e Ampliada Eidtora UFV 295p

Morettin, P. A., Toloi, C. M. (2006). *Análise de Séries Temporais,* Segunda edição 2006 429-433 Editora Edgar Blücher LTDA., CDD-519.232

Oliveira-Filho, A. T. , Ratter J. A., (2002). *The Cerrados of Brazil: ecology and natural history of a neotropical savannah,* Columbia University Press Publishers, New York Chischester, West Sussex pp. 91-121

Oliveira-Filho, A. T. (2011) TreeAtlan 1.0 - Flora Arbórea da Mata Atlântica e Domínios Adjacentes: Um Banco de dados envolvendo Geografia Diversidade e Conservação Available at : http://www. treeatlan.dcf.ufla.br/ access date : 17-09-2011

Oliveira, L.T.; (2004) *Fusão De Imagens De Sensoriamento Remoto E Mineração De Dados Geográficos Para Mapear As Fitofisionomias Do Bioma Cerrado.* 131p. (MSC thesis). UFLA. Lavras - MG, Brazil

Roerink, G. J.; Menenti, M.; Verhoef, W. (2000), *Reconstructing Cloudfree NDVI composites using Fourier analysis of time series International Journal of Remote Sensing* print/ISSN 1366-5901 online © 2000 Taylor & Francis Ltd, vol. 21, no. 9, pp. 1911-1917. ISSN 0143-1161

Sakamoto, T. ; Yokozawa, M. ; Toritani, H. ; Shibayama M. ; Ishitsuka N.; Ohno H. (2005). *A Crop Phenology Detection Method Using Time-Series Modis Data; Remote Sensing of Environment* 96, pp. 366 - 374

Sano et al. (2008). *Mapeamento semidetalhado do uso da terra do Bioma Cerrado,* Pesq. agropec. bras., Brasília, vol.43, no.1, pp.153-156

Silveira et al. (2006). *The Assesment of Vegetational Seasonal Dynamics Using Multitemporal NDVI and EVI Images Derived from MODIS* , CERNE , vol.4, n.2, pp. 177-184. ISSN 0104-7760.

Tan B. et al.(2010). *An Enhanced TIMESAT Algorithm for Estimating Vegetation Phenology Metrics From MODIS Data,* IEEE Journal Of Selected Topics In Applied Earth Observations And Remote Sensing, pp. 01-11

Verhoef, W. (1996). *Application of Harmonic Analysis of NDVI Time Series (HANTS). In Fourier Analysis of Temporal NDVI in the Southern African and American Continents,* edited by S. Azzali and M. Menenti, DLO Winand Staring Centre, Wageningen, TheNetherlands, Report 108, pp. 19–24,

Wang Q. et al. (2004). *Similarities in ground and satellite-based NDVI time series and their relationship to physiological activity of a Scots pine forest in Finland.* Remote sensing of environment., vol. 93, no1-2, pp. 225-237

Yu, Xinfang; et al. (2004). *Forest Classification Based on MODIS Time Series and Vegetation Phenology*. International Geoscience And Remote Sensing Symposium. VOL 4, pp. 2369-2372

Zhang, F. et al. (2003). *An Advanced Tool for Real-Time Crop Monitoring in China,* In IEEE International Geoscience and Remote Sensing Symposium, Toulouse, France, vol IV: 2242 - 2244. July 21-25,

Zhang, X. et al.(2003). *Monitoring Vegetation Phenology Using Modis.* Remote Sensing of Environment 84 471 - 475, Sensing of Environment Vol 84 pp. 471 - 475

Zhang, X. et al. (2006) *Global vegetation phenology from Moderate Resolution Imaging Spectroradiometer (MODIS): Evaluation of global patterns and comparison with in situ measurements*. Journal Of Geophysical Research Vol 111 No G04017 pp. 1-14,

15

Long-Term Detection of Global Vegetation Phenology from Satellite Instruments

Xiaoyang Zhang[1], Mark A. Friedl[2], Bin Tan[3], Mitchell D. Goldberg[4] and Yunyue Yu[5]
[1]*Earth Resources Technology Inc. at NOAA/NESDIS/STAR, College Park,*
[2]*Department of Geography and Environment Boston University, Boston,*
[3]*Earth Resources Technology Inc. at NASA Goddard Space Flight Center, Greenbelt,*
[4]*NOAA/NESDIS/STAR, Camp Springs,*
[5]*NOAA/NESDIS/STAR, College Park,*
USA

1. Introduction

Vegetation phenology is the expression of the seasonal cycles of plant processes and their connections to climate change (temperature and precipitation). The timing of phenological events can be used to document and evaluate the effects of climate change on both individual plant species and vegetation communities. Thus, vegetation phenology (including shifts in the timing of bud burst, leaf development, senescence, and growing season length) is considered as one of the simplest and most effective indicators of climate change (IPCC, 2007). Long-term observing and recording of changes in plant phenology support efforts to understand trends in regional and global climate changes, to reconstruct past climate variations, to explore the magnitude of climate change impacts on vegetation growth, and to predict biological responses to future climate scenarios.

Field phenological observations and calendars provide details of timing of seasonal development for specific plant species. The attributes of field phenophase observations include timing of flower bud or inflorescence appearance, first bloom, 50% bloom, end of blooming, fruit or seed maturing, fruit or seed shedding, first leaf unfolding, bud burst, 50% leaf unfolding, first leaf coloration, full leaf coloration, first defoliation, and end of defoliation. Such field observations have a history extending back for thousands of years in China (Zhu and Wan, 1963), and as far back as the early 1700s in Europe (e.g., Sparks and Carey, 1995), and the 1800s in Japan (Lauscher, 1978). Recently, several networks of field phenological observations have been established worldwide. The most notable of these networks are PlantWatch in Canada (http://www.naturewatch.ca/english/plantwatch/), the National Phenology Network (NPN, http://www.usanpn.org/) in the USA (United States of America), the European Phenology Network, the Japan Phenological Eyes Network (PEN, http://pen.agbi.tsukuba.ac.jp/), and the UK (United Kingdom) Phenology Network (http://www.phenology.org.uk/). The PlantWatch network is part of the Canadian national nature watch series of volunteer monitoring programs designed to help identify ecological change. Plants chosen for the network are perennial, easy-to-identify, broadly-distributed, and naturally occurring species that bloom every spring in response to changing temperature. The European Phenology Network (EPN) involves various universities and research centers, and is supported by the International Society for Biometeorology, Commission on Vegetation Dynamics, Climate and Biodiversity. The USA NPN was established with support from USGS (US Geological Survey) in 2007, and is an interdisciplinary effort involving botanical gardens, academia, and government agencies, with the goal of systematically collecting and analyzing phenological data. This network observes phenology in about 2000 evenly distributed field sites across the USA.

An advanced technique for qualifying seasonality of plant canopy in the field is to take measurements using a digital webcam. The camera is generally mounted in a high tower (around 30 m tall) and is connected to a local wireless network and a personal computer running camera image-capture software (Richardson *et al.*, 2007). The imagery from digital cameras is capable of monitoring plant canopy seasonality (Richardson *et al.*, 2009), crop growth (Goddijn and White, 2006), and timing and duration of flowering (Adamsen *et al.*, 2000).

Long-term field observations of species-level phenophases have been successfully used to reveal local and regional climatic variations occurring for several decades (Fitter *et al.*, 1995; Kramer, 1996; Rötzer and Chmielewski, 2000; Chen *et al.*, 2005). Specifically, long-term records of budburst and flowering dates have been associated with inter-annual variation in air temperature. Previous studies revealed that warmer spring temperature has advanced flowering dates by about 4 days/°C (Fitter *et al.*, 1995) and leaf unfolding by about 3.2–3.6 days/°C in Europe (Kramer, 1996; Rötzer and Chmielewski, 2000). On average, springtime phenological events have changed globally by 2.3 days per decade (Parmesan and Yohe, 2003). Similarly, the growing-season length (GSL) of deciduous broadleaf forests during the period from 1900–1987 increased by about five days as a result of a one-degree increase in mean annual temperature in the eastern United States (White *et al.*, 1999). Moreover, phenological records have been used to model historical climate change. Indeed, the record of grape harvest dates for the period 1523–2007 in the area around Vienna, Austria, reveals that temperature was as warm in the 16th century as in the 1990s; the mean May to July temperature then started to fall, with the coldest decade of the record from 1771 to 1780; and a constant temperature increase from the 1970s to the present seems to be unprecedented during the last 470 years (Maurer *et al.*, 2009).

During the last three decades, remote sensing has become a widely-used mechanism for monitoring the activity of vegetation at large spatial scales. The satellite-derived vegetation indices, commonly termed normalized difference vegetation index (NDVI), provides an indication of the canopy "greenness" of vegetation communities, which is a composite property of leaf chlorophyll content, leaf area, canopy cover and structure. Therefore, the time series of NDVI data derived from the Advanced Very High Resolution Radiometer (AVHRR) have been used extensively for monitoring vegetation phenology (Lloyd, 1990; Reed *et al.*, 1994; White *et al.*, 1997; Zhang *et al.*, 2007). More recently, the VEGETATION instrument onboard the SPOT 4 spacecraft, and the Moderate Resolution Imaging Spectroradiometer (MODIS) onboard NASA's Terra and Aqua spacecraft, have provided a new era of global remote sensing observations. MODIS data produce time series of vegetation indices at spatial resolutions of 250 m, 500 m, and 1 km globally, with substantially improved geometric and radiometric properties (Huete *et al.*, 2002).

Various phenology products have been developed from satellite data at regional and global scales. These products include: (1) the MODIS Land Cover Dynamics Product (MCD12Q2) derived from MODIS NBAR (nadir bidirectional reflectance distribution function adjusted reflectance) EVI (enhanced vegetation index) (500m–1000m), which is the only global product that is produced on an operational basis from 2001 to present (Zhang *et al*, 2006; Ganguly *et al*, 2010); (2) the MODIS-based product generated at NASA-GSFC (Goddard Space Flight Center) in support of the North American Carbon Program, which was produced using MODIS data at a spatial resolution of 250m–500m (Morisette *et al.*, 2009; Tan *et al.*, 2011); (3) the MODIS phenology product being generated for the contiguous United States (CONUS) by the US Forest Service (Hargrove *et al.*, 2009); (4) the USGS long-term 1-km AVHRR phenology product for CONUS (1989–present; Reed *et al.*, 1994); (5) the NOAA 4-km GVIx phenology over North America from 1982–2006 (Zhang *et al.*, 2007); (6) the global 4.6 km product for 2005 from the Medium Resolution Imaging Spectrometer (MERIS) Terrestrial Chlorophyll Index (MTCI) (Dash *et al.*, 2010); and (6) the global product based on FPAR (Fraction of Photosynthetically Active Radiation) developed by the European Space Agency (Verstraete *et al.*, 2008).

Satellite-derived phenology demonstrates recent climate change at a large spatial coverage. Using AVHRR NDVI between 1981 and 1991, Myneni *et al.* (1997) have estimated an advance of 8 ± 3 days in the onset of spring and an increase of 12 ±4 days in GSL in northern latitudes (45–70°N). An extended comparison of average AVHRR-NDVI values from July 1981 to December 1999 has shown that the duration of growing seasons increased by as much as 18 days in Europe and Asia, and by 12 days in northern North America (Zhou *et al.*, 2001). Furthermore, analysis of phenology derived from AVHRR NDVI between 1981 and 2006 across North America indicates that vegetation greenup onset advanced by 0.32 days/year in cold and temperate climate regions because of spring warming temperatures, while it changed progressively from an early trend (north region) to a later trend (south region) in subtropical regions because the shortened winter chilling days were insufficient to fulfill vegetation chilling requirements (Zhang *et al.*, 2007). However, little significant phenological trend has been found using the phenology detection capabilities of AVHRR NDVI during 1982–2006 over North America (White *et al.*, 2009).

Monitoring of vegetation phenology from remote sensing remains a significant challenge, although this technique has been demonstrated to be a robust tool. This is because satellite observations are frequently interfered with various abiotic factors, and a satellite footprint covers a large vegetation community at landscape scales. This chapter briefly introduces current methods in phenology detection from satellite data, and further presents long-term variation in satellite-derived vegetation phenology at the scale of global coverage.

2. Overview of phenology detection from satellite data

2.1 Vegetation index for phenology detection

Vegetation index (VI) derived from satellite data has been widely applied to monitor vegetation properties. The most commonly used vegetation index is the Normalized

Difference Vegetation Index (NDVI). It was first formulated by Rouse *et al.* (1973) using the following formula:

$$NDVI = \frac{\rho_{NIR} - \rho_{red}}{\rho_{NIR} + \rho_{red}} \quad (1)$$

where ρ_{NIR} and ρ_{red} stand for the spectral reflectance measurements acquired in the near-infrared and red regions.

The NDVI derived from satellite data has been proved to be a robust tool for retrieving local and global vegetation properties, including vegetation type, net primary product, leaf area index, foliage cover, phenology, photosynthetically active radiation absorbed by a canopy (FPAR), evapotranspiration (ET), and biomass (e.g. Tucker *et al.*, 1986; Unganai and Kogan, 1998; Loveland *et al.*, 1999; Myneni *et al.*, 2002, Friedl *et al.*, 2002). More importantly, a long time series of AVHRR NDVI data has been widely applied for exploring global climate change reflected by variation of inter-annual vegetation phenology (Read *et al.*, 1994; Myneni *et al.*, 1997; Zhou *et al.*, 2001; Nemani *et al.*, 2003; Zhang *et al.*, 2007). Although NDVI provides researchers with a way to monitor vegetation characteristics, the use of NDVI across a variety of vegetation types may be limited by sensitivity to background reflectance (soil background brightness and moisture condition) (Huete *et al.*, 1985; Bausch, 1993), the attenuation caused by highly variable aerosols (Kaufman and Tanré, 1992; Miura *et al.*, 1998; Ben-Ze'ev *et al.*, 2006), and the saturation at densely vegetated areas (Huete *et al.*, 2002; Gitelson, 2004).

The enhanced vegetation index (EVI) has been developed to improve the quantification of vegetation activity (Huete *et al.*, 2002). EVI reduces sensitivity to soil and atmospheric effects, and remains sensitive to variation in canopy density where NDVI becomes saturated (Huete *et al.*, 2002). It is calculated from reflectance in blue, red and near-infrared bands, using the formula:

$$EVI = G\frac{\rho_{NIR} - \rho_{red}}{\rho_{NIR} + C_1\rho_{red} - C_2\rho_{blue} + L} \quad (2)$$

where ρ_{blue}, ρ_{red} and ρ_{NIR} are values in the blue, red, and near-infrared bands, respectively, L (=1) is the canopy background adjustment, C_1 (=6) and C_2 (=7.5) are aerosol resistance coefficients, and G (=2.5) is a gain factor.

As described in the above equation, EVI requires information on reflectance in blue wavelengths, which is not available on some satellite instruments, including SPOTVGT, SeaWiFS, ENVISAT-MERIS, GLI, and AVHRR. To overcome this limitation, a two band EVI (EVI2) has been proposed (Huete *et al.*, 2006; Jiang *et al.*, 2008), which is described as:

$$EVI2 = G\frac{\rho_{NIR} - \rho_{red}}{\rho_{NIR} + C_3\rho_{red} + L} \quad (3)$$

where C_3 is a coefficient (2.4).

The two-band adaptation of EVI2 is fully compatible with EVI (Huete *et al.*, 2006; Jiang *et al.*, 2007). The EVI2 remains functionally equivalent to the EVI, although slightly more prone to aerosol noise, which is becoming less significant with continuing advancements in atmosphere correction. Similar to EVI, EVI2 is less sensitive to background reflectance, including bright soils and non-photosynthetically active vegetation (i.e. litter and woody tissues) (Rocha *et al.*, 2008). Thus, it could be used to monitor vegetation phenology and activity across a variety of ecosystems (Rocha and Shaver, 2009).

There are several other vegetation indices in vegetation phenology detections. These include Normalized Difference Water Index (NDWI) (Delbart *et al.*, 2005), FPAR (Verstraete *et al.*, 2008), and LAI (Obrist *et al.*, 2003).

2.2 Algorithm of phenology detection

Phenology detections from time series of satellite data are commonly composed of two steps: modeling of the temporal VI trajectory and identification of the timing of phenological phases. Modeling (or smoothing) of the temporal VI trajectory is to reduce non-vegetative information (noise) in the satellite observations. The noise in an annual time series is mainly caused by environmental impacts: cloud cover, atmospheric effects, and snow cover. To minimize cloud and atmospheric contamination, the maximum value composite (MVC) (Holben, 1986) and best index slope extraction (BISE) (Viovy *et al.*, 1992) are commonly applied to create weekly, biweekly, or monthly composites. To further reduce noise, time series of VI data are often smoothed using a variety of different methods including Fourier harmonic analysis (Moody and Johnson, 2001), asymmetric Gaussian function-fitting (Jonsson and Eklundh, 2002), piece-wise logistic functions (Zhang *et al.*, 2003), Savitzky-Golay filters (Chen *et al.*, 2004), degree-day based quadratic models (de Beurs and Henebry, 2004), and polynomial curve fitting (Bradley *et al.*, 2007). In mid- and high latitudes, vegetation signals are also contaminated by snow cover during winter. To reduce snow contamination, which generally results in a dramatically steep drop in NDVI and irregular variation in EVI (Zhang *et al.*, 2006), snow cover observations are explicitly removed or replaced. This is done using nearest non-snow observations in a temporal VI trajectory after winter periods are determined using ancillary data of land surface temperature and snow detection (Zhang *et al.*, 2004a; Tan *et al.*, 2011) or high values of NDWI (Delbart *et al.*, 2005).

For long-term VI data record, the noises also result from instrumental uncertainties related to sensor decay and inconsistency among multi-sensors. A variety of studies have simulated VI values across different sensors to investigate the uncertainty caused by various impact factors and to establish VI translation equations. Generally, the VI values from various instruments are continued using a set of linear or quadric equations (Steven *et al.*, 2003; Fensholt and Sandholt, 2005; Miura *et al.*, 2006).

The modeled annual time series of VI data is not necessary for the accurate reflection of seasonal vegetative signals because of the complex abiotic influences. The degree of vegetation representation is strongly dependent on the model approaches used. The uncertainty in the temporal VI trajectory is generally the main source of errors in the detection of vegetation phenologic metrics, which is currently lack of detailed investigations.

A number of methods have been developed to identify the timing of phenological phases (or metrics) from the modeled/smoothed temporal VI trajectory at regional and global scales.

The commonly used methods are the threshold-based technique which is divided into absolute VI threshold (e.g., Lloyd, 1990; Fischer, 1994; Myneni *et al.*, 1997; Zhou *et al.*, 2001) and relative threshold (e.g., White *et al.*, 1997; Jonsson and Eklundh, 2002; Delbart *et al.*, 2005; Karlsen *et al.*, 2006; Dash *et al.*, 2010), moving average (Reed *et al.*, 1994), spectral analysis (Jakubauskas *et al.*, 2001; Moody and Johnson, 2001), and inflection point estimation in the time series of vegetation indices (Moulin *et al.* 1997; Zhang *et al.* 2003; Tan *et al.*, 2011). Various approaches in detecting phenological timing, particularly the greenup onset, are compared using the same dataset (de Beurs and Henebry, 2010; White *et al.*, 2009). Evidently, most of the methods work well at local and regional scales, or for specific vegetation types. However, they are difficult to implement globally since empirical constants are involved and generally do not account for ecosystem specific characteristics of vegetation growth.

3. Long-term satellite detection of global vegetation phenology

3.1 Global vegetation phenological metrics

Phenology observed from satellite data is usually defined as land surface phenology (de Beurs and Henebry, 2004; Friedl et al., 2006) because an annual cycle of satellite data reflects seasonal variation composed of vegetation, atmosphere, snow cover, water conditions, and other land disturbance. However, vegetation seasonal dynamics are generally the parameters of interest to retrieve, whereas the abiotic signals in the temporal satellite data are considered to be noise. As a result, long-term global satellite-based phenological metrics in this chapter are defined according to vegetation seasonal cycles. Briefly, a seasonal cycle of vegetation growth consists of a greenup phase, a maturity phase, a senescent phase, and a dormant phase (Figure 1, Zhang *et al.*, 2003). These four phases are characterized using four phenological transition dates in the time series of VI data: (1) greenup onset (leaf-out): the date of onset of VI increase; (2) maturity onset: the date of onset of VI maximum; (3) senescence onset: the date of onset of VI decrease; and (4) dormancy onset: the date of onset of VI minimum. Furthermore, the time series of VI data provides the integrated VI for the growing season (the sum of daily VI values varying from greenup onset to dormancy onset), maximum and minimum VI values during a growing season, and the length of the vegetation growing season.

During a senescent phase, foliage senescent development consists of several coloration statuses (Zhang and Goldberg, 2011). Fall foliage coloration is a phenomenon occurring in many deciduous trees and shrubs worldwide. Fall foliage status is a function of the colored leaves on the plant canopy. With the spread of colored foliage, the percentage of fallen leaves increases. Their difference represents relative variation in colored leaves on plant canopy, which can be quantified using a temporally-normalized brownness index. The occurrence of the maximum relative variation derived from the brownness index is considered to be a critical point in foliage coloration status, this being the onset timing of peak foliage coloration. Prior to this point, foliage status is generally defined using the categories of little/no change, low coloration, moderate coloration, and near-peak coloration. Following the critical point, it is divided into peak coloration phase and post-peak coloration phase (Figure 2).

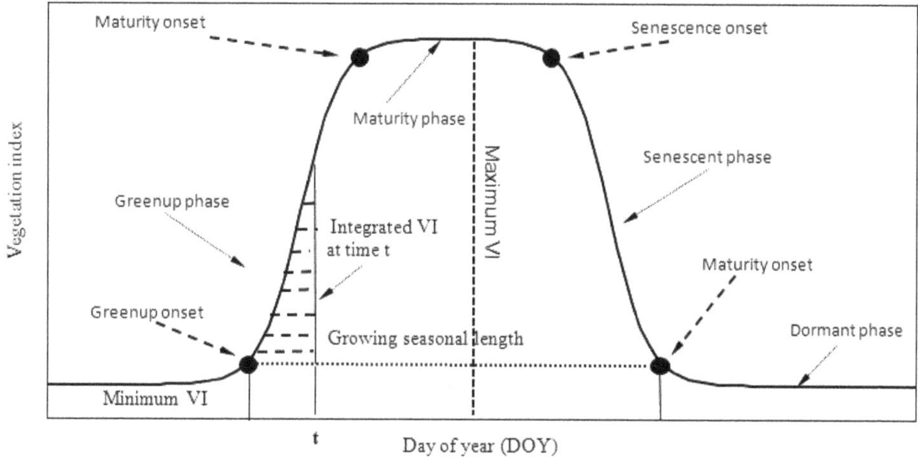

Fig. 1. Key phenological metrics in an annual trajectory of satellite vegetation index.

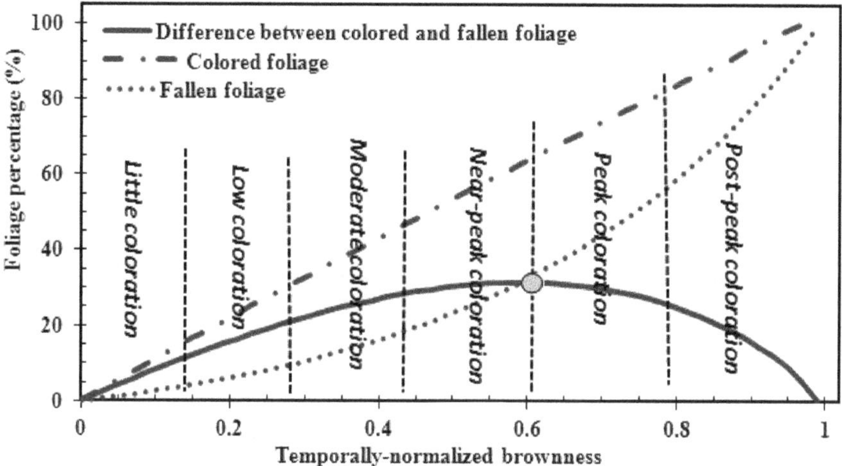

Fig. 2. Correlation of the temporally-normalized brownness index with colored leaves and fallen leaves, separately, and determination of foliage coloration status. The grey dot indicates the critical point when colored foliage reaches maximum on a plant canopy.

More than one set of vegetation phenological metrics could occur within a one-year period because of the complexity of phenological cycles across the globe. Vegetation growth can undergo one or more cycles, and may include an incomplete cycle (truncated at the beginning or end) during a year (Figure 3). The simplest case is illustrated in Figure 3a, where a single and complete growth cycle centers near the mid-point of a 12-month period. Two partial cycles are recorded in Figure 3b, 3c, and 3d. Figure 3e illustrates the situation where two complete growth cycles are finished, which leads to two complete sets of phenological metrics. Figure 3f–3h shows examples of two incomplete cycles and one complete cycle. To capture vegetation phenological timing properly from the complex cycles within a given one-year period, the satellite data should be extended by periods of a half-year prior to and following the period of interest, separately.

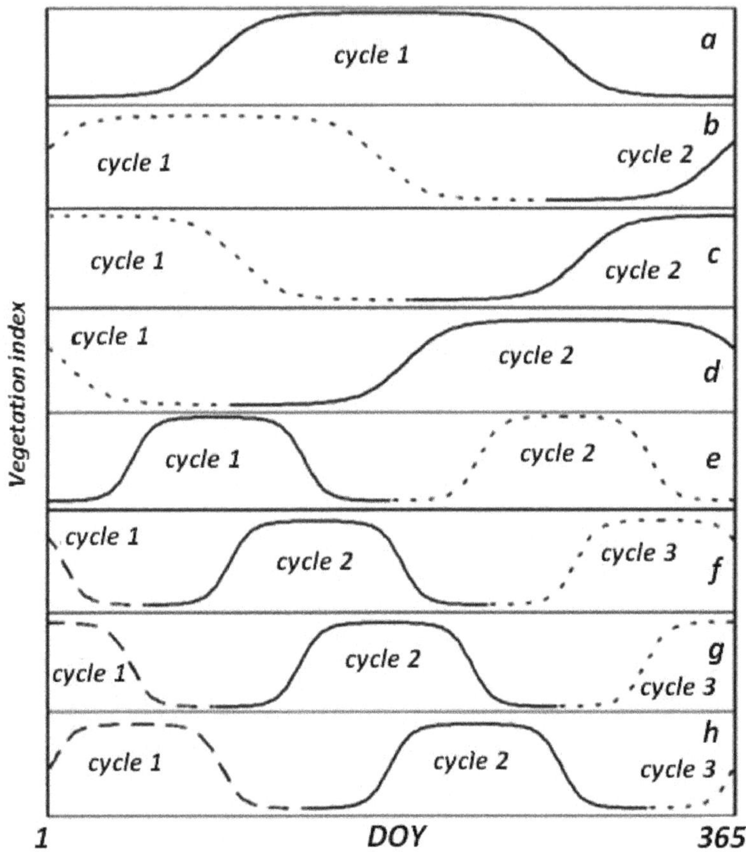

Fig. 3. Various hypotheses of vegetation phenological cycles across the globe.

3.2 Detection of global vegetation phenology

To determine the global phenological metrics described above, the following approaches are conducted. Temporal VI data are first preprocessed to remove or reduce the impacts of clouds, atmosphere, snow cover, etc. Specifically, the data gaps caused by clouds–creating isolated missing values–are filled by linear interpolation using neighbor good quality data. The time series of VI data at each pixel is then smoothed using a Savitzky-Golay and running local median filter. The background VI value at each pixel, which represents the minimum VI of soil and vegetation in an annual time series (Zhang *et al.*, 2007), is identified and it is used to replace VI values in the time series flagged as snow contaminations.

Vegetation growth cycle is identified using a moving slope along the VI time series. The periods with sustained VI increase and decrease at each pixel are determined using a five-point moving slope technique, where transitions from periods of increasing VI to periods of decreasing VI are identified by changes from positive to negative slope, and vice versa. Because slight decreases or increases in VI can be caused by local or transient processes unrelated to vegetation-growth cycles, two heuristics are applied to exclude such variation: (1) the change in VI within any identified period of VI increase or decrease must be larger than 35% of the annual range in VI for that pixel; and (2) the ratio of the local maximum VI to the annual maximum VI should be at least 0.7. This approach screens out short-term variation unrelated to growth and senescence cycles in VI data, while at the same time identifying multiple growth cycles within any 12-month period.

VI time series in the growing phases (VI consistent increase) and senescent phases (VI consistent decrease) is modeled using a sigmoidal vegetation growth function (Zhang et al., 2003). The specific sigmoid function used to model temporal VI dynamics is the logistic function of vegetation growth:

$$y(t) = \frac{c}{1 + e^{a+bt}} + d \tag{4}$$

where t is time in days, $y(t)$ is the VI value at time t, a and b are free parameters that are fitted using a non-linear least sqaures approach, c is the amplitude of VI variation and d is the initial background VI value. The advantages of the sigmoidal model are that: (1) it provides a simple, bounded, continuous function for modeling growth and decay processes; and that (2) each parameter can be assigned a biophysical meaning related to vegetation growth or senescence.

This sigmoidal model has been demonstrated to be effective in depicting seasonality of vegetation growth as a function of time (or cumulative temperature) in various ecosystems and data measurements. It was originally developed for monitoring crop growth based on field measurements (e.g., Richards, 1959; Ratkowsky, 1983) and adopted to simulate temporal satellite vegetation index (Zhang *et al.*, 2003). It has then been applied to investigate seasonal vegetation growth using webcam data (Richardson *et al.*, 2006; Kovalskyy *et al.*, 2012), Landsat TM data (e.g., Fisher *et al.*, 2006; Kovalskyy *et al.*, 2011), AVHRR data (e.g., Zhang *et al.*, 2007), and MODIS data (e.g., Zhang *et al.*, 2003, 2006; Ahl *et al.*, 2006; Liang *et al.*, 2011). Moreover, studies have shown that the sigmoidal model performance is superior to both Fourier functions and asymmetric Gaussian functions for dictping remotely sensed phenology (Beck *et al.*, 2006). Thus, the physically-based sigmoidal model is applicable for the detection of global vegetation phenology.

Phenological transition dates within each growth or senescence phase are identified using the rate of change in the curvature of the modeled sigmoidal curves (Zhang et al., 2003; Figure 1). Specifically, transition dates correspond to the day-of-year (DOY) on which the rate of change in curvature in the VI data exhibits local minima or maxima. These dates indicate when the annual cycle makes a transition from one approximately linear stage to another. Formally, at any time t, the curvature (K) for the sigmoidal function given above is:

$$K = \frac{d\alpha}{ds} = \frac{y''}{\left(1 + y'^2\right)^{\frac{3}{2}}} = -\frac{b^2 c e^{a+bx}\left(1 - e^{a+bx}\right)\left(1 + e^{a+bx}\right)^3}{\left[\left(1 + e^{a+bx}\right)^4 + b^2 c^2 e^{2(a+bx)}\right]^{\frac{3}{2}}} \tag{5}$$

where α is the angle (in radians) of the unit tangent vector at time t along a differential curve, and s is the unit length of the curve. Setting $z = e^{a+bt}$, the rate of change of curvature (K') is:

$$K' = b^3cz\left\{\frac{3z(1-z)(1+z)^3\left[2(1+z)^3+b^2c^2z\right]}{\left[(1+z)^4+(bcz)^2\right]^{\frac{5}{2}}}-\frac{z^2\left(1+2z-5z^2\right)}{\left[(1+z)^4+(bcz)^2\right]^{\frac{3}{2}}}\right\} \tag{6}$$

During the growth period, when vegetation transitions from a dormant state to a growth phase, three extreme points in a VI curve can be identified using the equation 6 (Zhang *et al.*, 2003). The two maximum values correspond to the onset of greenup (onset of VI increase) and the onset of maturity (onset of VI maximum), respectively (Figure 1). Similarly, the extreme points during the senescent phase represent the transition dates of the senescent onset (onset of VI decrease) and the dormancy onset (onset of VI minimum).

To determine foliage coloration status, a temporally-normalized brownness index is derived from the relative percentage dynamics of the fraction of colored foliage (Zhang and Goldberg, 2011). This brownness index is described as:

$$\mathrm{TNBI}_{\mathrm{b(t)}} = \frac{F_{c(t)}-F_{c\min}}{F_{c\max}-F_{c\min}} = \frac{F_{cb(t)}-F_{cb\min}}{F_{cb\max}-F_{cb\min}} \tag{7}$$

where $F_{cbmin}=F_{cmin}+F_b$; $F_{cbmax}=F_{cmax}+F_b$; $F_{cb(t)}=F_{c(t)}+F_b$; $\mathrm{TNBI}_{\mathrm{b(t)}}$ is defined as the temporally-normalized brownness at time t; F_b is the exposed surface background; $F_{c(t)}$ and $F_{cb(t)}$ are the fraction of colored foliage on plant canopy and total brown material at time t, separately; F_{cmin} and F_{cmax} are the maximum and minimum fractions of colored leaf cover; and F_{cbmin} and F_{cbmax} are the minimum and maximum fractions of brown material during the senescent phase, separately.

The temporally-normalized brownness index is directly linked to the temporal trajectory of vegetation index (Zhang and Goldberg, 2011). Specifically, the colored foliage is determined after the modeled temporal VI trajectory during the senescent phase is further combined with a linear mixture model of surface components consisting of green (or photosynthetic) vegetation, colored (or non-photosynthetic) vegetation, and exposed surface background (bare soil and rock). As a result, the temporally-normalized brownness index is deduced as:

$$\mathrm{TNBI}_{\mathrm{b(t)}} = 1-\frac{1}{1+e^{a+bt}} \tag{8}$$

The temporally-normalized brownness index represents relative changes in colored foliage, and varies with time in each pixel individually. It is independent of the surface background, vegetation abundance, and species composition. Thus, it is robust to divide the foliage coloration status, as displayed in Figure 2, into separate categories of little coloration, low coloration, moderate coloration, near-peak coloration, peak coloration, and post-peak coloration.

3.3 Vegetation index for global phenology detection

A long-term dataset of global EVI2 has been generated from the daily land surface reflectance from the AVHRR Long-Term Data Record (LTDR) and the MODIS Climate Modeling Grid (CMG) records. AVHRR LTDR provides daily surface spectral reflectance at a spatial resolution of 0.05 degrees from various AVHRR sensors from 1981-1999 (Vermote and Saleous, 2006), which is available at the NASA funded REASoN project web site (http://ltdr.nascom.nasa.gov/). The MODIS CMG dataset provides Terra and Aqua MODIS daily CMG surface reflectance (Collection 5.0) at a spatial resolution of 0.05 degrees, covering the period from 2000 to 2010, which is available at the USGS for EROS DAAC (http://edcdaac.usgs.gov/main.asp). From these daily surface spectral reflectance, the long-term daily EVI2 has been calculated and available for the last 30 years (http://vip.arizona.edu/viplab_data_explorer).

4. Results in global vegetation greenup onset

Global vegetation phenological metrics during the last three decades were detected from global daily EVI2 using a series of piece-wise logistic models. Here, only the greenup onset is presented and discussed because it is the most important parameter in a vegetation seasonal cycle.

4.1 Spatial pattern in the timing of greenup onset

Figure 4 sets out the average onset of vegetation greenup in the 1980s, 1990s, and 2000s. If there were multiple seasonal cycles in a given calendar year, the first occurrence of greenup onset was selected. As expected, the spatial pattern in the three periods is very similar. However, the spatial variation in phenological transition dates reflects both broad-scale patterns in controlling mechanisms related to climate, and more local factors related to land cover and human activities.

Several spatially distinctive properties of greenup onset are evident. Changes in phenology with latitude are apparent in most of the northern hemisphere, from 30°N northwards (Figure 4). Greenup onset occurs in early March in the southern USA (south of 40°N), April in the northern USA, and at the end of June in northern Canada. Zonal patterns in the timing of greenup onset indicate that the transition date of greenup varies at a rate of about 2-3 days per degree of latitude in North America, Europe, and Asia (Figure 5a, 5b). This latitude dependence is assumed to be a function of temperature variation (Myneni *et al.*, 1997; Zhang *et al.*, 2004a).

The dependence on latitude is spatially variable because of the spatial complexity in elevation and human activities. For example, the timing shift in greenup onset is about one and half months from bottom to top of the Carpathian Mountains and Dinaric Alps in Europe (Figure 5b). This reflects that the timing of greenup onset is also a function of elevation in mountains, because temperature decreases with increasing elevation. Moreover, agricultural land use is one of the most geographically extensive land cover types on the Earth. Their phenological behavior is frequently distinct from that of surrounding natural vegetation because of controls applied by human management. It is highly evident in central North America, where the onset of greenup occurs much later in the Mississippi River valley and the mid-western agricultural heartland, relative to the surrounding natural vegetation (Figure 4). This pattern depends strongly on crop type and human management. Moreover, urban lands advance greenup onset relative to rural areas surrounding the urban regions because of the urban heat island effects (Zhang *et al.*, 2004b), although this is not clearly visualized on the 0.05 degree maps.

In dry climate (arid and semi-arid regions), the spatial pattern in vegetation greenup onset is very complex because it is generally controlled by water availability. In Mediterranean climates and the southwestern United States, the start of vegetation growth occurs mainly in winter and early spring and, in some cases, during the summer monsoon season. Outside of the humid tropical regime in sub-Saharan Africa, Australia, and southern South America, the dominant vegetation types are grasses, shrubs and savannas. The onset of vegetation greenup in these vegetation types generally depends on timing of the rainy season.

Inspection of the greenup onset in dry climates reveals several regular patterns in local regions. The most notable pattern is present in northern Africa (the Sahelian and sub-Sahelian region). The timing of greenup onset shifts smoothly from early March, at around 6.5°N, to mid-October in the boundary between the Sahel and the Sahara desert (17.9°N, Figure 5b). The shift rate is about 20 days per degree of latitude, which is about 10 times slower than that in temperate North America and Eurasia. This pattern reflects the start of the rainy season, which triggers the onset of vegetation growth in this region (Zhang *et al.*, 2005), which is in turn controlled by the migration of the Intertropical Convergence Zone (ITCZ). In contrast, the phenological pattern found in southern Africa is much more complex (roughly 1°S southward), although greenup onset shows a regular delayed shift from 1°S to 22°S and an advanced shift of 22°S southwards (Figures 4 and 5b). In the eastern part of this region, vegetation growth generally starts between September and November, whereas it tends to occur in February and March in southwestern Africa (west of the Kalahari Desert). In the Great Horn of Africa, two cycles of vegetation growth are evident, which reflects the bimodal precipitation regime in this region. These irregular patterns coincide strongly with patterns evident in the arrival of the rainy season (Zhang *et al.*, 2005).

In South America, four different phenological regions follow the variation in the onset of vegetation greenup. Greenup onset occurs in the boreal winter, with no obvious gradient in the northern Andes mountainous region. In southern South America, green leaves emerge in the boreal summer and gradually push northward at a rate of about three days per latitude (Figure 5c). However, a remarkable phenological trend exists along the Brazilian Highlands (in the direction from 60°W and 39°S to 35°W and 5°S), where the greenup onset shifts from July to next February at a rate of about 0.12 days/km. In contrast, the timing of greenup onset is very irregular in the Amazon rainforest, where the values are of poor quality because of high frequencies of cloud cover and weak seasonality in vegetation index. Overall, the complex phenological pattern is likely to be associated with precipitation and latitude-elevation-dependent temperature.

Phenological variation in Australia divides into three distinct regions. Greenup onset occurs in the late boreal autumn and winter in northern areas, in the boreal summer in southern areas, and in the boreal spring, or with no clear phenology, in central Australia. Although

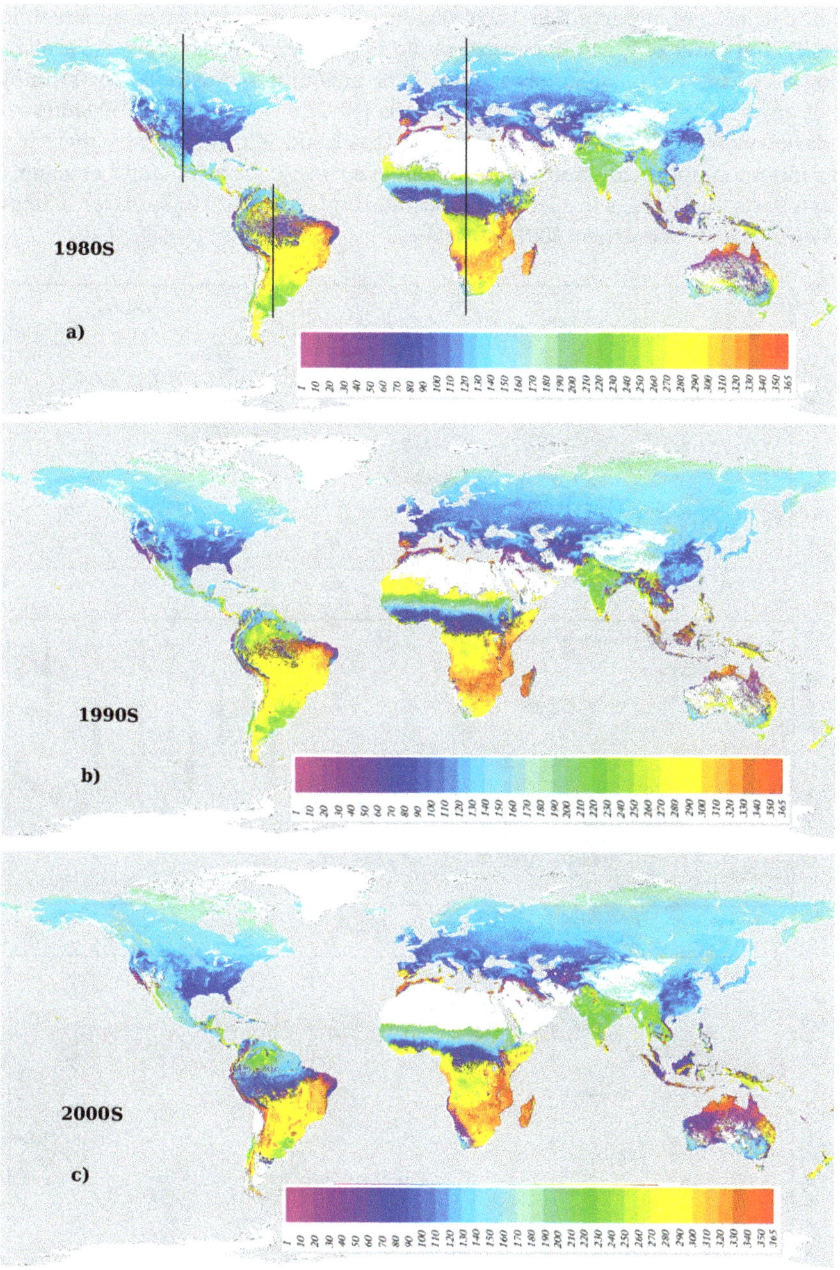

Fig. 4. Average timing of greenup onset in three periods: a) 1980s (1982--1989), b) 1990s (1990--1999), and c) 2000s (2001--2009), separately. The color legend is the day of year. Three vertical lines on a) present the locations of profiles in Figure 5.

irregular patches are observed in each region, a regular gradient is apparent locally. Specifically, the onset of greenup occurs mainly in January over northern Australia, while phenological phases occur about six months later in southern Australia. For example, the timing of greenup onset in central north Australia (13–21.5°S and 128–140°E) shifts at a rate of 0.1 days/km from October to late January. This trend is controlled by the Australian summer monsoon and extra-monsoonal rainfall events (e.g., Hendon and Lebmann, 1990), and also reflects the changes in species composition and a decrease in both tree biomass and diversity (Cook and Heerdegen, 2001).

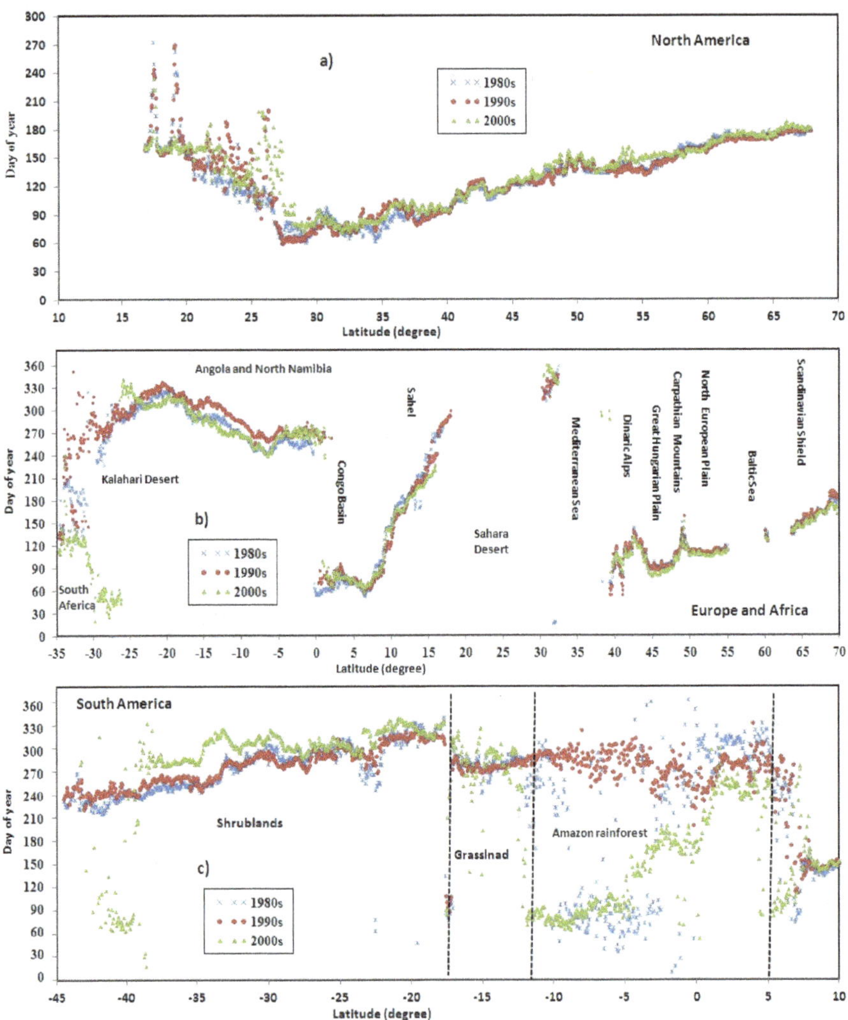

Fig. 5. Profiles of the shift of greenup onset: a) along a meridian of 100°W in North America, b) along a meridian of 20°E in Europe and Africa, and c) along a meridian of 65°W in South America. The geographic locations are displayed on Figure 4a.

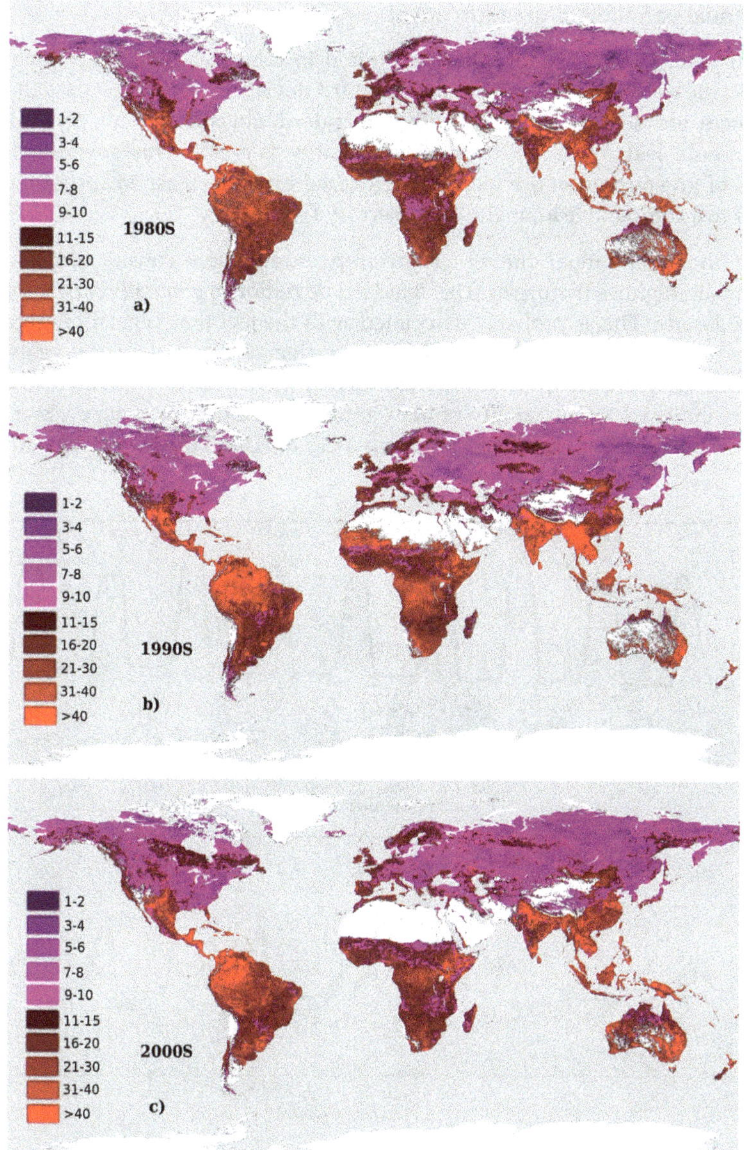

Fig. 6. Inter-annual variation (standard variation) in the timing of greenup onset in a) 1980s, b) 1990s, and c) 2000s, separately. The color legend is the number of days.

Note that the detected phenology metrics are of poor quality for evergreen vegetation (tropical rainforests and boreal forests) in many areas. This is because the annual variation in EVI2 is too subtle to retrieve phenology effectively. Moreover, there is no vegetation growth in tropical desert areas and polar regions of permanent snow cover.

4.2 Inter-annual variation in greenup onset

Inter-annual variation in greenup onset is limited in temperate and cold climate regimes (Figure 6). In the northern hemisphere, the standard deviation is generally less than 10 days, although there are several locations with a standard deviation of about 11-15 days in evergreen needle leaf forest where EVI2 seasonality is weak. This suggests that spring occurrences of greenup onset are regularly triggered by an increase in spring temperature, which leads to a comparable annual EVI2 trajectory (Figure 7a).

In contrast, the inter-annual timing of greenup onset varies considerably in arid and semiarid climate regimes (Figure 6). The standard deviation is generally larger than 15 days within each decade. This is probably associated with the fact that vegetation greenup onset strongly tracks rainy season occurrence, which can change greatly between years (Zhang *et al.*, 2005). For example, a temporal EVI2 trajectory in shrubland in the southwestern United States clearly indicates the variability of inter-annual vegetation growing cycles (Figure 7b), with the timing of greenup onset varying from DOY 85 to 213 during the period from 2001 to 2009.

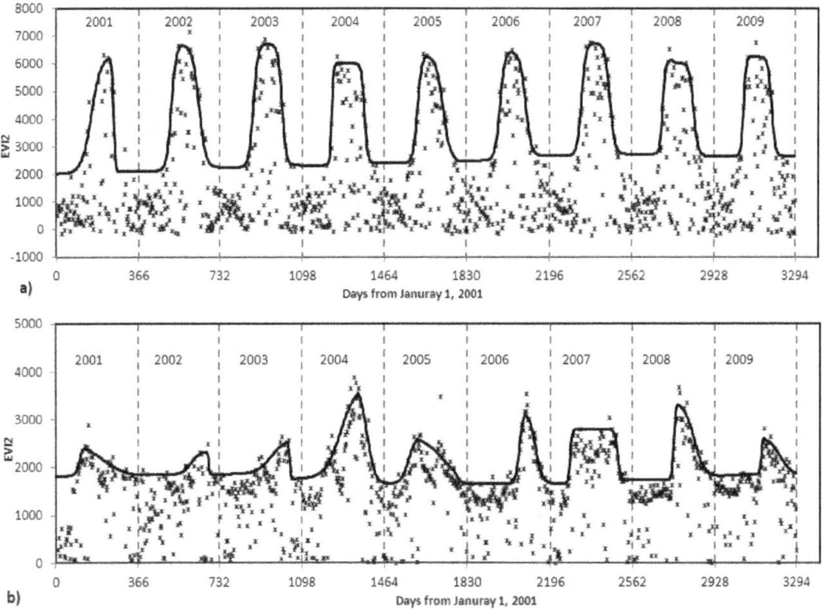

Fig. 7. Time series of daily EVI2 from 2001-2009 in two sample pixels. Solid line is the modeled vegetative EVI2 while the asterisks are the raw EVI2. a) Deciduous forests in northeastern North America and b) shrubland in the semiarid region of southwestern North America.

4.3 Shift in greenup onset

Figure 8 sets out the shift of greenup onset during the past three decades. From the 1980s to the 1990s, the onset of vegetation greenup became advanced in most of the northern hemisphere, South America, and the Sahelian and sub-Sahelian regions. However, delayed shifts appeared in relatively small regions in each continent, except for the southern semiarid region in Africa.

From the 1990s to the 2000s, shifts in greenup onset were basically opposite to those in the previous period in large parts of South America, Africa, and North America. In contrast, persistent trends during the three decades occurred in relatively small regions. In particular, an advanced trend was evident in most of Eurasia.

It is worth noting that the vegetation greenup occurred in a much larger area across the Sahel in the 1990s than in the 1980s. This trend agrees with the result derived by Tucker and Nicholson (1999) and Olsson *et al.* (2005). However, the region with greenup occurrence was reduced during 2000s, which is probably associated with retreat of the ITCZ migration.

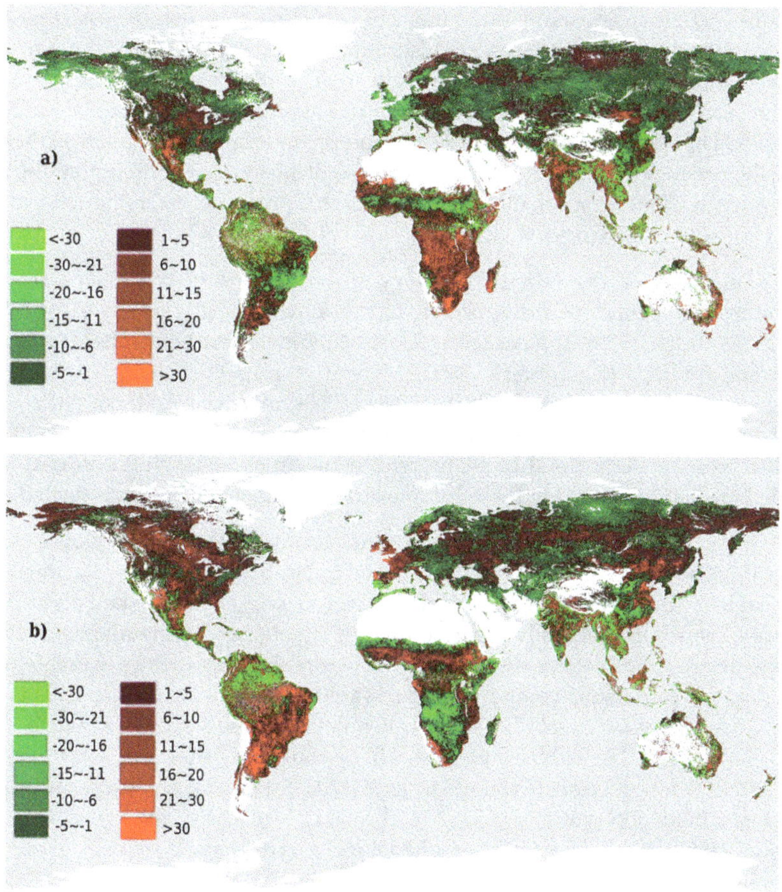

Fig. 8. Shift in greenup onset. a) The difference between the 1990s and 1980s; b) the difference between the 2000s and 1990s. The green color indicates the number of advanced days, while the red color shows delayed days.

5. Discussion and conclusions

This chapter provides an overview of methods and results in the detection of vegetation phenology using satellite data. Various vegetation indices derived from satellite data reflect seasonal dynamics in vegetation growth with reasonable accuracy, and a variety of methods have been developed for detecting vegetation phenological metrics. In the detection of long-term global vegetation phenology, EVI2 from AVHRR and MODIS data has advantages over NDVI and EVI (Rocha and Shaver, 2009) and a series of pieces-wise sigmoidal models of vegetation growth provide a flexible, repeatable, and realistic means to monitor seasonal and inter-annual dynamics in vegetation using remote sensing data across the globe.

At global scales, vegetation greenup onset during the past three decades suggests that AVHRR and MODIS-derived estimates are geographically and ecologically realistic. In particular, patterns in the timing of greenup onset are strongly dependent on latitude (temperature patterns) in temperate and cold climate regimes across the northern hemisphere, although these patterns are also impacted by elevation and human activities locally. Their inter-annual variance is relatively small, with a value generally less than 10 days within each decade. In contrast, greenup onset in arid and semiarid climate regions is very complex. The regular spatial gradient only occurs in local regions, such as the Sahelian and sub-Sahelian region. The inter-annual variance of phenological timing could be larger than one month. This is probably in response to precipitation regimes and rainfall seasonality migrations (Zhang *et al.*, 2005).

The long-term shifts of vegetation phenology in most parts of the globe are generally episodic rather than persistent in response to climate changes. Early trends of greenup onset from the 1980s to the 1990s appear across most of the northern hemisphere, which agrees with previous findings (Zhou *et al.*, 2001; Myneni *et al.*, 1997; Zhang *et al.*, 2007). The consistent advanced trends from the 1980s–1990s–2000s only occur in large parts of Eurasia and small parts of North America. In most regions of South America, the timing of greenup onset shifted from an early trend to a late trend while an opposite shift occurred in Africa. The detailed mechanisms driving these complex trends will be further investigated.

Finally, it is critical to provide the quality and accuracy of satellite global vegetation phenology detections. Without this, trends derived to predict the response to climate change are less reliable. The quality of phenological detection is strongly dependent on the temporal VI trajectory, which is generally affected by the frequency of cloud cover and snow appearance, and by the model efficiency in removing abiotic noise. To validate accuracy, sufficient field measurements comparable to a satellite footprint are required. This requires field data to reconcile with satellite-based phenological observations, which is currently extremely challenging. The validation effort will become more practical, with the inclusion of observations from webcam (Richardson *et al.*, 2009) and the landscape measurements upscaled from field observations (Liang *et al.*, 2011). Currently, the effort to assess the quality and accuracy of global vegetation phenology is underway.

6. Acknowledgements

This work was partially supported by NASA MEaSUREs contract NNX08AT05A. The authors wish to express their thanks to Kamel Didan and Armando Barreto for help in long-term EVI2 data. The views, opinions, and findings contained in this study are those of the author(s) and should not be interpreted as an official NOAA or US Government position, policy, or decision.

7. References

Adamsen, F.J., Coffelt, T.A., and Nelson, J.M., Barnes, E.M., and Rice, R.C., 2000. Method for using image from a color digital camera to estimate flower number. *Crop Science,* 44: 704–09.

Ahl, D. E., Gower, S. T., Burrows, S. N., Shabanov, N. V., Myneni, R. B., and Knyazikhin, Y., 2006. Monitoring spring canopy phenology of a deciduous broadleaf forest using MODIS. *Remote Sensing of Environment,* 104(1): 88-95.

Bausch, W. C., 1993. Soil background effects on reflectance-based crop coefficients for corn. *Remote Sensing of Environment,* 46: 213–222.

Beck, P.S.A., Atzberger, C., Høgda, K.A., Johansen, B., and Skidmore, A.K., 2006. Improved monitoring of vegetation dynamics at very high latitudes: A new method using MODIS NDVI. *Remote Sensing of Environment,* 100: 321 - 334.

Ben-Ze'ev, E., Karnieli, A., Agam, N., Kaufman, Y., and Holben, B., 2006. Assessing vegetation condition in the presence of biomass burning smoke by applying the aerosol-free vegetation index (AFRI) on MODIS. *International Journal of Remote Sensing,* 27: 3203–3221.

Bradley, B. A., Jacob, R. W., Hermance, J. F., and Mustard, J. F., 2007. A curve fitting procedure to derive inter-annual phenologies from time series of noisy satellite NDVI data. *Remote Sensing of Environment,* 106: 137–145.

Chen, J., Jonsson, P., Tamura, M., Gu, Z.H., Matsushita, B., and Eklundh, L., 2004. A simple method for reconstructing a high-quality NDVI time-series data set based on the Savitzky-Golay filter. *Remote Sensing of Environment, 91*: 332-344.

Chen, X., Hu, B., and Yu, R., 2005. Spatial and temporal variation of phenological growing season and climate change impacts in temperate eastern China. *Global Change Biology,* 11: 1118–1130.

Cook, G. D., and Heerdegen, R. G., 2001. Spatial variation in the duration of the rainy season in monsoonal Australia. *International Journal of Climatology, 21*: 1723-1732.

Dash, J., Jeganathan, C., and Atkinson, P.M., 2010. The use of MERIS Terrestrial Chlorophyll Index to study spatio-temporal variation in vegetation phenology over India. *Remote Sensing of Environment, 114*: 1388-1402.

de Beurs, K.M., and Henebry, G.M. 2010. Spatio-temporal statistical methods for modeling land surface phenology. In *Phenological Research - Methods for Environmental and Climate Change Analysis,* Irene L. Hudson and Marie R. Keatley (Eds), Spring, New York, pp. 177-208.

de Beurs, K.M., and Henebry, G.M., 2004. Land surface phenology, climatic variation, and institutional change: Analyzing agricultural land cover change in Kazakhstan. *Remote Sensing of Environment, 89*: 497-509.

Delbart, N., Kergoat, L., Le Toan, T., Lhermitte, J., and Picard, G., 2005. Determination of phenological dates in boreal regions using normalized difference water index. *Remote Sensing of Environment, 97*: 26-38.

Fensholt, R. and Sandholt, I., 2005. Evaluation of MODIS and NOAA AVHRR vegetation indices with in situ measurements in a semi-arid environment. *International Journal f Remote Sensing,* 26(12): 2561-2594.

Fisher, A., 1994. A model for the seasonal variations of vegetation indices in coarse resolution data and its inversion to extract crop parameters. *Remote Sensing of Environment*, 48: 220-230.

Fisher, J.I., Mustard, J.F., and Vadeboncoeur, M.A., 2006. Green leaf phenology at Landsat resolution: scaling from the field to the satellite. *Remote Sensing of Environment*, 100: 265–279.

Fitter, A.H., Filtter, R.S.R., Harris, I.T.B., and Williamson, M.H., 1995. Relationship between first flowering date and temperature in the flora of a locality in central England. *Functional Ecology*, 9: 55-60.

Friedl, M.A, Henebry, G., Reed, B. Huete, A., White, M., Morisette, J., Nemani, R., Zhang, X., and Myneni, R., 2006. Land surface phenology: a community white paper requested by NASA. ftp://ftp.iluci.org/Land_ESDR/Phenology_Friedl_whitepaper.pdf.

Friedl, M. A., McIver, D. K., Hodges, J. C. F., Zhang, X. Y., Muchoney, D.,Strahler, A. H., Woodcock, C. E., Gopal, S., Schneider, A., Cooper, A., Baccini, A., Gao, F., and Schaaf, C., 2002. Global land cover mapping from MODIS: algorithms and early results. *Remote Sensing of Environment*, 83: 287-302.

Ganguly, S., Friedl, M.A., Tan, B., Zhang, X., and Verma, M., 2010. Land Surface Phenology from MODIS: Characterization of the Collection 5 Global Land Cover Dynamics Product. *Remote Sensing of Environemt*, 114 (8): 1805-1816, doi:10.1016/j.rse.2010.04.005.

Gitelson, A. A., 2004. Wide dynamic range vegetation index for remote quantification of biophysical characteristics of vegetation. *Journal of Plant Physiology*, 161, 165–173.

Goddijn, L.M. and White, M. 2006. Using a digital camera for water quality measurements in Galway Bay. *Estuar Coast Shelf S*, 66: 429–36.

Hargrove, W.W., Spruce, J.P., Gasser, G.E., and Hoffman, F.M., 2009. Toward a National Early Warning System for Forest Disturbances Using Remotely Sensed Canopy Phenology. *Photogrammetric Engineering and Remote Sensing, 75*: 1150-1156.

Hendon, H. H. and Lebmann, B., 1990. A composite study of the Australian summer monsoon. *Journal of Atmospheric Science*, 47: 2227-2240.

Holben, B.N., 1986. Characteristics of maximum value composite images from temporal AVHRR data. *International Journal of Remote Sensing*, 7: 1417-1434.

Huete, A. R., Didan, K., Miura, T., Rodriguez, E. P., Gao, X., and Ferreira, L. G., 2002. Overview of the radiometric and biophysical performance of the MODIS vegetation indices. *Remote Sensing of Environment*, 83: 195–213.

Huete, A. R., Didan, K., Shimabukuro, Y. E., Ratana, P., Saleska, C. R., Hutyra, L. R., Yang, W., Nemani, R.R., and Myneni, R., 2006. Amazon rainforests green-up with sunlight in dry season. *Geophysical Research Letters*, 33, L06405. doi:10.1029/2005GL025583.

Huete, A. R., Jackson, R. D., and Post, D. F., 1985. Spectral response of a plant canopy with different soil backgrounds. *Remote Sensing of Environment*, 17: 37–53.

IPCC 2007 *Climate Change 2007: Impacts, Adaptation, and Vulnerability.* Contribution of Working Group II to the Fourth Assessment Report of the Intergovernment Panel on Climate Change, ed M L Parry, O F Canziani, J P Palutikof, P J van der Linden and C E Hanson (Cambridge: Cambridge University Press) 976pp.

Jakubauskas, M.E., Legates, D.R., and Kastens, J.H., 2001. Harmonic analysis of time-series AVHRR NDVI data. *Photogrammetric Engineering and Remote Sensing*, 67: 461-470.

Jiang, Z., Huete, A.R., Didan, K., and Miura, T., 2008. Development of a two-band enhanced vegetation index without a blue band. *Remote Sensing of Environment*, 112: 3833-3845.

Jönsson, P., Eklundh, L., 2002. Seasonality extraction by function fitting to time-series of satellite sensor data. *IEEE Geosciences and Remote Sensing*, 40: 1824–1831.

Karlsen, S.R., Elvebakk, A., and Hogda, K.A. et al., 2006. Satellite-based mapping of the growing season and bioclimatic zones in Fennoscandia. *Global Ecology Biogeography*, 15: 416–430.

Kaufman, Y. J., and Tanré, D., 1992. Atmospherically resistant vegetation index (ARVI) for EOS-MODIS. *IEEE Transactions on Geoscience and Remote Sensing*, 30: 261–270.

Kovalskyy, V., David, P. Roy, D.P., Zhang, X., and Ju, J., 2012. The suitability of multi-temporal web-enabled Landsat data NDVI for phenological monitoring - a comparison with flux tower and MODIS NDVI. *Remote Sensing Letters*, 3(4): 325–334.

Kramer, K., 1996. *Phenology and growth of European trees in relation to climate change*. Thesis Landbouw Universiteit Wageningen.

Kramer, K., Leinonen, I., and Loustau, D., 2000. The importance of phenology for the evaluation of impact of climate change on growth of boreal, temperate and Mediterranean forests ecosystems: An overview. *International Journal of Biometeorology*, 44: 67-75.

Lauscher, F, 1978. Neue Analysen ältester und neuerer phänologischer Reihen. *Arch. für Meteorologie, Geophysik und Klimatologie (Ser. B)*, 26: 373-385.

Liang, L., Schwartz, M. D., and Fei, S., 2011. Validating satellite phenology through intensive ground observation and landscape scaling in a mixed seasonal forest. *Remote Sensing of Environment*, 115: 143-157.

Lloyd, D., 1990. A phenological classification of terrestrial vegetation cover using shortwave vegetation index imagery. *International Journal of Remote Sensing*, 11: 2269-2279.

Loveland, T. R., Zhu, Z. L., Ohlen, D. O., Brown, J. F., Reed, B. C., and Yang, L. M., 1999. An analysis of the IGBP global land-cover characterization process. *Photogrammetric Engineering and Remote Sensing*, 65: 1021-1032.

Miura, T., Huete, A. R., and van Leeuwen, W. J. D., 1998. Vegetation detection through smoke-filled AVHRIS images: An assessment using MODIS band passes. *Journal of Geophysical Research*, 103(D24): 32, 001–32,011.

Miura, T., Huete, A., and Yoshioka, H., 2006. An empirical investigation of cross-sensor relationships of NDVI and red/near-infrared reflectance using EO-1 hyperion data. *Remote Sensing of Environment*, 100 (2): 223-236.

Moody, A., and Johnson, D.M., 2001. Land-surface phenologies from AVHRR using the discrete Fourier transform. *Remote Sensing of Environment*, 75: 305–323.

Morisette, J.T., Richardson, A.D., Knapp, A.K., Fisher, J.I., Graham, E.A., Abatzoglou, J., Wilson, B.E., Breshears, D.D., Henebry, G.M., Hanes, J.M., and Liang, L., 2009. Tracking the rhythm of the seasons in the face of global change: phonological research in the 21st century. *Front Ecology Environment*, 7: 253-260, doi:10.1890/070217

Moulin, S., Kergoat, L., Viovy, N., and Dedieu, G.G., 1997. Global-scale assessment of vegetation phenology using NOAA/AVHRR satellite measurements. *Journal of Climate*, 10: 1154-1170.

Myneni, R. B., Keeling, C. D., Tucker, C. J., Asrar, G., and Nemani, R. R., 1997. Increased plant growth in the northern high latitudes from 1981- 1991. *Nature*, 386: 698–702.

Myneni, R.B., S. Hoffman, Y. Knyazikhin, J. L. Privette, J. Glassy, Y. Tian, Y. Wang, X. Song, Y. Zhang, G. R. Smith, A. Lotsch, M. Friedl, J. T. Morisette, P. Votava, R. R. Nemani and S. W. Running, 2002. Global products of vegetation leaf area and fraction absorbed PAR from year one of MODIS data. *Remote Sensing of Environment*, 83: 214-231.

Nemani, R.R., Keeling, C.D., Hashimoto, H., Jolly, W.M., Piper, S.C., Tucker, C.J., Myneni, R.B., and Running, S.W., 2003. Climate-driven increases in global terrestrial net primary production from 1982 to 1999. *Science*, 300(5625):1560-1563.

Olsson, L., Eklundh, L., and Ardö, J., 2005. Greening of the Sahel - trends, patterns and hypotheses. *Journal of Arid Environments*, 63: 556-566.

Parmesan, C., and Yohe, G., 2003. A globally coherent fingerprint of climate change impacts across natural systems. *Nature*, 421: 37-42.

Ratkowsky, D. A., 1983. *Nonlinear regression modeling – A unified practical approach*. New York: Marcel Dekker. (pp. 61- 91).

Reed, B. C., Brown, J. F., VanderZee, D., Loveland, T. R., Merchant, J. W., and Ohlen, D. O., 1994. Measuring phenological variablity from satellite imagery. *Journal of Vegetation Science*, 5: 703-714.

Richards, F.J., 1959. A flexible growth function for empirical use. *Journal of Experimental Botany*, 10, 290-300.

Richardson, A. D., Braswell, B. H., Hollinger, D., Jenkins, J. P., and Ollinger, S. V., 2009. Near-surface remote sensing of spatial and temporal variation in canopy phenology. *Ecological Applications*, 19(6): 1417–1428.

Richardson, A.D., Bailey, A.S., Denny, E.G., Martin, C.W., and O'Keefe, J., 2006. Phenology of a northern hardwood forest canopy. *Global Change Biology*, 12: 1174–1188.

Richardson, A.D., Jenkins, J. P., and Braswell, B.H., *et al*. 2007. Use of digital webcam images to track spring green-up in a deciduous broadleaf forest. *Oecologia*, 115522: 323–34.

Rocha, A.V., Potts, D.L., Goulden, M.L., 2008. Standing litter as a driver of interannual CO2 exchange variability in a freshwater marsh. *Journal of Geophysical Research*, 113, G04020, doi:10.1029/2008JG000713.

Rocha, A.V., and Shaver, G.R., 2009. Advantages of a two band EVI calculated from solar and photosynthetically active radiation fluxes. *Agricultural and Forest Meteorology*, 149(9): 1560-1563.

Rötzer, T., and Chmielewski, F.M., 2000. Phenological maps of Europe. *Agrarmeteorologische Schriften, H6*, 1-12.

Rouse, J.W., Haas, R.H., Schell, J.A., and Deering, D.W., 1973. Monitoring vegetation systems in the Great Plains with ERTS. In *3rd ERTS Symposium, NASA* SP-351 I, pp. 309–317.

Smith, M. O., Ustin, S. L., Adams, J. B., and Gillespie, A. R., 1990. Vegetation in deserts: I. A regional measure of abundance from multispectral images. *Remote Sensing of Environment*, 31: 1-26.

Sparks, T.H., and Carey, P.D., 1995. The responses of species to climate over 2 centuries-an analysis of the Marsham phonological record, 1736-1947. *Journal of Ecology*, 83: 321-329.

Steven, D. M., Malthus, J. T., Baret, F., Xu, H., and Chopping, J. M., 2003. Intercalibration of vegetation indices from different sensor systems. *Remote Sensing of Environment*, 88: 412–422.

Tan, B., J.T. Morisette, R.E. Wolfe, F. Gao, G.A. Ederer, J. Nightingale, and J.A. Pedelty. 2011. An enhanced TIMESAT algorithm for estimating vegetation phenology metrics from MODIS data. *IEEE Journal of Selected Topics in Applied Earth Observations and Remote Sensing*, 4(2): 361-371.

Tucker, C.J., and Nicholson, S.E., 1999. Variations in the size of the Sahara desert from 1980 to 1997, *Ambio*, 28(7): 587-591.

Tucker, C.J., Fung, I.Y., Keeling, C.D., and Gammon, R.H., 1986. Relationship between atmosphere CO2 variations and a satellite-derived vegetation index. *Nature*, 319:195-199.

Unganai, L. S., and Kogan, F. N. 1998. Drought monitoring and corn yield estimation in Southern Africa from AVHRR data. *Remote Sensing of Environment*, 63: 219–232.

Vermote, E.F., and Saleous, N., 2006. Calibration of NOAA16 AVHRR over a desert site using MODIS data. *Remote Sensing of Environment*, 105: 214-220.

Verstraete, M.M., Gobron, N., Aussedat, O., Robustelli, M., Pinty, B., Widlowski, J.L., and Taberner, M., 2008. An automatic procedure to identify key vegetation phenology events using the JRC-FAPAR products. *Advances in Space Research*, 41(11): 1773-1783.

Viovy, N., Arino, O., and Belward, A.S., 1992. The Best Index Slope Extraction (Bise) - a Method for Reducing Noise in Ndvi Time-Series. *International Journal of Remote Sensing*, 13: 1585-1590.

White, M. A., Thornton, P. E., and Running, S. W., 1997. A continental phenology model for monitoring vegetation responses to interannual climatic variability. *Global Biogeochemical Cycles*, 11: 217- 234.

White, M.A., de Beurs, K.M., Didan, K., Inouye, D.W., Richardson, A.D., Jensen, O.P., O'Keefe, J., Zhang, G., Nemani, R.R., van Leeuwen, W.J.D., Brown, J.F., de Wit, A., Schaepman, M., Lin, X.M., Dettinger, M., Bailey, A.S., Kimball, J., Schwartz, M.D., Baldocchi, D.D., Lee, J.T., and Lauenroth, W.K., 2009. Intercomparison, interpretation, and assessment of spring phenology in North America estimated from remote sensing for 1982-2006. *Global Change Biology*, 15: 2335-2359.

White, M.A., Running, S.W., and Thornton, P.E., 1999. The impact of growing-season length variability on carbon assimilation and evapotranspiration over 88 years in the eastern US deciduous forest. *International Journal of Biometeorology*, 42: 139-145.

Zhang, X. and Goldberg, M, 2011. Monitoring Fall Foliage Coloration Dynamics Using Time-Series Satellite Data. *Remote Sensing of Environment*, 115(2): 382-391.

Zhang, X., Friedl, M. A., Schaaf, C. B., and Strahler, A. H., 2004a. Climate controls on vegetation phenological patterns in northern mid- and high latitudes inferred from MODIS data. *Global Change Biology*, 10: 1133–1145.

Zhang, X., Friedl, M. A., Schaaf, C. B., Strahler, A.H., Hodges, J. C. F., Gao, F., Reed, B. C., and Huete, A., 2003. Monitoring vegetation phenology using MODIS. *Remote Sensing of Environment*, 84, 471-475.

Zhang, X., Friedl, M.A., and Schaaf, C.B., 2006. Global vegetation phenology from MODIS: Evaluation of global patterns and comparison with in situ measurements. *Journal of Geophysical Research*, 111, G04017, doi:10.1029/2006JG000217.

Zhang, X., Friedl, M.A., Schaaf, C. B., Strahler, A.H., and Schneider, A., 2004b. The footprint of urban climates on vegetation phenology. *Geophysical Research Letter*, 31, L12209, doi:10.1029/2004GL020137.

Zhang, X., Friedl, M.A., Schaaf, C.B., Strahler, A.H., and Liu, Z., 2005. Monitoring the response of vegetation phenology to precipitation in Africa by coupling MODIS and TRMM instruments. *Journal of Geophysical Research-Atmospheres,* 110:D12103, doi:10.1029/2004JD005263.

Zhang, X., Tarpley, D., and Sullivan, J., 2007. Diverse responses of vegetation phenology to a warming climate, *Geophysical Research Letters,* 34: L19405, doi:10.1029/2007GL031447.

Zhou, L., Tucker, C.J., Kaufmann, R.K., Slayback, D., Shabanov, N.V., and Myneni, R.B., 2001. Variation in northern vegetation activity inferred from satellite data of vegetation index during 1981 to 1999. *Journal of Geophysical Research,* 106(D17): 20069-20083.

Zhu, K., and Wan, M., 1963. A productive science - Phenology, *Public Science (Chinese),* No 1.

16

Correlation of Vegetation and Air Temperature Seasonal Profiles – Spatial Arrangement and Temporal Variability

Simoniello Tiziana, Lanfredi Maria, Coppola Rosa,
Imbrenda Vito and Macchiato Maria
*National Research Council of Italy,
Institute of Methodologies for Environmental Analysis, CNR - IMAA,
University of Basilicata, Dep. Environmental Engineering and Physics, DIFA,
University of Naples "Federico II", Dep. of Physics, DSF,
Italy*

1. Introduction

The existence of a strict link between land cover annual cycles and seasonality in the meteorological variability is well known. Joint observations have been exploited for thousand of years as an efficient basis for the empirical land management in agriculture. In Europe, the first attempt to catalogue detailed records of seasonal weather and vegetation phenological events was made by the Marsham family in early 1700s but only during the 19th century this activity become more systematic in the attempt to collect records for wide scientific utility (e.g., Sparks & Carey, 1995).

More recently, the large diffusion of satellite observations has enabled direct monitoring of the land surface at repetitive time intervals and over extended areas. They have proven to be cost effective tools and are expected to contribute to a wide array of local/global change-related applications (vegetation and ecosystem dynamics, hazard monitoring, land surface climatology, hydrology, land cover change etc.) (see e.g. EU Global Monitoring for Environment and Security - GMES, 2010). Mainly thanks to these observations, today's researchers can investigate many problems related to atmosphere-biosphere interactions and physiological responses of vegetation to climate change. In particular, vegetation cover phenology is attracting much attention both for the intrinsic interest in studying collective behaviours of plant communities in specific geographic and anthropic contexts (e.g., Ting et al., 2008) and for understanding their link with climatic variability (e.g., Viña and Henebry, 2005; Maignan et al., 2008; Normand et al., 2009; Maselli et al., 2009; Ma and Zhou, 2011).

Changes in climate can alter land cover phenology both directly, by modifying timing and amplitude of phenological profiles (e.g., Menzel et al., 2001; 2006; Penuelas and Filella, 2001; Cleland et al., 2007; Prieto et al., 2009) and indirectly, by affecting the plant community distribution and dynamics (Feehan et al., 2009). In turn, modifications in the vegetation phenological cycle imply alteration in the exchanges with the atmosphere and particularly in the Carbon balance (e.g., Piao et al., 2008).

Unfortunately, heterogeneity over a wide range of scales severely limits the characterization of biosphere-atmosphere interactions (e.g. Göckede et al., 2009; Zhao et al., 2009). The approximation of ecosystems by considering the prevailing plant functional type (PFT) is too rough for many applications since a high percentage of the terrestrial biosphere is actually made by highly intermixed PFT compositions (Bhreshears, 2006) and new aggregation concepts are required for describing functional categories (Williams et al, 2009; Ustin and Gamon, 2010). Therefore, it is necessary to reduce details without destroying information of dynamical value, especially in regions characterized by a mosaic of human land use and natural vegetation whose role in local and global dynamics is crucial, as shown by recent studies on the repercussions of land-use/land-cover modifications on the climate system (e.g., Pielke, 2005; Douglas et al, 2009; Zhao et al., 2009; Mahmood et al., 2010).

In this Chapter we approach the estimation of the monthly correlation between land surface phenological cycles and seasonal air temperature patterns in order to investigate how this correlation is distributed in space and how this distribution varies in the passage from a year to another one. We refer to "land surface phenology" rather than to "vegetation phenology", according to the remote sensing community (deBeurs & Henebry, 2005), because the variability observed at the pixel detail is the integrated effect of a mixing of signals, derived from extended plant aggregations and non vegetated traits. However, it is widely recognised that satellite data, by means of suited vegetation indexes, are able to capture collective behaviours of plant communities adapted to the specific climatic, structural, and ecological characteristics of the geographical area.

In order to investigate the spatial and temporal variability of vegetation and temperature in a typical, very complex, Mediterranean environment, we analyze monthly NDVI (Normalized Difference Vegetation Index) and air temperature time series over Southern Italy, core of Mediterranean Basin, where the responsiveness of vegetation to temperature increase is the greatest in the European continent (e.g. Estrella et al., 2009).

Heterogeneity is reduced by looking for phenological alikeness, as it appears at the sensor resolution, since this reflects analogies in the collective functionality of the sub-pixel components. Profiles of monthly NDVI, acquired by the MODIS sensor (250m), are elaborated by performing a multitemporal clustering on annual basis to aggregate areas with similar phenological cycles. In the same way, areas with common phenological and temperature seasonal patterns are analyzed by classifying temperature and NDVI profiles jointly. In order to evaluate the spatial variability in the phase-relation between vegetation and temperature, temporal correlations are estimated per pixel and gathered in a correlation map (r-map), which is created per each year. Sign and magnitude of the correlation coefficient inform us about the phase relation between the annual patterns: high positive values of r characterize areas where phenology and temperature are nearly in phase; whereas high negative values characterize areas where they have about opposite phases. Thus, the r-map provides a measure of the similarity between phenology and temperature profiles and a representation of the spatial distribution of such a similarity. This should point out leading spatial structures dominated by analogous biogeographical features. In addition, the comparison between r-maps estimated for different years should enhance the influence of inter-annual meteorological fluctuations on such structures thus discriminating stable regions from more variable areas where the dependence of phenology on temperature seasonality varies with meteorology.

2. Study area

The area under study belongs to the Mediterranean biogeographical region and is located in the Southern part of the Italian peninsula (Fig. 1). Its territory is mainly mountainous: the Apennine chain runs along the peninsula from NW to SE, splitting it in two parts and forming a broad irregular mountain system with generally narrow coastal plains on both sides (Fig. 2a). In particular, the East side, overlooking the Adriatic Sea, is characterized by level areas such as most of the Apulia region and the plains lying in the far south-east. The west side, overlooking the Tyrrhenian Sea, is characterized by an orography more complex than the other: uplands and plains stretch along the west coast between the Apennines and the sea. This region includes a number of major cities, among which there are Naples and Palermo. The Strait of Messina separates Sicily, the largest island in the Mediterranean Sea, from mainland Italy. Much of the northern portion of the island is mountainous, and Mount Etna, one of the largest active volcanoes in the world, reaches 3000 m in altitude.

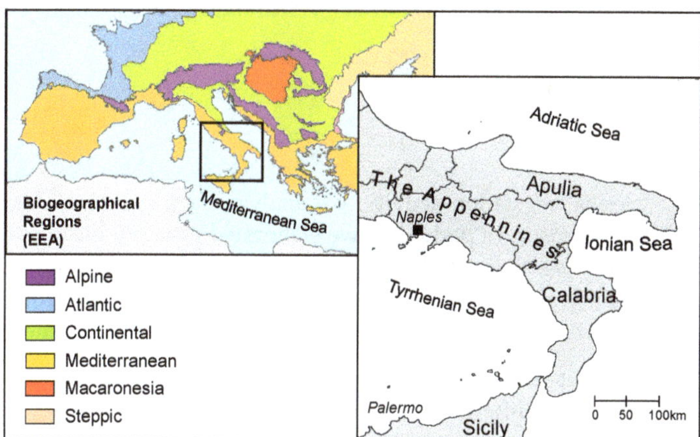

Fig. 1. Location of the study area within the map of biogeographical regions (source European Environment Agency-EEA) and the administrative limits of the investigated areas with the principal placenames.

Altogether, Southern Italy enjoys a mild climate thanks to its position in the middle of the Mediterranean, to the influence of air masses coming from North Africa, and to the presence of the barrier formed by the Alps chain that shelters from the northern cold winds. Nevertheless, local significant differences are present, especially due to latitude variability (~6^0), to orography, and to the maritime climate of the coasts. In particular, its rich orographic variability makes this territory a good test site for assessing the effect of altitude on phenology. In the inland areas, where the sea effects are negligible and altitude is high, climate is typically mountainous with cold winters. On the Adriatic side, the thermoregulatory action of the sea is reduced by the eastern cold and humid winds. The seaboard and the southernmost zones have a typical Mediterranean climate, characterized by rainy autumns and winters, and very warm dry summers. Gathered rain events often trigger floods and erosion. The areas of Calabria well synthesize the high basic heterogeneity of the territory. Here, we can find typically sub-tropical coastal environments opposed, at a distance of few tens of kilometers, to typical alpine mountain environments, such as those of the Sila National Park. Agriculture is the main economic activity in Southern Italy. Cereal monocultures are prevalent and farmlands as well as pastures are widespread on the territory (Fig. 2b), also at relatively high altitude. Industrial activities and urban settlements are mainly distributed according to orography: main cities and industrial areas are prevalently located along the coast. Inland, the industrial sector is less developed, so there are only small factories. Although natural vegetation is present along the coast (pine forests and maquis prevalently), it predominates at high altitude, where broad leaved forests and maquis are densely distributed.

Degradation signatures are evident in the southernmost part of the territory, where any climatic change involving increase of drought duration and/or intensity of rare rainfall events can favor vulnerability to degradation (Lanfredi et al., 2003). Since also agricultural practices often are obsolete and not accompanied by suited soil preservation strategies, desertification could be expected in some areas (Sciortino et al., 2000; Liberti et al., 2009; De Santis et al., 2010).

3. Data

3.1 Satellite data

To evaluate the cycles of vegetation activity, we analyzed a vegetation index time series acquired by the MODIS sensor. In particular, we elaborated monthly mean values of NDVI (Normalized Difference Vegetation Index) at full spatial resolution (250m) for the period 2005-2008. NDVI data, available as 16-day composite, were gathered from the MODIS dataset by NASA LP DAAC (Land Processes Distributed Active Archive Center).

The selected index is related to the structural and functional characteristics of vegetation cover, since it is based on the difference of reflectance responses in the red and near infrared bands. The index ranges between -1 and +1; negative values are linked to water and thick clouds, low positive values to bare soils, high positive values represent dense and photosynthetically active vegetation. NDVI is adopted as proxy for photosynthetic activity to characterize vegetation dynamics. In particular, MODIS-NDVI time series are widely analyzed to study land surface phenology in both natural environment and cultivated covers (e.g., Butt et al., 2011; Jönsson et al., 2010; Karlsen et al., 2008).

3.2 Orographic data

The analysis of the dependence of phenology on altitude was performed by using a Digital Elevation Model (DEM) derived by the NASA SRTM (Shuttle Radar Topographic Mission). The original data available as 3 arc second (~90m pixel resolution) were reprojected to satellite data projection and resampled using an average filter to 250m (Fig. 2a) to make them comparable with MODIS data.

3.3 Temperature maps

We used temperature data collected by national networks distributed over Southern Italy. In particular, maps of air temperature at 250m resolution were obtained from ground station measurements by means of a combined deterministic-stochastic model suitably devised for the study region on the basis of latitude, elevation (from the re-sampled SRTM DEM), and sea distance as independent variables (Coppola et al., 2006). Data to implement the spatialization model were obtained from 15 years daily temperature time series (1994-2008) collected at 35 stations of the Central Office of Agrarian Ecology - National Agronomic Network (UCEA-RAN). The model coefficients were used to elaborate the maps of monthly mean temperature (hereafter T) for the period 2005-2008.

3.4 Land cover data

Land cover information was extracted from the Corine map downloaded from the High Institute for Environment Protection and Research (ISPRA, former APAT). For the purpose

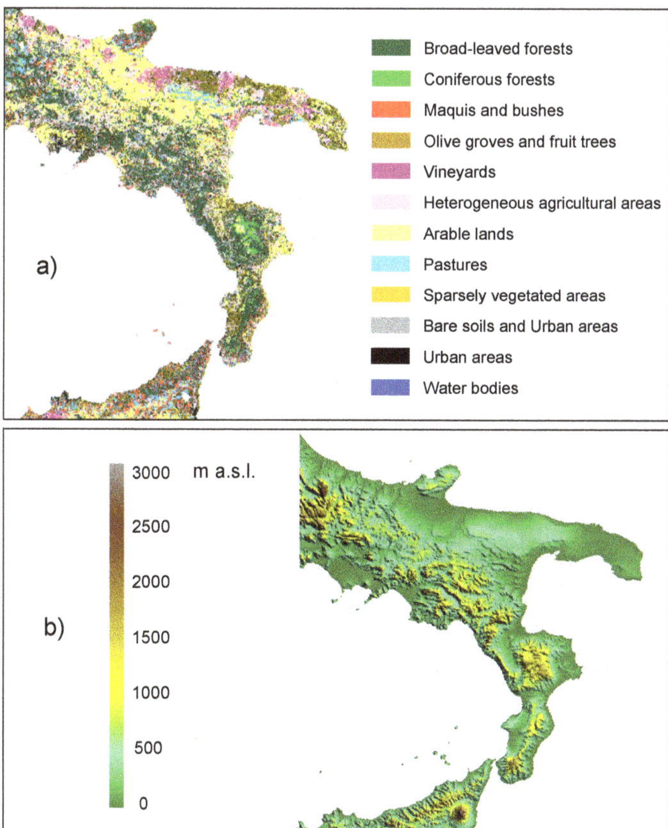

Fig. 2. Land cover map (a) obtained by recoding the Corine classes; and Digital Elevation Model (b) derived from SRTM.

of this study, the original Corine level 1-3 codes (http://sia.eionet.europa.eu/CLC2000/classes) were reclassified to group artificial surfaces and water bodies and to preserve detailed information on vegetated covers (Fig. 2b); the adopted map has the following 12 classes: Broad-leaved and mixed forests, Coniferous forests, Maquis and bushes; Sparsely vegetated natural areas, Pastures, Arable lands, Vineyards, Olive groves and fruit trees, Heterogeneous agricultural areas, Bare soils, Urban and Industrial Areas, and Water bodies. The Corine data in vector format were rasterized in a map at the same resolution (250m) of vegetation indices.

4. Classification procedure for the identification of phenological classes

In order to group pixels on a phenological basis, we adopted an unsupervised classification since no a priory knowledge was available for pattern identification. We classified NDVI and T images at the sensor resolution (250m) by means of fuzzy k-means unsupervised algorithm, which represents an extension of the widely adopted k-means by evaluating the probability of a point to belong to a cluster instead of hard cluster identification (e.g., Richards & Jia, 2006). We selected the fuzzy approach since we do not expect abrupt variations between one phenological cluster and another one; instead, especially for natural vegetation, we can expect gradual differences among phenological patterns. In this context, the fuzziness concept can be more appropriate than an abrupt partition (e.g., Simoniello et al., 2008).

In general, this clustering algorithm groups the multispectral data according to statistical properties of their spectra for obtaining land cover/land use information (e.g., Richards & Jia, 2006). Thus, each image pixel is represented as a point in the n-dimensional space, where the n dimensions are the spectral bands and the coordinates are the brightness values in each band. Iteratively, the algorithm evaluates the probability of each point to belong to a cluster (degree of membership) starting from the primary cluster prototypes. For each iteration, the cluster prototypes are rearranged and the membership degree recomputed until the movements of cluster prototypes for successive iteration is less than a predetermined threshold. Finally, a data point is assigned to the cluster with the highest membership degree.

In this study, we applied the fuzzy k-means algorithm to the temporal domain by classifying NDVI from April up to September (period of maximum phenological variability) to avoid cluster aggregations driven by residual clouds or hardly filtered NDVI data. This seasonal period was selected just to minimize non vegetated effects on the identification of the main phenological areas; the successive analyses were instead obtained by considering the complete annual profiles. We used the Euclidean distance for computing the membership degree among patterns and fixed to 1% the minimum threshold of centroid movement, which means a variation on the second decimal figure of the NDVI value.

Although phenology is mainly driven by land cover type, we also included temperature patterns in the analysis because climatic conditions can significantly affect the phenological development for the same cover type. Thus, we jointly classified NDVI and T; for such an implementation, each pixel was represented as a point in a 12-dimensional space, where the 12 dimensions are the 6+6 monthly images of T and NDVI and the coordinates are the values of two variables. The results of two clustering procedures (NDVI alone and NDVI-T) were compared in order to evaluate the ability of NDVI to account also for temperature.

5. Spatial structure analysis

The spatial variability of phenological aggregates was evaluated by analyzing some peculiar indices commonly adopted in Landscape Ecology analysis (e.g., Diaz-Varela et al., 2009; Fischer and Lindenmayer, 2007; McGarigal et al., 2002). We evaluated the indices described in the following on the four landscapes represented by the phenological classifications related to the investigated years (2005-2008). In particular, we analyzed:

- Number of patches, which is a simple measure of the extent of subdivision or fragmentation in a landscape.
- Mean patch size, which is the average area of patches comprising a landscape mosaic; it is perhaps the single most important and useful piece of information contained in the landscape having a great deal of ecological utility.
- Aggregation Index (AI) is equals the number of like adjacencies involving the corresponding class, divided by the maximum possible number of like adjacencies involving the corresponding class, which is achieved when the class is maximally clumped into a single, compact patch; multiplied by 100 (to convert to a percentage). AI increases as the landscape is increasingly aggregated from 0 to 100, on the basis of the following formula

$$AI = \frac{g_{ii}}{max - g_{gg}} 100 \quad (1)$$

where, g_{ii} = number of like adjacencies (joins) between pixels of patch type (class) i; and max-g_{ii} = maximum number of like adjacencies between pixels of patch type.

- Intersperion and Justapposition index (IJI) evaluates the observed interspersion over the maximum possible interspersion for the given number of patch types. IJI approaches 0 when the distribution of adjacencies among unique patch types becomes increasingly uneven. IJI = 100 when all patch types are equally adjacent to all other patch types,(i.e., maximum interspersion and juxtaposition):

$$IJI = \frac{-\sum_{i=1}^{m}\sum_{k=i+1}^{m}\left[\left(\frac{e_{ik}}{E}\right)\ln\left(\frac{e_{ik}}{E}\right)\right]}{\ln\left(0.5\left[m(m-1)\right]\right)} 100 \quad (2)$$

where, e_{ik} = total length of edge in landscape between classes i and k; E = total length of edge in landscape, excluding background; and m = number of classes present in the landscape.

- Shannon's diversity index is a popular measure of diversity in community ecology, applied to landscapes. ShDI = 0 when the landscape contains only 1 patch (i.e., no diversity). ShDI increases as the number of different patch types (i.e., patch richness,

PR) increases and/or the proportional distribution of area among patch types becomes more equitable.

$$ShDI = -\sum_{i=1}^{m}(PlnP) \tag{3}$$

where, Pi = proportion of the landscape occupied by class i.

6. Results

In order to evaluate the optimal number of clusters n to represent the heterogeneity of the investigated territory, we considered the trade-off between the efficacy of ecosystem representation and local fragmentation effects. A low number of classes enabled us just to distinguish natural from managed covers; whereas a high number of classes ($n>10$) put into evidence very small clusters ($n<2\%$ of the total pixels) arisen from the fragmentation of the edges of the dominant clusters, making the identified cluster less representative of collective dynamics. These additional clusters are not representative of extended collective dynamics. They are expected to account for that heterogeneity we would like to reduce. On this basis, we selected $n=10$ to obtain a phenological segmentation of the territory from the classification of the NDVI temporal profiles.

A second sequence of clustering maps (n from 2 to 30) was elaborated by jointly classifying NDVI and temperature profiles. We obtained very similar arrangements, with the differences between the NDVI and the NDVI-T clustering maps that tend to increase with n. Such differences are however very small; in particular, for the selected 10 classes we found no variations in the mean phenological profiles and differences in their spatial arrangement less than 0.5% (Tab. 1).

On the whole, NDVI and NDVI-T provide the same territorial segmentation. Such a similarity confirms the strong role of temperature in the vegetation development and the capability of NDVI to account for such temperature-vegetation interaction.

Cluster	1	2	3	4	5	6	7	8	9	10
Difference (%)	0.08	0.21	0.12	-0.46	0.33	-0.13	-0.39	-0.01	0.17	0.08

Table 1. Percentage difference (on the total number of pixels) between the number of pixels obtained by classifying in 10 clusters the NDVI profiles alone or the NDVI and temperature profiles together.

6.1 Characterization of phenological classes

The phenological clusters obtained by classifying on annual basis the NDVI profiles show a very coherent structure in both the spatial and temporal domains (Fig. 3). Their temporal patterns are organized in a continuum of phenological curves, which start from the early peaked curves of clusters 1 and 2, then become ever smoother up to the quite flat profile of cluster 5, and finally reach the highest summer values of clusters 10.

The spatial arrangement strongly reflects the land cover spatial distribution and the orography iso-level curves (compare Fig.2). From the lowest to the highest cluster number, there is a reduction of covers related to anthropic activities and an increase of natural vegetation (Fig. 4). In particular, the first two clusters are dominated by arable lands generally non-irrigated; then, clusters 3 and 4 also include a high percentage of annual crops but they are mainly associated with olive groves and complex cultivation patterns, with a higher percentage in the first one. In clusters 5, olive groves and vineyards are prevalent similarly to cluster 6, where also fruit trees are present. Starting from cluster 7, the presence of natural vegetation increases and also in lands principally occupied by agriculture there are significant areas of natural vegetation (Corine code 2.4.3). The cluster sequence 8-9-10 reflects the typical spatial coherence and organization of natural ecosystems in this area, i.e. the presence of transitional vegetation around the core of forested areas. In particular, the core of forests (cluster 10) is largely composed by broad-leaved trees (more than 75% of the cluster).

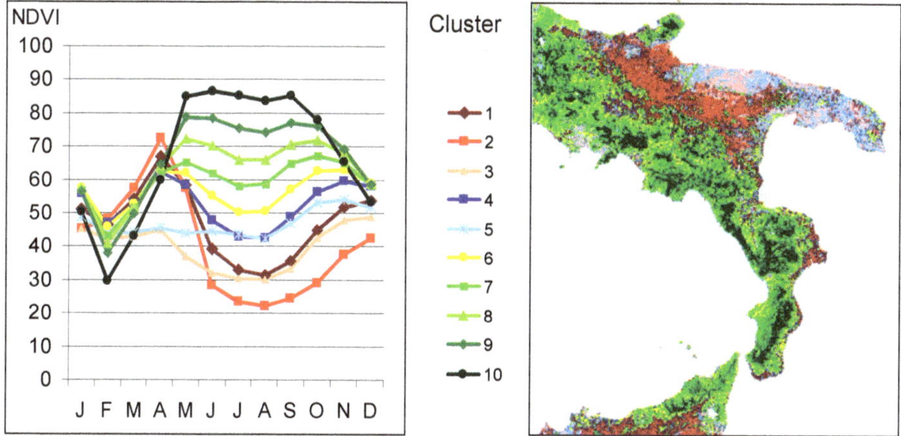

Fig. 3. Phenological profiles obtained by the fuzzy k-means classification on 2005 MODIS-NDVI data, values are multiplied by 100 (left) and their corresponding spatial patterns (right).

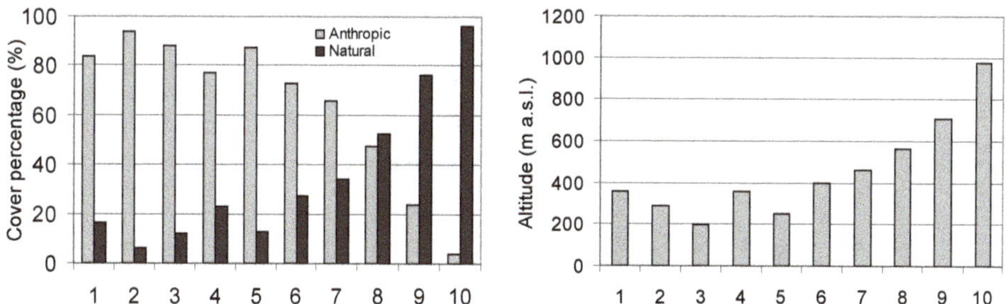

Fig. 4. Relative distribution of natural and anthropic covers (left); mean altitude value per phenological cluster (right). Shown data are related to 2005 as representative also of the other three years.

The distribution of the clusters along the altitude gradient starts from the central hilly areas of the first two clusters, goes downwards to the coastal aggregations, and then gradually increases from cluster 5 up to the mountainous areas of cluster 10. The only exception is represented by cluster 4 that is more spread over the territory being also composed by scrub and herbaceous vegetation associations for the natural portion (high presence of sclerophyllous vegetation, often left to delimit cultivation fields).

6.2 Spatial and temporal variability of vegetation patterns

The clustering analysis of the four investigated years does not highlight dramatic changes in the spatial aggregation of the phenological clusters (Fig. 5). The distribution of cluster cores seems to be very similar in the map sequence, particularly for the clusters with a high percentage of natural covers; whereas a slightly higher variability appears at the cluster borders.

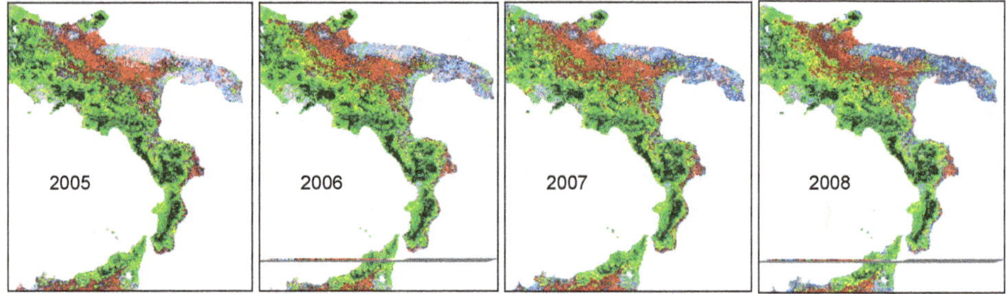

Fig. 5. Sequence of phenological maps obtained by the fuzzy classification.

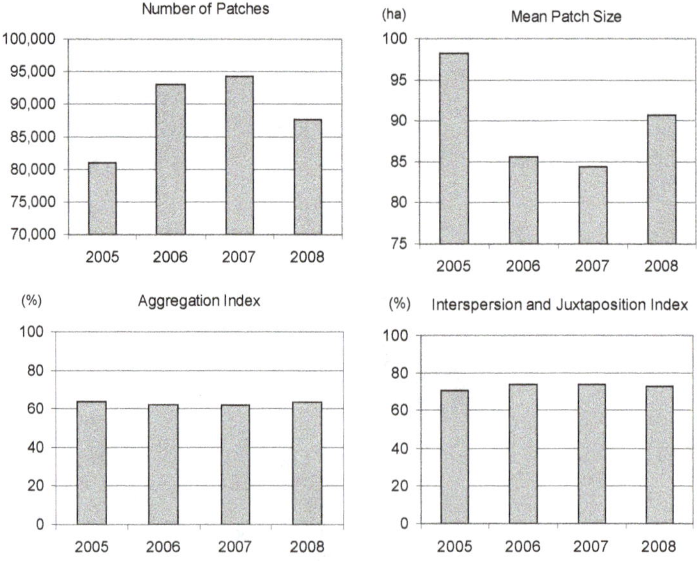

Fig. 6. Spatial characteristics of the phenological cluster aggregates.

To quantify the spatial variability of phenological aggregates, we analyzed the spatial structure of the maps (Fig. 6). Even if in 2006 and 2007 there is a higher number of patches (about 15%) having smaller dimensions, the level of cluster aggregation is quite high (~62%) and fairly constant. It was evaluated as the ratio between the number of like adjacencies (joins) inside the clusters and the greatest possible joins' number related to a maximally clumped shape with the same dimension of the given cluster. Similarly the level of interspersion and juxtaposing of patches belonging to different clusters is preserved in time. The Shannon's diversity index, commonly adopted in ecology to assess the richness of species (number and proportion), also shows quite high values (Fig. 7, top). In our case, being the cluster number constant, it highlights the level of uniformity in the proportional area distribution among the clusters.

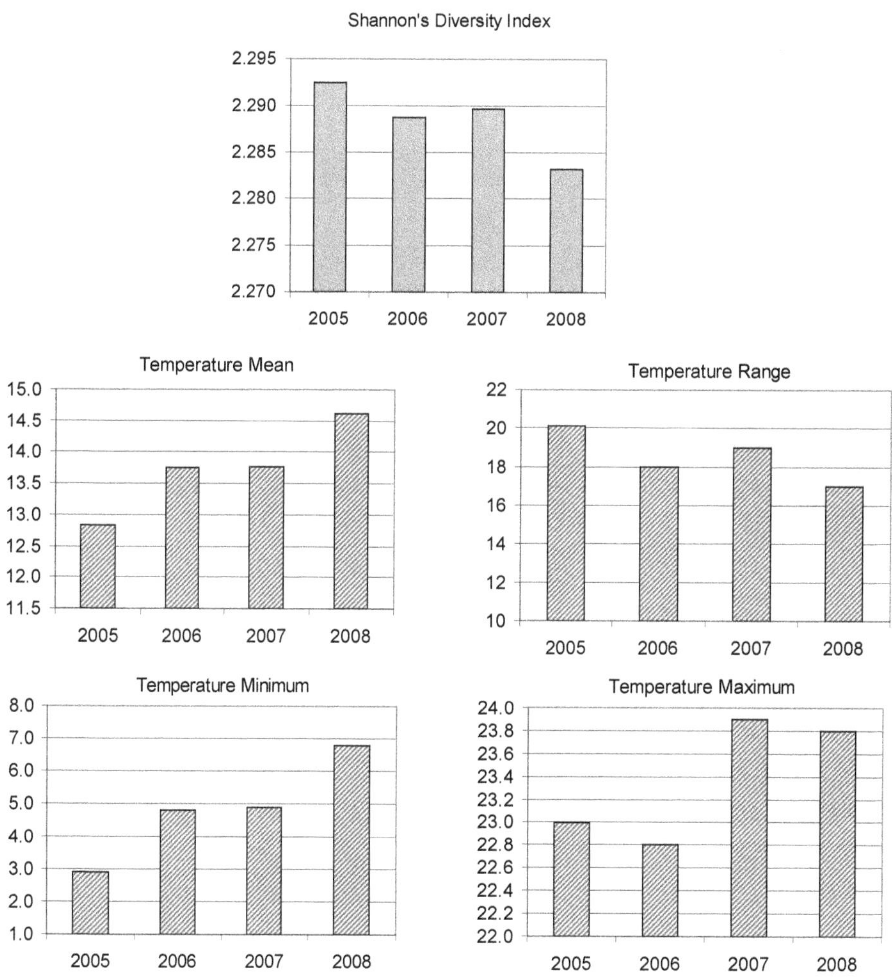

Fig. 7. Shannon's diversity index evaluated on the phenological maps (top) and characteristic parameters of mean temperature profiles for the whole study areas.

Among the four years, 2005 shows the most equitable cluster distribution; instead in 2008 the phenological aggregates appear to some extent less uniformly distributed. At broad scale level, such variation in the spatial heterogeneity evidenced by the Shannon's index appears to follow the behavior of temperature features (Fig. 7, bottom). In particular, it seems to be positively related to the annual temperature range (R 0.95) and negatively affected by the temperature mean and minimum values (R -0.97). The maximum of temperature does not appear to have a direct link with the spatial uniformity of phenological aggregates.

The NDVI profiles corresponding to the spatial clusters show a great variability, particularly in the growing and greening seasons with differences in both amplitude and length (Fig. 8). Such marked interannual differences in NDVI values jointly with the relative stability of the phenological aggregates suggest the effect of external forcing, such as the variability in temperature seasonality, capable to generate collective modifications of phenological responses.

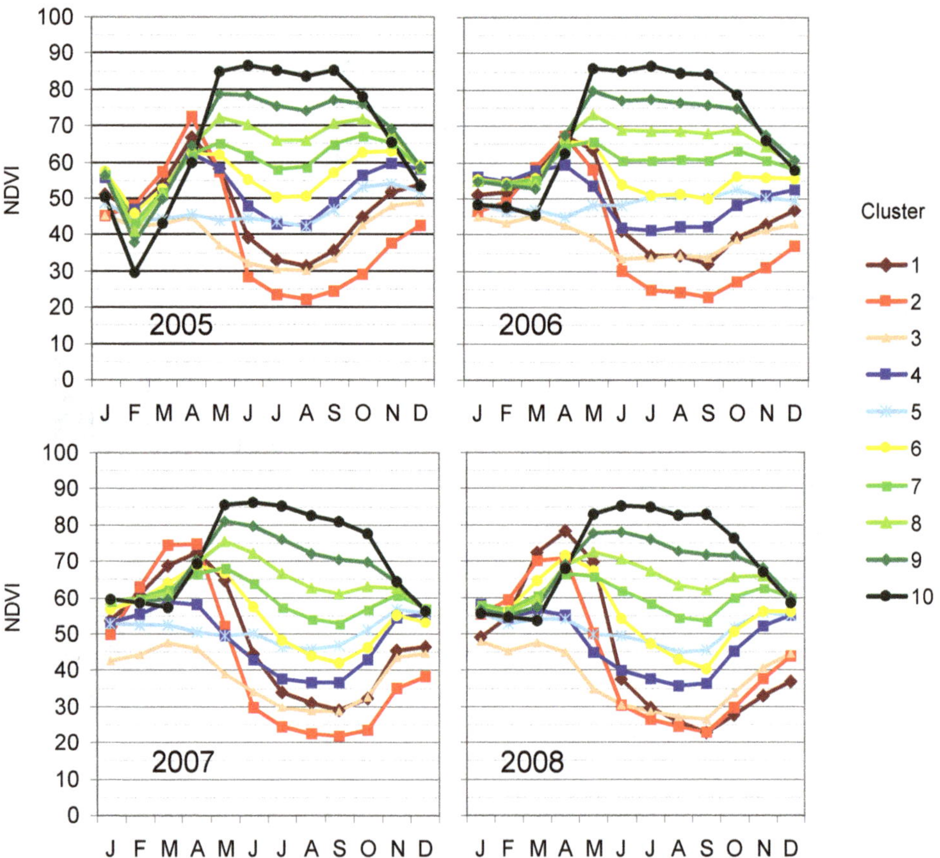

Fig. 8. Annual profiles corresponding to the phenological clusters in Fig. 8. The cluster colors in the maps correspond to the colors of the profiles.

The profiles of mean air temperature for the investigated area (Fig.9) seem to support such a hypothesis. In particular, 2005 appears colder than the other years during the first months to which low NDVI values correspond, especially in February. The relative low summer temperature in 2006, jointly with a gradual summer-autumn reduction, seems to characterize the flat behavior of vegetation response during the greening season of this year. The warmer initial period of 2007 and 2008 correspond to higher NDVI values during the first moths of both the years; conversely the warmer temperature in summer is related to a reduced vegetation activity response, except for the cluster at high altitude (cluster10). The behavior of the two years diverge from September, when 2007 show lower temperatures and slightly higher NDVI value reductions for the clusters with a large presence of natural covers.

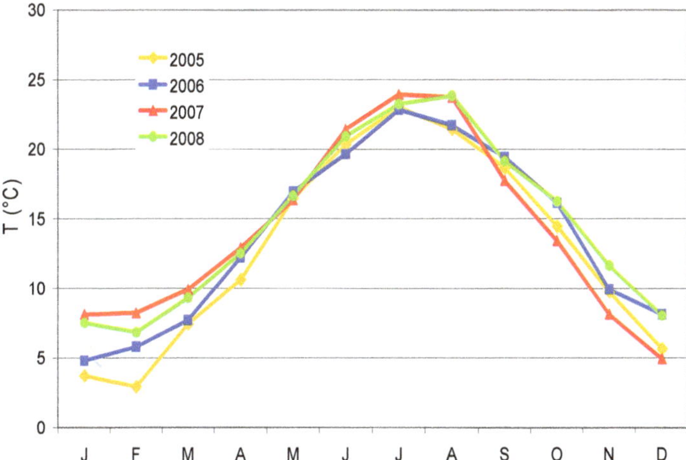

Fig. 9. Mean monthly temperature over the study area.

6.3 Correlation between vegetation and temperature patterns: r- maps

In order to investigate the relationship between temperature and vegetation development in detail, we analyzed the correlation between NDVI and T profiles at full spatial resolution by estimating the correlation per pixel (r-maps) (Fig.10).

The r-maps point out some leading spatial structures: there is a clear separation between the level areas of the Adriatic side (high negative correlation) and the Apennine areas of the Tyrrhenian side (high positive correlation). These aggregates are separated by a central transition area with low correlation magnitude.

The prevalent presence of cultivated lands in the level areas largely explains the anticorrelation in the phase-relation between phenology and temperature seasonality; whereas the high positive correlation on the Apennine mainly results from the diffuse presence of forested covers. In particular, broad-leaved forests and cultivations with olive and fruit trees represent the two extremes in the correlation range of the land cover classes: high positive the first and high negative the second (Fig. 11). The mixing and prevalence of peculiar covers determine the overall spatial structure of the correlation between vegetation activity and temperature cycles. Where natural vegetation prevails, the correlation is positive and fairly high, where man managed covers are dominant, there is an opposite NDVI-T phase-relation and a reduced correlation magnitude. The spatial structure of correlation is then largely preserved in time: only small variations of the spatial patterns are present in the r-map sequence.

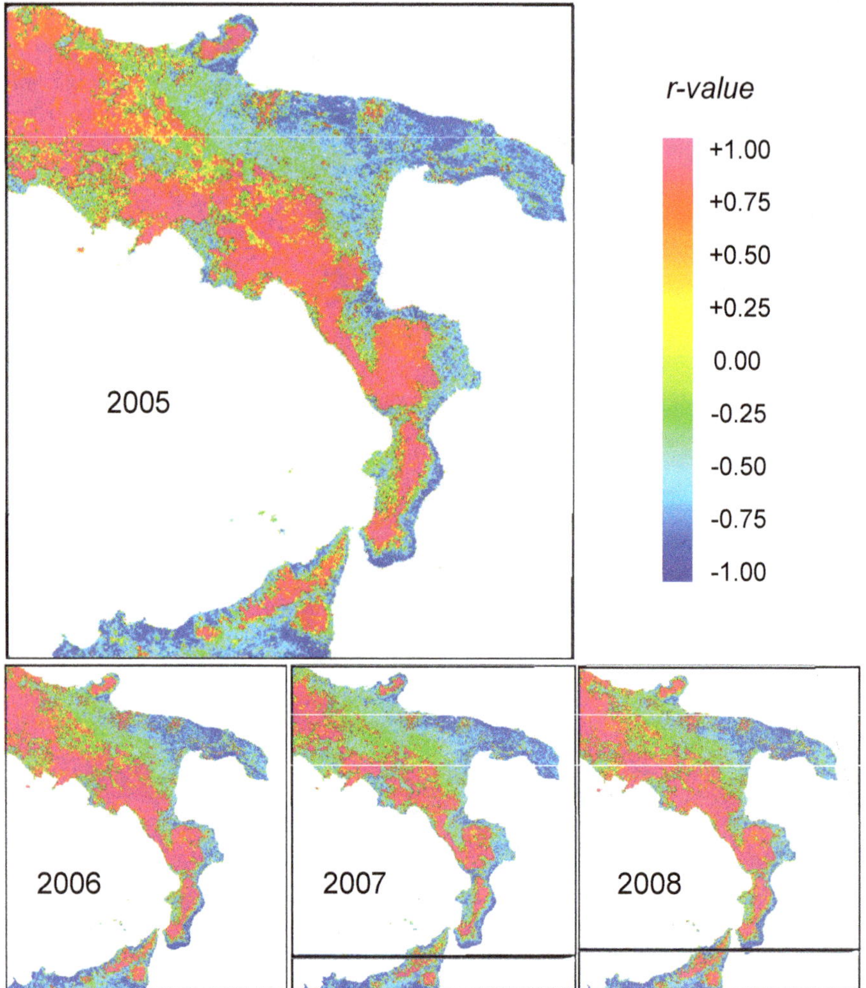

Fig. 10. Maps of correlation coefficient (r-map) between the NDVI and T seasonal profiles.

The strong link between land cover allocation and orography in the study area largely characterizes the distribution of correlation values along the altitude gradient (Fig. 12).

The correlation increases with the elevation and such a behaviour is preserved even if its strength is variable from one year to another. In particular, the variation in the correlation strength is higher in 2007 when there is a shift of the mean temperature profile, mainly characterized by a warm initial period and colder autumnal-winter months (see Fig. 9). The different mean effects per altitude range (lower at high elevation) suggest different responses of the present vegetation to the peculiar 2007 temperature pattern.

Such different responses induced a large deviation toward anticorrelation in the distribution of vegetation-temperature phase-relation (Fig. 13), with a particularly evident population reduction for the highest correlation class in 2007.

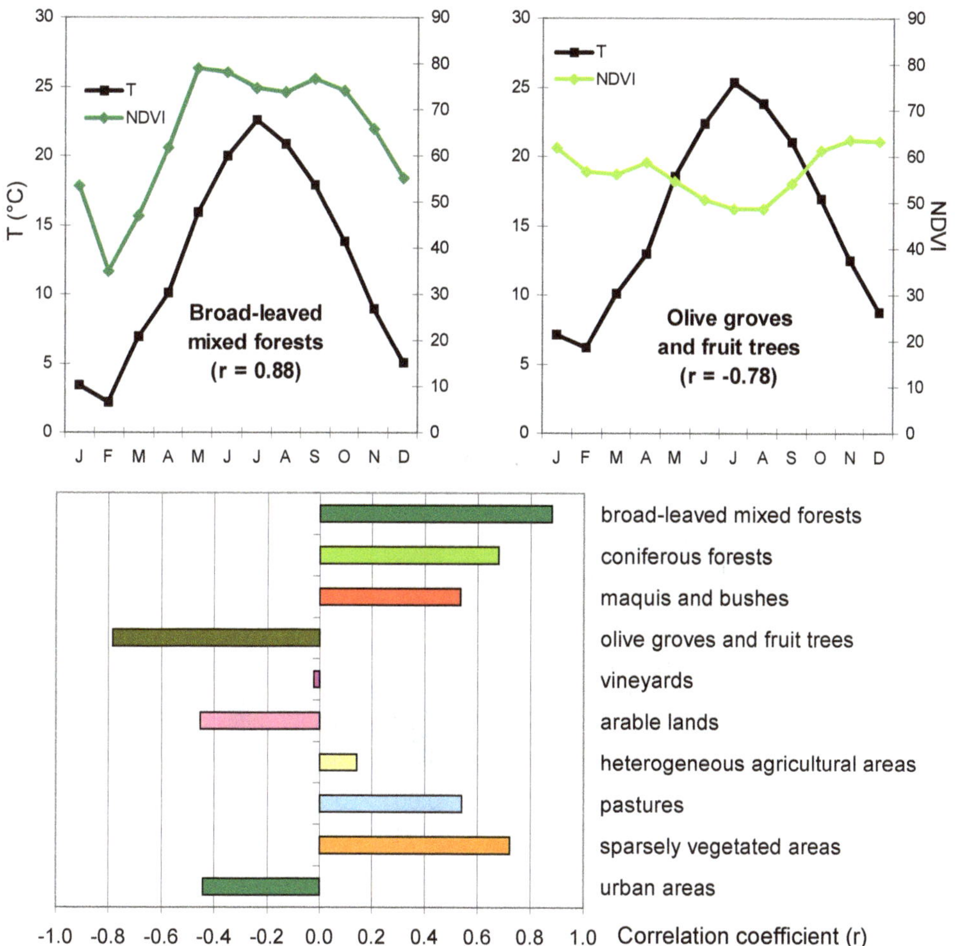

Fig. 11. Top: mean NDVI and temperature seasonal profiles for Broad leaved forests (high positive correlation) and Olive grove and fruit trees (high negative correlation); bottom: values of correlation between NDVI and temperature per land cover (data for 2005).

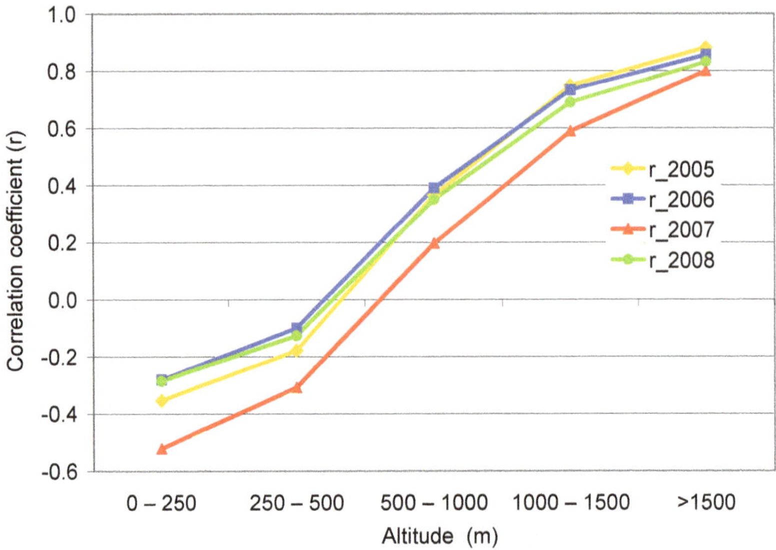

Fig. 12. NDVI and temperature correlation values per altitude range.

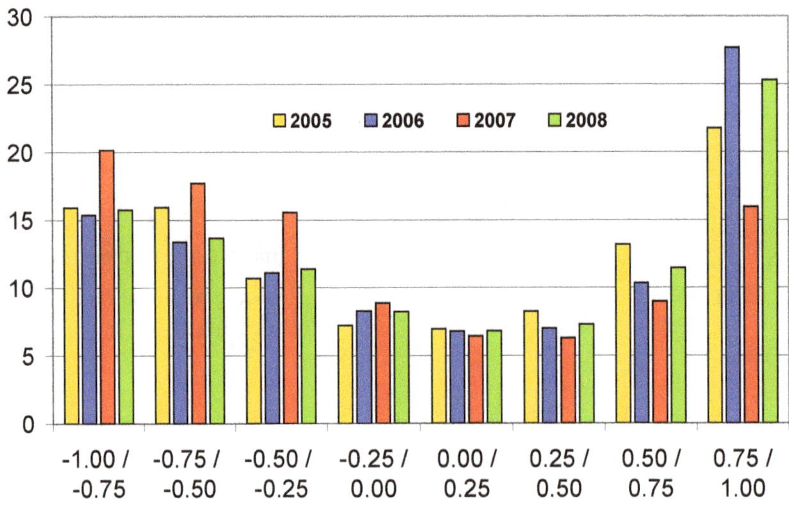

Fig. 13. Percentage distribution of correlation coefficients between NDVI and T profiles.

6.4 Spatial variability of phenological classes vs. temperature-vegetation temporal correlation

In order to evaluate the capability of the clustering procedure to follow the different responses of vegetation to temperature variability, we analyzed the changes in the cluster arrangements vs the correlation between phenology and temperature seasonality. We compared the distribution of the correlation values for pixels that permanently belong to a given cluster during the investigated period (stable areas) with the distribution of pixels moving from one cluster to an other (unstable areas) (Fig. 14). For stable areas, even if there is some variability among correlation classes, the configuration of their distribution is preserved in time. For unstable areas, the shape of correlation distribution shows an evident modification in 2007.

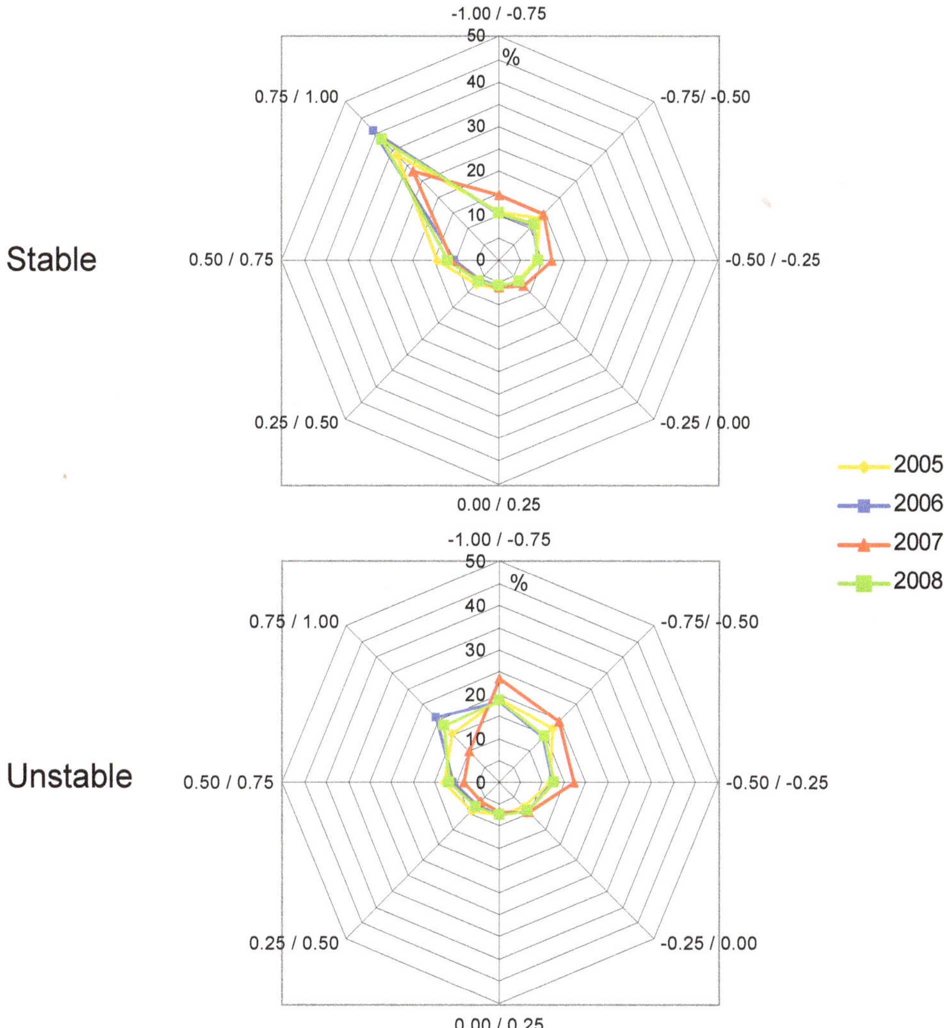

Fig. 14. Comparison of NDVI and T correlation value distributions for stable areas (persisting in the same cluster) and for unstable areas (changing clusters) for the whole investigated region. In the charts, the percentage distributions are represented in correlation bins between -1 and 1 at a 0.25 step.

To better analyze such different behaviours and to evaluate their relationship with temperature patterns, we focused on clusters with less anthropic influence. Then, we identified pixels moving from the cluster having the highest percentage of natural covers (cluster 10) during the investigated period. About 13% of pixels belonging to cluster 10 in

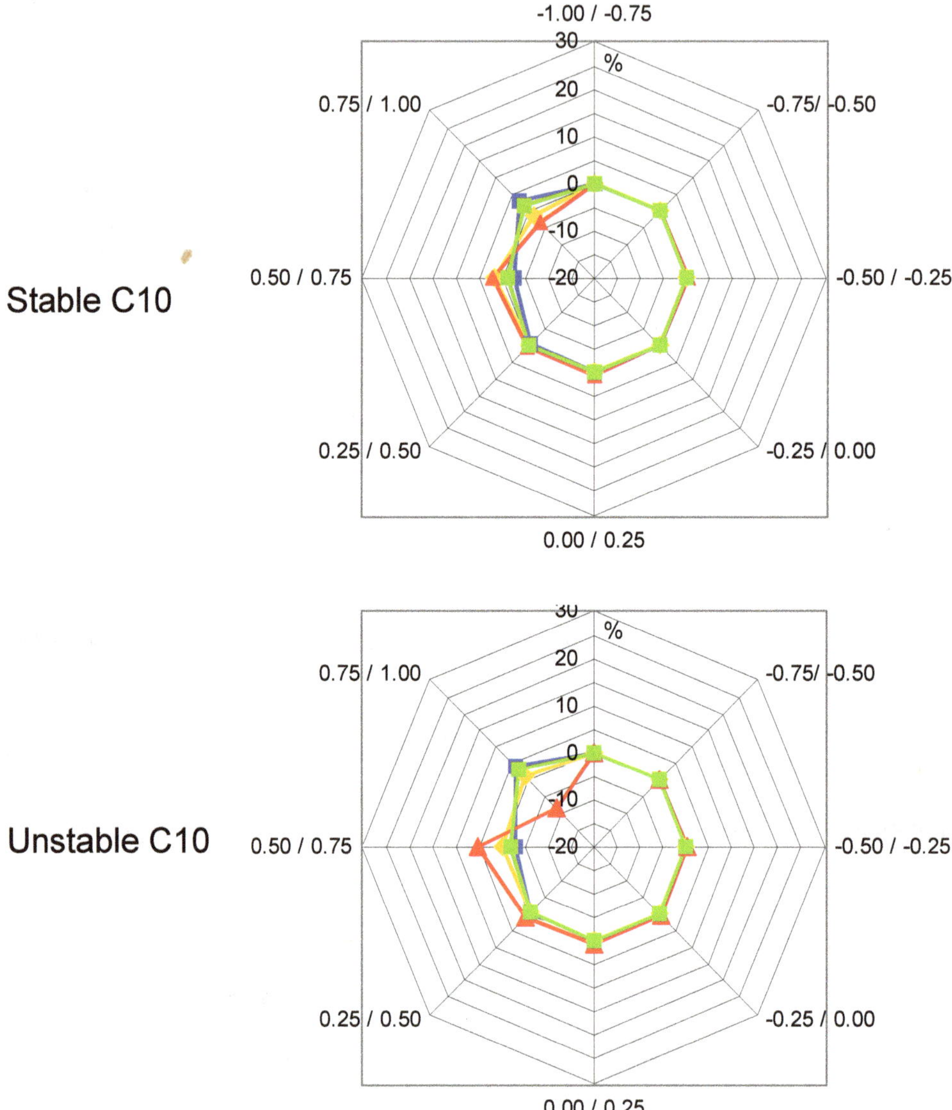

Fig. 15. Comparison of NDVI and T differences in correlation distribution for pixels stable in cluster 10 and for pixels transited to cluster 9 in 2007 (unstable C10). In the charts, the percentage distributions are represented in correlation bins between -1 and 1 at a 0.25 step.

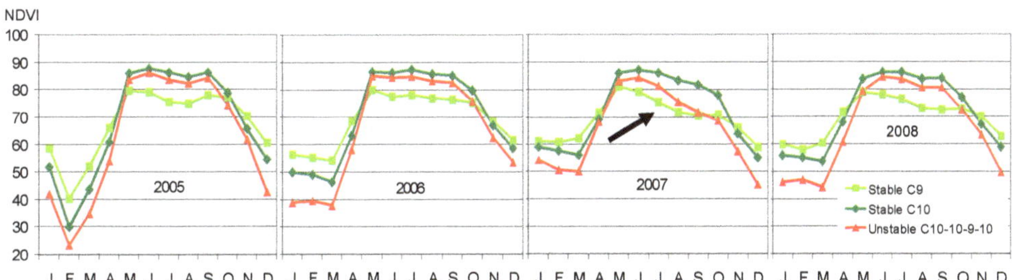

Fig. 16. NDVI profiles for pixels stable in cluster 9 and cluster 10 and for pixels transited from cluster 10 to cluster 9 only in 2007.

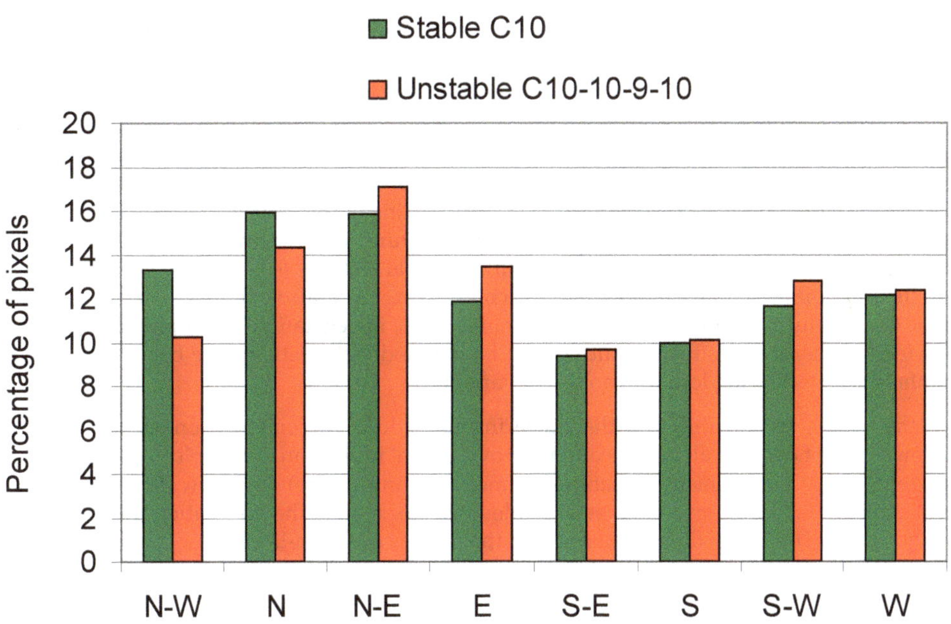

Fig. 17. Aspect distribution for pixels stable in cluster 10 and for pixels transited from cluster 10 to cluster 9 only in 2007.

2005, 2006 and 2008 were associated by the clustering procedure to cluster 9 in 2007. The quasi totality of such pixels is covered by forested areas (84%), mainly by broad-leaved trees, and is located at an average altitude around 1000m, similarly to stable pixels. By analyzing the differences in their distribution per correlation class (Fig. 15), we can note a marked configuration change in 2007. Conversely, pixels that are stable in cluster 10 during the four years considerably preserved the correlation distribution.

Such a variation in the correlation between vegetation and temperature cycles highlights a different phenological response of pixels transited to cluster 9 in 2007 compared to pixels persisting in cluster 10. As shown by NDVI cycles, the profiles of transited pixels, indeed, is very similar to those of cluster 10 in 2005, 2006 and 2008; whereas during 2007 it is more similar to the behaviour of cluster 9, in particular, in the summer period (Fig. 16). By analyzing the peculiarities of such transited pixels, we found a higher exposition to sunlight, mainly towards East - Northeast and Southwest, whereas stable pixels are prevalently exposed to North - Northwest (Fig. 17). The aspect factor could have exacerbated the effects of the warm summer and drought condition prolonged up to autumn in 2007.

7. Concluding remarks

Our results point out a significant degree of interannual variability of phenological responses, already evidenced by other studies (e.g. Hanes et al., 2010; Bradley et al. 2008), which interests not only herbaceous species but also structured vegetation, such as broad-leaved forests. The spatial distribution of phenological aggregates roughly reflects the macroscopic characteristics of the biogeographical variability (mainly altitude and land cover). Their spatial heterogeneity is characterized by a background structure, determined by the acclimatization setting, and by fluctuating components which aggregate to one or to another core according to different meteorological scenarios.

The joint evaluation of the cluster sequences and maps of correlation with temperature show that pixels persisting in the same clusters generally keep correlation even if both phenology and meteorology exhibit slightly different characteristic curves. Conversely, larger correlation differences are detected for pixels fluctuating among clusters, which could agree with the presence of non-linear phenology-temperature relationship (Hudson, 2010). This relationship is also influenced by some additional local features (e.g., surface aspect), whose effects emerge in mutated meteorological conditions. Since also rainfalls, jointly with the basic temperature variable, can have a relevant control in vegetation development especially in Mediterranean areas, future studies will be devoted to investigate the combined role of rainfalls in determining local phenological variability.

On the whole, the capability of the clustering procedure implemented on annual basis to follows the phenological modifications provides an indication for exploring the use of cluster spatial arrangement variations (dimension and shape) jointly with temperature-phenology correlation maps in studies devoted to model the vegetation responses to meteorological/climatic fluctuations also in the presence of non-linearity.

8. Acknowledgment

This work was developed in the framework of "TeRN" project (Tecnologie per le Osservazioni della Terra e i Rischi Naturali/Technologies for Earth Observation and Natural Hazards) supported by the Italian Ministry of University and Research.

9. References

Breshears, D.D. (2006). The grassland-forest continuum: trends in ecosystem properties for woody plant mosaics?. *Frontiers in Ecology and the Environment.* vol. 4, n.2, (March 2006), pp. 96-104, ISSN 1540-9309

Butt, B., M. D. Turner, A. Singh & L. Brottem (2011). Use of MODIS NDVI to evaluate changing latitudinal gradients of rangeland phenology in Sudano-Sahelian West Africa. *Remote Sensing of Environment*, in press, ISSN: 0034-4257

Cleland, E.E.; Chuine, I.; Menzel, A.: Mooney, H.A. & Schwartz, M.D. (2007). Shifting plant phenology in response to global change. *Trends in Ecology and Evolution*, vol. 30, (July 2007), pp. 357-365, ISSN: 0169-5347

Coppola, R.; Liberti, M.; D'Emilio, M.; Lanfredi, M.; Simoniello, T. & Macchiato, M. (2006). Combined approach for air temperature spatialization using DEM, latitude and sea distance: variability of monthly data in Southern Italy. *Proceedings of the workshop Spatial Data Methods for Environmental and Ecological Processes*, p. 4, ISBN 88-8459-078-7, Baia delle Zagare (Foggia), Italy, September 4-15, 2006.

de Beurs, K.& Henebry, G.M. (2005). Land surface phenology and temperature variation in the International Geosphere–Biosphere Program high-latitude transects, *Global Change Biology*, vol. 11, (April 2005), pp. 779–790, ISSN: 1365-2486

De Santis, F.; Giannossi, M.L.; Medici, L.; Summa, V. & Tateo, F. (2010). Impact of physico-chemical soil properties on erosion features in the Aliano area (Southern Italy), *Catena*, vol. 81, (May 2010), pp. 172-181, ISSN 0341-8162

Diaz-Varela, E. R., Marey-Pérez, M. F. & Alvarez-Alvarez, P. (2009). Use of simulated and real data to identify heterogeneity domains in scale-divergent forest landscapes. *Forest Ecology and Management*, Vol.258, n.11, pp. 2490-2500.

Douglas, E.M.; Beltrán-Przekurat, A.; Niyogi, D.; Pielke, Sr. R.A. & Vörösmartyet, C.J. (2009) The impact of agricultural intensification and irrigation on land-atmosphere interactions and Indian monsoon precipitation – A mesoscale modeling perspective. *Global Planetary Change*, vol. 67, (May 2009), pp. 117-128, ISSN 0921-8181

Estrella N., Sparks T.H., & Menzel A. (2009). Effects of temperature, phase type and timing, location, and human density on plant phenological responses in Europe. *Climate Research*, Vol.39, n. 3, (September 2009), pp. 235-248, ISSN 1616-1572

Feehan, J.; Harley, M. & van Minnen, J. (2009) Climate change in Europe. 1. Impact on terrestrial ecosystems and biodiversity. A review. *Agronomy for Sustainable Development*, vol. 29, (July-September 2009), pp. 409-421, ISSN 1773-0155.

Fischer, J. & Lindenmayer, D.B. (2007). Landscape modification and habitat fragmentation: a synthesis. *Global Ecology and Biogeography*, Vol. 16, n.3, pp. 265-280, ISSN 1466-8238.

GMES (2010). Global Monitoring for Environment and Security Regulation of the European parliament and of the council (EU) n.911/2010 (available at: www.gmes.info/).

Göckede, M., A. M. Michalak, D. Vickers, D. P. Turner & B. E. Law (2010). Atmospheric inverse modeling to constrain regional-scale CO2 budgets at high spatial and temporal resolution, *Journal of Geophysical Research*, 115(D15), (August 2010), pp. D15113, , ISSN: 0094-8276.

Jönsson, A. M.; Eklundh L.; Hellström M.; Bärring L. & Jönsson P. (2010). Annual changes in MODIS vegetation indices of Swedish coniferous forests in relation to snow dynamics and tree phenology, *Remote Sensing of Environment*, vol. 114, (November 2010), n. 11, pp. 2719-2730, ISSN: 0034-4257

Karlsen, S.R.; Tolvanen A.; Kubin E.; Poikolainen, J; Høgda, K. A.; Johansen, B., Danks, F. S.; Aspholm, P.; Wielgolaski, F. E. & Makarova, O. (2008), MODIS-NDVI-based mapping of the length of the growing season in northern Fennoscandia, *International Journal of Applied Earth Observation and Geoinformation*, vol. 10, n. 3, (September 2008), pp. 253-266, ISSN: 0303-2434

Lanfredi, M.; Lasaponara, R.; Simoniello, T.; Cuomo, V. & Macchiato, M. (2003). Multiresolution spatial characterization of land degradation phenomena in Southern Italy from 1985 to 1999 using NOAA-AVHRR NDVI data. *Geophysical Research Letters*, vol. 30, pp.1069, doi:10.1029/2002GL015514, ISSN: 0094-8276

Liberti, M.; Simoniello, T.; Carone, M.T.; Coppola, R.; D'Emilio, M. & Macchiato, M. (2009). Mapping badland areas using LANDSAT TM/ETM satellite imagery and morphological data. *Geomorphology*, vol. 106, (May 2009), pp. 333-343, ISSN 0169-555X

Ma, T.& Zhou, C. (2011). Climate-associated changes in spring plant phenology in China, *International Journal of Biometeorology*, in press, ISSN: 0020-7128

Mahmood, R.; Pielke, Sr. R.A.; Hubbard, K.G.; Niyogi, D.; Bonan, G.; Lawrence, P.; Baker, B.; McNider, R.; McAlpine, C.; Etter, A.; Gameda, S.; Qian, B.; Carleton, A.; Beltran-Przekurat, A.; Chase, T.; Quintanar, A.I.; Adegoke, J.O.; Vezhapparambu, S.; Conner, G., Asefi, S., Sertel, E., Legates, D.R., Wu, Y., Hale, R., Frauenfeld, O.W., Watts, A. ; Shepherd, M.; Mitra, C.; Anantharaj, V.G.; Fall, S.; Lund, R., Nordfelt, A.; Blanken, P.; Du, J.; Chang, H.-I.; Leeper, R.; Nair, U.S.; Dobler, S.; Deo, R. & Syktus, J. (2010). Impacts of land use land cover change on climate and future research priorities. *Bulletin of American Meteorological Society*, vol. 91, n. 1, (January 2010), pp. 37-46, *ISSN* 1520-0477

Maignan, F.; Bréon, F.M.; Bacour, C.; Demarty, J. & Poirson, A. (2008). Interannual vegetation phenology estimates from global AVHRR measurements. Comparison with in situ data and applications. *Remote Sensing of Environment*, vol. 112, n.2, February 2008), pp. 496–505, ISSN: 0034-4257

Maselli, F.; Di Gregorio, A.; Capecchi V. & Breda, F. (2009). Enrichment of land-cover polygons with eco-climatic information derived from MODIS NDVI imagery. *Journal of Biogeography*, vol. 36, n. 4, (April 2009), pp. 639-650, ISSN: 1365-2699

McGarigal, K., Cushman, S. A., Neel, M. C. & Ene, E. (2002). FRAGSTATS: Spatial Pattern Analysis Program for Categorical Maps. Computer Software Program Produced by the Authors at the University of Massachusetts, Amherst. Available at the following web site: www.umass.edu/landeco/research/fragstats/fragstats.html.

Menzel et al., European phenological response to climate change matches the warming pattern, (2006*), Global Change Biology*, vol. 12, n. 10, (October 2006), pp. 1969–1976, ISSN: 1365-2486

Menzel, A.; Estrella, N. & Fabian, P. (2001). Spatial and temporal variability of the phenological seasons in Germany from 1951-1996. *Global Change Biology*, vol. 7, pp. 657-666, (August 2001), ISSN 1365-2486

Normand, S.; Treier, U.A.; Randin, C.; Vittoz, P.; Guisan, A. & Svenning J.-C. (2009). Importance of abiotic stress as a range-limit determinant for European plants: insights from species responses to climatic gradients *Global Ecology and Biogeography*, vol. 18, n. 4, (April 2009), pp. 437-449, ISSN 1466-8238

Penuelas, J. & Filella, I. (2001). Response to a Warming World. *Science*, vol. 294, October 2001), pp. 793, ISSN 1095-9203

Piao, S.; Ciais, P.; Friedlingstein, P.; Peylin, P.; Reichstein, M.; Luyssaert, S.; Margolis, H.; Fang, J.; Barr, A.; Chen, A.; Grelle, A; Hollinger, D.Y.; Laurila, T.; Lindroth, A.; Richardson A.D. & Vesala, T. (2008). Net carbon dioxide losses of northern ecosystems in response to autumn warming. *Nature*, vol. 451, (January 2008), pp. 49-53, ISSN 0028-0836

Pielke, Sr. R.A. (2005) Land use and Climate change. Science, vol. 310, (December 2005), pp. 1625-1626, , ISSN 1095-9203

Prieto, P.; Peñuelas, J.; Niinemets, U.; Ogaya, R.; Schmidt, I.K.; Beier, C.; Tietema, A.; Sowerby, A.; Emmett, B.A.; Láng, E.K.; Kröel-Dulay, G.; Lhotsky, B.; Cesaraccio, C.; Pellizzaro, G.; de Dato, G.; Sirca, C. & Estiarte, M. (2009). Changes in the onset of spring growth in shrubland species in response to experimental warming along a north–south gradient in Europe. *Global Ecology and Biogeography*, vol. 18, n.4, (April 2009), pp. 473-484, ISSN 1466-8238

Richards, J.A. & Jia, X. (2006). *Remote Sensing Digital Image Analysis: An Introduction*. 4th Edn., Sprinter, ISBN 978-3-540-25128-6

Sciortino, N.; Colonna, N.; Ferrara, V.; Grauso, S.; Iannetta M. & Svalduz A. (2000) La lotta alla desertificazione in Italia e nel bacino del Mediterraneo. *Energia, Ambiente e Innovazione*, vol. 2, pp. 35–47, ISSN 1124-0016

Simoniello, T.; Lanfredi, M.; Liberti, M.; Coppola, R. & Macchiato, M. (2008). Estimation of vegetation cover resilience from satellite time series. *Hydrology and Earth System Sciences*, vol. 12, n.4, (July 2008), pp. 1053-1064, ISSN 1607-7938

Sparks, T.H. & Carey, P.D, (1995) The responses of species to climate over 2 centuries - an analysis of the Marsham phonological record, 1736–1947. *Journal of Ecology*, vol. 83, pp. 321–329., ISSN 1365-2745

Ting,, S.; Hartley, S. & Burns, K.C. (2008) Global patterns in fruiting seasons. *Global Ecology and Biogeography*, vol. 17, n. 5, (Septeber 2008), pp. 648-657, ISSN 1466-8238

Viña, A. & Henebry, G. (2005). Spatio-temporal change analysis to identify anomalous variation in the vegetated land surface: ENSO effects in tropical South America *Geophysical Research Letters*, vol. 32, pp. L17403, ISSN 0094-8276

Williams, M., Richardson, A.D., Reichstein, M., Stoy, P.C., Peylin, P., Verbeeck, H., Carvalhais, N., Jung, M., Hollinger, D.Y., Kattge, J., Leuning, R., Luo, Y., Tomelleri, E., Trudinger, C.M., & Wang, Y.P. (2009). Improving land surface models with FLUXNET data, *Biogeosciences*, vol. 6, (July 2009), 1341-1359

Zhang, X.Y.; Friedl, M.A.; Schaaf, C.B.; Strahler, A.H.; Hodges, J.C.F.; Gao, F.; Reed, B.C. & Huete, A., (2003). Monitoring vegetation phenology using MODIS. *Remote Sensing of Environment*, vol. 84, n. 3, (March 2003), pp. 471–475, ISSN 0034-4257

Zhao, B.; Yan, Y.; Guo, H.; He, M.; Gu, Y. & Li, B., (2009). Monitoring rapid vegetation succession in estuarine wetland using time series MODIS-based indicators: an application in the Yangtze River Delta area. *Ecological Indicators,* vol. 9, n.2, (March 2009), pp. 346–356, ISSN 1470-160X

Permissions

The contributors of this book come from diverse backgrounds, making this book a truly international effort. This book will bring forth new frontiers with its revolutionizing research information and detailed analysis of the nascent developments around the world.

We would like to thank Dr. Xiaoyang Zhang, for lending his expertise to make the book truly unique. He has played a crucial role in the development of this book. Without his invaluable contribution this book wouldn't have been possible. He has made vital efforts to compile up to date information on the varied aspects of this subject to make this book a valuable addition to the collection of many professionals and students.

This book was conceptualized with the vision of imparting up-to-date information and advanced data in this field. To ensure the same, a matchless editorial board was set up. Every individual on the board went through rigorous rounds of assessment to prove their worth. After which they invested a large part of their time researching and compiling the most relevant data for our readers. Conferences and sessions were held from time to time between the editorial board and the contributing authors to present the data in the most comprehensible form. The editorial team has worked tirelessly to provide valuable and valid information to help people across the globe.

Every chapter published in this book has been scrutinized by our experts. Their significance has been extensively debated. The topics covered herein carry significant findings which will fuel the growth of the discipline. They may even be implemented as practical applications or may be referred to as a beginning point for another development.

The editorial board has been involved in producing this book since its inception. They have spent rigorous hours researching and exploring the diverse topics which have resulted in the successful publishing of this book. They have passed on their knowledge of decades through this book. To expedite this challenging task, the publisher supported the team at every step. A small team of assistant editors was also appointed to further simplify the editing procedure and attain best results for the readers.

Our editorial team has been hand-picked from every corner of the world. Their multi-ethnicity adds dynamic inputs to the discussions which result in innovative outcomes. These outcomes are then further discussed with the researchers and contributors who give their valuable feedback and opinion regarding the same. The feedback is then collaborated with the researches and they are edited in a comprehensive manner to aid the understanding of the subject.

Apart from the editorial board, the designing team has also invested a significant amount of their time in understanding the subject and creating the most relevant covers. They scrutinized every image to scout for the most suitable representation of the subject and create an appropriate cover for the book.

The publishing team has been involved in this book since its early stages. They were actively engaged in every process, be it collecting the data, connecting with the contributors or procuring relevant information. The team has been an ardent support to the editorial, designing and production team. Their endless efforts to recruit the best for this project, has resulted in the accomplishment of this book. They are a veteran in the field of academics and their pool of knowledge is as vast as their experience in printing. Their expertise and guidance has proved useful at every step. Their uncompromising quality standards have made this book an exceptional effort. Their encouragement from time to time has been an inspiration for everyone.

The publisher and the editorial board hope that this book will prove to be a valuable piece of knowledge for researchers, students, practitioners and scholars across the globe.

List of Contributors

Pierangelo Crucitti
Società Romana di Scienze Naturali, SRSN, Roma, Italy

Mohamed Sghaïer Achouri
Unité de Recherche de Bioécologie Animale et Systématique évolutive, Faculté des Sciences de Tunis, Campus Universitaire Manar, Université Tunis El Manar, Tunis

David R. Gray
Natural Resources Canada, Canadian Forest Service, Canada

Shinsuke Koike
Tokyo University of Agriculture and Technology, Tokyo, Japan

Isao Arimoto, Hideto Morimoto and Chinatsu Kozakai
Tokyo University of Agriculture and Technology, Tokyo, Japan

Koji Yamazaki
Ibaraki Nature Museum, Ibaraki, Japan

Shinsuke Kasai
Tokyo University of Agriculture and Technology, Tokyo, Japan
Present address: Shireto Foundation, Hokkaido, Japan

Yusuke Goto
Tokyo University of Agriculture and Technology, Tokyo, Japan
Present address: Tateyama Caldera Sabo Museum, Toyama, Japan

María de las Mercedes Ciciarelli
Laboratorio de Estudios de Anatomía Vegetal Evolutiva y Sistemática (LEAVES), Facultad de Ciencias Naturales y Museo, Universidad Nacional de La Plata, Argentina

Lenka Hajkova and Jiri Nekovar
Czech Hydrometeorological Institute Prague, Charles University Prague, Czech Republic

Dasa Richterova and Vera Koznarova
Czech University of Life Sciences Prague, Czech Republic

Sona Sulovska, Ales Vavra, Alena Vondrakova and Vit Vozenilek
Palacky University Olomouc, Czech Republic

Anne Fournier
IRD-UMR PALOC, Paris, France

Frederic C. Do
IRD-UMR Eco&Sols, Campus SupAgro-INRA, France

Josiane Seghieri
IRD-UMR HydroSciences Montpellier, Université Montpellier, France

Jean-Louis Devineau
CNRS-Museum National d'Histoire Naturelle, Département Hommes, Natures, Sociétés, Eco-Anthropologie et Ethnobiologie, France

Rozilaine Aparecida Pelegrini Gomes de Faria, Maria de Fatima Barbosa Coelho, Maria Cristina de Figueiredo e Albuquerque and Rodrigo Aleixo Brito de Azevedo
Universidade Federal de Mato Grosso, Brasil

Yule Roberta Ferreira Nunes, Giovana Rodrigues da Luz and Lílian de Lima Braga
Universidade Estadual de Montes Claros, Brazil

Vasudeva Ramesh
University of Agricultural Sciences, Dharwad, Karnataka, India

Rajesh Gunaga
Dr. B.S.K. Krishi Vidyapeeth, Dapoli, Maharashtra, India

José Luiz Petri, Fernando José Hawerroth, Gabriel Berenhauser Leite, Marcelo Couto and Poliana Francescatto
Epagri/Estação Experimental de Caçador, Caçador, Santa Catarina, Embrapa Agroindústria Tropical, Fortaleza, Ceará, UFSC/Universidade Federal de Santa Catarina, Florianópolis, Santa Catarina, Brazil

Rodolfo López Gómez, and Luis María Suárez Rodríguez
Universidad Michoacana de San Nicolás de Hidalgo, Instituto de Investigaciones Químico-Biológicas, Edificio B-1. Francisco J Mujica S/N Col Felicitas del Rio, Mexico

Heriberto Martínez López
Centro Interdisciplinario de Investigaciones para el Desarrollo Integral Regional, Unidad Oaxaca, México

Norberto Gariglio, Marcela Weber
Damián Castro and Norma Micheloud, Facultad de Ciencias Agrarias, Universidad Nacional del Litoral, Argentina

Thomaz Oliveira, Luis Carvalho, Luciano Oliveira, Wilian Lacerda and Fausto Acerbi Junior
UFLA, Brazil

Xiaoyang Zhang
Earth Resources Technology Inc. at NOAA/NESDIS/STAR, College Park, USA

Mark A. Friedl
Department of Geography and Environment Boston University, Boston, USA

Bin Tan
Earth Resources Technology Inc. at NASA Goddard Space Flight Center, Greenbelt, USA

Mitchell D. Goldberg
NOAA/NESDIS/STAR, Camp Springs, USA

Yunyue Yu
NOAA/NESDIS/STAR, College Park, USA

Simoniello Tiziana, Lanfredi Maria, Coppola Rosa, Imbrenda Vito and Macchiato Maria
National Research Council of Italy, Institute of Methodologies for Environmental Analysis, CNR - IMAA, University of Basilicata, Dep. Environmental Engineering and Physics, DIFA, University of Naples "Federico II", Dep. of Physics, DSF, Italy

www.ingramcontent.com/pod-product-compliance
Lightning Source LLC
Chambersburg PA
CBHW070728190326
41458CB00004B/1088

* 9 7 8 1 6 3 2 3 9 5 1 0 8 *